Impact of Artificial Intelligence in Radiology

Implementation of artificial intelligence (AI) in radiology is an important topic of discussion. Advances in AI—which encompass machine learning, artificial neural networks, and deep learning—are increasingly being applied to diagnostic imaging. While some posit radiologists are irreplaceable, certain AI proponents have proposed to "stop training radiologists now." By compiling perspectives from experts from various backgrounds, this book explores the current state of AI efforts in radiology along with the clinical, financial, technological, and societal perspectives on the role and expected impact of AI in radiology.

AI in Clinical Practice Series

The Impact of Artificial Intelligence in Radiology

Edited by Adam E. M. Eltorai, Ian Pan, and H. Henry Guo

https://www.routledge.com/AI-in-Clinical-Practice/book-series/AICLINICAL

Impact of Artificial Intelligence in Radiology

Edited by
Adam E. M. Eltorai, Ian Pan, and H. Henry Guo

CRC Press
Taylor & Francis Group
Boca Raton London New York

CRC Press is an imprint of the
Taylor & Francis Group, an **informa** business

Designed cover image: provided by author

First edition published 2025
by CRC Press
2385 NW Executive Center Drive, Suite 320, Boca Raton FL 33431

and by CRC Press
4 Park Square, Milton Park, Abingdon, Oxon, OX14 4RN

CRC Press is an imprint of Taylor & Francis Group, LLC

© 2025 Taylor & Francis Group, LLC

Library of Congress Cataloging-in-Publication Data
Names: Eltorai, Adam E. M., editor. | Pan, Ian, editor. | Guo, Haiwei Henry, editor.
Title: Impact of artificial intelligence in radiology / edited by Adam E.M. Eltorai, Ian Pan and Haiwei Henry Guo.
Identifiers: LCCN 2024015263 (print) | LCCN 2024015264 (ebook) | ISBN 9780367558161 (hardback) | ISBN 9780367558154 (paperback) | ISBN 9781003095279 (ebook)
Subjects: MESH: Technology, Radiologic | Artificial Intelligence | Medical Informatics Applications | Radiology--methods
Classification: LCC RC78.7.D35 (print) | LCC RC78.7.D35 (ebook) | NLM WN 160 | DDC 616.07/570285--dc23/eng/20240625
LC record available at https://lccn.loc.gov/2024015263
LC ebook record available at https://lccn.loc.gov/2024015264

ISBN: 9780367558161 (hbk)
ISBN: 9780367558154 (pbk)
ISBN: 9781003095279 (ebk)

DOI: 10.1201/9781003095279

Typeset in Palatino
by Deanta Global Publishing Services, Chennai, India

Contents

Part I: Technology in Medicine—Disruptive Innovation

Part II: Radiology's Role in Medicine

Part III: What is AI?

Part IV: Current State of AI in Radiology

Part V: AI Applications in Development

Part VI: Potential of AI

Part VII: Expectations—Radiologists' Jobs, Job Satisfaction, Salary, and Role in Society

Part VIII: Attitudes—Implementation Feasibility

Part IX: Technology Determinism

Preface

By compiling perspectives from experts from various backgrounds, this book explores the current state of AI efforts in radiology along with the clinical, financial, technological, and societal perspectives on the role and expected impact of AI in radiology. Perceptions can become self-fulfilling and shape technology and societal development. Technology will continue to advance, in ways that often supersede human capabilities, which can help to bring improvement to human lives. As the future of AI in radiology unfolds, this book represents a historical timestamp of which group of disparate experts' perceptions were closer to reality and offer insights into future development opportunities.

Editor Biographies

Adam E. M. Eltorai, MD, PhD, completed his graduate studies in Biomedical Engineering and Biotechnology along with his medical degree from Brown University, followed by radiology residency at Brigham and Women's Hospital/Harvard Medical School. He is interested in the development and clinical implementation of AI applications. Dr Eltorai has published numerous journal articles and books.

Ian Pan, MD is currently a diagnostic radiology resident and former chief resident in the Brigham and Women's Hospital/Harvard Medical School Diagnostic Radiology Residency Program. He graduated from the Program in Liberal Medical Education at Brown University where he received concurrent bachelor's and master's degrees in applied mathematics, biology, and biostatistics in 2016, as well as his MD from the Warren Alpert Medical School in 2020. His expertise lies at the intersection of artificial intelligence and medical imaging, having won multiple international competitions sponsored by organizations such as the Radiological Society of North America and published over 30 peer-reviewed manuscripts in this domain.

H. Henry Guo, MD, PhD is a clinical professor in the department of radiology at the Stanford University School of Medicine. He received his MD and PhD in the department of pathology at the University of Washington, followed by radiology residency and fellowships in Nuclear Medicine and Thoracic Imaging at Stanford. Dr Guo is focused on cancer and lung diseases in his clinical practice and research, co-authoring over 80 research articles, book chapters, and web-based educational resources, and is a recognized expert in interpretation of thoracic CTs and PET-CTs. Dr Guo is translating the use of quantitative CT and AI-enabled tools in clinical practice, and collaborates with faculty members in Computer Science, Engineering, and Medicine on applications of AI to topics including interstitial lung diseases, pulmonary hypertension, and early cancer detection.

Contributors

Ibrahim Abdalla
University of Minnesota Medical School
Delaware, Minneapolis, MN

Madeline Ahern
University of Minnesota Medical School
Delaware, Minneapolis, MN

Sinibaldo Rafael Romero Arocha
University of Minnesota Medical School
Delaware, Minneapolis, MN

Mohammad Mirza-Aghazadeh-Attari
Department of Radiology
Tabriz University of Medical Sciences
Tabriz
Iran Department of Diagnostic and
    Interventional radiology
Urmia University of Medical Sciences
Urmia, Iran

Dina Belhasan
University of Minnesota Medical School
Delaware, Minneapolis, MN

Abhishta Bhandari
School of Medicine and Dentistry
James Cook University

Christian Bluethgen
Institute for Diagnostic and Interventional
    Radiology
University Hospital Zurich

Benard Ohene Botwe
Department of Radiography
School of Biomedical & Allied Health Sciences
College of Health Science
University of Ghana

Ken Chang
Stanford University
Stanford, CA

Leonid L. Chepelev
Joint Department of Medical Imaging
Toronto General Hospital
University of Toronto
Toronto

Youngmin Chu
Department of Radiology
State University of New York Upstate Medical
    University
New York

Gian Marco Conte
Department of Radiology
Mayo Clinic
Rochester, MN

Kate Dembny
University of Minnesota Medical Scientist
    Training Program
Minneapolis, MN

Lalitha S. Denduluri
University of Minnesota
Twin Cities, MN

Rajiv Dharnipragada
University of Minnesota Medical School
Delaware, Minneapolis, MN

Ruiwen Ding
Department of Bioengineering
Samueli School of Engineering
University of California
Los Angeles, CA

Florian Dubost
Senior Machine Learning Engineer at Google

Mina Estafanos
Cardiothoracic Surgery Resident
University of Minnesota
Minneapolis, MN

Eduardo Farina
Diagnósticos da América AS (Dasa)
São Paulo, Brazil

Sara J. Fardin
Toronto Sick Kids Hospital
Tufts University School of Medicine and Tufts
    Medical Center
Department of Radiology
Boston, MA

Christian Federau
AI Medical AG
University of Zürich

Suely Fazio Ferraciolli
Diagnósticos da América AS (Dasa) and
    Universidade de São Paulo
São Paulo, Brazil

Christopher G. Filippi
MD, Toronto Sick Kids Hospital
Tufts University School of Medicine and Tufts
    Medical Center
Department of Radiology
Boston, MA

Tyler Gathman
University of Minnesota Medical School
Delaware, Minneapolis, MN

Bilwaj Gaonkar
Department of Neurosurgery
University of California, Los Angeles (UCLA)
Los Angeles, CA

Ribhav Gupta
University of Minnesota Medical School
Minneapolis, MN
Stanford University School of Medicine

Ronit Gupta
University of California
Berkeley, CA

Jordan Helmer
University of Minnesota Medical School
Delaware, Minneapolis, MN

Elisa Hofmeister
University of Minnesota Medical School
Delaware, Minneapolis, MN

Madeleine Howard
University of Minnesota Medical School
Delaware, Minneapolis, MN

William Hsu
Department of Radiological Sciences
David Geffen School of Medicine at UCLA
Los Angeles, CA

Alexander E. Jacobs
University of Minnesota Medical School
Minneapolis, MN

Jayashree Kalpathy-Cramer
Professor of Ophthalmology
Chief of the Division of Artificial Medical
    Intelligence
University of Colorado Anschutz Medical
    Campus
and
Director of Health Informatics
Colorado Clinical and Translational Sciences
    Institute (CCTSI)
Aurora, CO

Devasenathipathy Kandasamy
Radio Diagnosis and Interventional Radiology
All India Institute of Medical Sciences
New Delhi, India

Thomas Kane
University of Minnesota Medical School
Delaware, Minneapolis, MN

Saurin Kantesaria
University of Minnesota Medical School
Minneapolis, MN

Felipe Campos Kitamura
Diagnósticos da América AS (Dasa) and
    Universidade Federal de São Paulo
São Paulo, Brazil

Timothy L. Kline
Mayo Clinic
Rochester, MN

Megan Kollitz
University of Minnesota Medical School
Minneapolis, MN

Panagiotis Korfiatis
Radiology Informatics
Mayo Clinic in Rochester
Minnesota, MN

Osvaldo Landi
Diagnosticos da America AS (Dasa)
Sao Paulo, Brazil

Adrianna M. Rivera-León
University of Minnesota Medical Scientist
    Training Program
Minneapolis, MN

Chien-Hung Liao
Chang Gung Memorial Hospital
Taoyuan City, Taiwan

Ranson Liao
Department of Trauma and Emergency Surgery
    Chang Gung Memorial Hospital Linkou
Taiwan

Charles Lu
Massachusetts Institute of Technology
Cambridge, MA

Luke Macyszyn
Board-Certified Neurosurgeon
DISC Sports & Spine Center

Christina Malamateniou
Director of Postgraduate Programme in
    Radiography
University of London (HESAV)
Switzerland

Joseph A. Maldjian
Department of Radiology
University of Texas Southwestern Medical
    Center
Dallas, TX

Sophia Mavrommatis
University of Minnesota Medical Scientist
    Training Program
Minneapolis, MN

Raphael Nicholas Mayeden
Department of Radiology
School of Medicine
University of Health and Allied Sciences
Ghana

Andrew J. Missert
Department of Radiology
Mayo Clinic
Rochester, MN

André Lupp Mota
Cardiovascular Imaging Postdoctoral Research
    Fellow - MGH/Harvard
Cardiovascular Surgeon at the Federal
    University of São Paulo

Léo Max Feuerschuette Neto
Diagnósticos da América AS (Dasa)
São Paulo, Brazil

Mireia Crispin Ortuzar
University of Cambridge
Cambridge

Christian Park
Penn State Health
Mechanicsburg, PA

Amy Patel
The Breast Care Center at Liberty Hospital
The University of Missouri-Kansas City School
    of Medicine
Liberty, MO

Jay B. Patel
Athinoula A. Martinos Center for Biomedical
    Imaging
Charlestown, MA

Mike Prairie
University of Minnesota Medical School
Minneapolis, MN

Christopher Prokosch
University of Minnesota Medical School
Minneapolis, MN

Krithika Rangarajan
Radio Diagnosis and Interventional Radiology
All India Institute of Medical Sciences
IIT Delhi
New Delhi, India

Katelyn Rypka
University of Minnesota Medical School
Minneapolis, MN

Edson Saito
Diagnósticos da América AS (Dasa)
São Paulo, Brazil

Aaron Schumacher
University of Minnesota Medical School
Minneapolis, MN

Bhavya R. Shah
Department of Radiology
University of Texas Southwestern Medical
    Center
Dallas, TX

Heena Shah
University of Minnesota Medical School
Minneapolis, MN

Praveer Singh
University of Colorado School of Medicine
Colorado, CN

Amy Song
University of Minnesota Medical School
Minneapolis, MN

Ali S. Tejani
Department of Radiology
University of Texas Southwestern Medical
    Center
Dallas, TX

Ranveer Vasdev
University of Minnesota Medical School
Minneapolis, MN

Joshua D. Warner
Department of Radiology
University of Wisconsin School of Medicine
    and Public Health
Wisconsin, WI

David J. H. Wu
University of Minnesota Medical School
Minneapolis, MN

Charlene Liew Jin Yee
Duke-NUS Medical School
Department of Radiology
Changi General Hospital
Singapore

Fang F. Yu
Department of Radiology
University of Texas Southwestern Medical
    Center
Dallas, TX

Chandan G. B. Yogananda
Department of Radiology
University of Texas Southwestern Medical
    Center
Dallas, TX

Armin Zarrintan
Department of Radiology
Tabriz University of Medical Sciences
Tabriz, Iran

PART I

TECHNOLOGY IN MEDICINE—DISRUPTIVE INNOVATION

1 Clinical View

Ranson Liao

Experienced clinicians and radiologists excel at collecting, classifying, and analyzing clinical information and images to reach a diagnosis and decide on treatment options [1]. However, in the data-intensive environment of today's clinical practice, physicians must cope with a relentless flow of information, some of it helpful, most of it not [2]. The ability to catalog, correlate, and classify these variables continuously lies well beyond the capabilities of even the most knowledgeable and perceptive physicians.

We have witnessed a massive increase in the ability to perform complex calculations in past decades, with exponential increases in computational power enabled by relentless downsizing of integrated circuits of microprocessors. Personal computers have become universal appliances. Moreover, the Internet facilitated the dissemination of software and further provided the impetus for computer scientists to develop robust algorithms to simulate human intelligence.

Artificial intelligence (AI) refers to a system "endowed with the intellectual processes characteristic of humans, such as the ability to reason, discover meaning, generalize, or learn from experience." In the last decade, a relatively new approach to artificial intelligence called deep learning has produced significant breakthroughs and is now used on billions of digital devices for complex tasks such as speech recognition, image interpretation, and language translation. Deep learning has the potential to affect major changes in clinical medicine and healthcare delivery.

Accelerating advancements in technology development and knowledge acquisition, such as the Internet of things, big data, smartphones, and novel AI algorithms are increasingly influencing our daily lives. In contrast, the pace of information technological change has been slower in the healthcare sector, in part reinforced by tradition and closed systems. The threat of the viral pandemic has compelled healthcare providers to accelerate acceptance of new technologies and new practice styles, enabling time savings and greater focus on the more challenging tasks of clinical care.

AI has been used in diverse fields in medicine: including in drug discovery, personalized diagnostics and therapeutics, molecular biology, bioinformatics, and medical imaging [3, 4]. AI applications can also discern disease patterns by searching through and analyzing massive amounts of digital information. The US Food and Drug Administration states that "Artificial intelligence-based technologies have the potential to transform healthcare by deriving new and important insights from the vast amount of data generated during the delivery of healthcare every day" [5].

Disruptive innovations often arise from cumulative experience. Computer vision and deep learning have significantly impacted the healthcare sector in the past years, particularly in medical imaging. A Pubmed search using the keyword "deep learning" reveals a dramatic increase in publication numbers, from 107 in 2010 to 9,373 in 2020. If we expand the keyword to "artificial intelligence," publications listed in Pubmed rose from 4,471 in 2010 to 21,147 by 2020. An increasing number of algorithms or devices have received Food and Drug Agency approval during these years. To fulfill the rules and requirements of this new technology, the FDA has announced the SaMD policy to facilitate physician and medical device manufacturers' understanding of the guidelines in developing the latest tools for broadening medical usage.

Application of AI can help cope with information overload. Deep learning algorithms have been used to analyze data stored in electronic health records (EHR) to diagnose disease. Numerous retrospective studies have demonstrated early identification and stratification of patients, representing the early low-hanging fruit in AI applications [6]. Radiological image analysis was one of the rapidly developing fields of novel medical technology. Since the 1990s, radiological images have been increasingly digitized and stored in picture archiving and communication systems (PACS) in medical institutes across the world, which provide materials for developing machine learning algorithms. Traditionally, researchers used "handcrafted" feature extraction methods to train various machine learning algorithms. However, after the deep learning algorithms attained success in image recognition, end-to-end convolutional neural networks have been used to generate "learned" features from the data itself. Moreover, increasing usage of graphics processing units (GPU) vastly accelerated the computational speed of deep learning algorithms, facilitating rapid development. The application of deep learning to radiological image analysis has grown explosively since 2017. Institutions have released large image datasets such as chest X-rays (CXR), brain computed tomography (CT)/magnetic resonance image (MRI), extremity X-rays, mammography,

DOI: 10.1201/9781003095279-2

and CT images of cancers, enabling academic and industrial researchers to scale up algorithm development.

Current published works mainly focus on lung nodule identification in chest CT, chest X-ray pathologies classification, fracture detections in plain radiographs, brain CT/MRI lesion detection, disease classification, prognosis prediction, and breast cancer screening in mammography. For example, in the respiratory disease field, the landmark dataset Chest X-ray [7] released by NIH contains over 100,000 anonymized chest X-ray images and corresponding labels. The algorithm competition platform Kaggle also released several valuable medical image datasets; one of the most famous is LUNA16 [8]. Various researchers in the computer vision and medical domain utilize these datasets with or without private data to develop algorithms for CXR and chest CT applications [9]. More and more researchers also use private data from home institutions to develop algorithms, although these studies can be difficult to replicate due to limited access to the data and codes.

Existing EHR systems can integrate with deep learning models [10], facilitating the building of EHR systems that allow voice input to generate clinical notes, infer diagnosis codes, and automatically enter data into deep learning algorithms. Radiology reports could be semi-structured, and a portion of the contents could be generated by automatic medical image analysis through deep learning models [11]. For example, researchers have developed deep learning algorithms to automatically generate chest X-ray reports from image analysis, including with impressions, findings, and medical text tags, which radiologists can then review for quality control.

The United States FDA approved the first deep learning algorithm-driven device in 2016 [12] three in 2017, and with gradually increased numbers over the following years. The applied medical domains include cardiology, endocrinology, radiology, neurology, emergency medicine, and oncology. However, although the algorithm and devices are approved, most of them still need prospective studies to demonstrate clear benefit in clinical practice. Convincing medical personnel to utilize AI in daily routine requires a paradigm shift that has largely yet to take place. Most existing medical AI systems are focused on accelerating or improving the quality of medical care, for which the current lack of cost–benefit data can be a barrier to investing in such products.

While AI has already demonstrated capabilities in image recognition, a more difficult task, but one with greater potential, is developing an intelligent algorithm that can recognize all of the lesions shown in particular images, similar to the human's response in interpretation of the images [13]. The development of such an intelligent algorithm can provide most of the care currently only delivered by humans. Furthermore, the deployment of such a clinical workflow is still a challenging task for scientists and doctors. Their research speaks to the existing artificial intelligence chasm, defined as "the gulf between developing a scientifically sound algorithm and its use in any meaningful real-world applications" [14]. Even technologic giants like IBM and Google are struggling to deploy and integrate high-performance AI into current clinical practice, which requires close collaboration between data scientists, engineers, hospital administrators, and healthcare providers, with differing expectations and different potential benefits for various stakeholders. While many occupations have a negative perception of AI due to fears of being replaced by AI, most physicians don't [15, 16]. Some physicians are looking forward to AI that will assist their patient care. Medicine is a domain that faces considerable uncertainty, and medical decision-making relies on numerous amounts of ambiguous information gathered from history-taking, physical examination, imaging studies, laboratory exams, etc. AIs currently trained using data from domain experts are less likely to exceed the performance of specialists. Therefore, AI will still make wrong decisions facing the most challenging scenarios in which humans will also make mistakes unless AI exceeds the specialist's performance. AI algorithms will not be adopted if not trusted by physicians or will increase the complexity of the workload.

Data and algorithm security is another important issue facing this burgeoning field. Because hackers can modify data to change the results of deep learning models, regulations may be needed to ensure model security, particularly as these techniques become more widely used. Whereas existing regulations focus on the privacy of medical data, new regulations should also protect analytical models [10].

The machine learning development lifecycle concept is also applicable toward medical deep learning development [17]. As Andrew Ng has described, each machine learning implementation can be divided into four parts: project scoping, data collection, model training, and deployment of the product. Project scoping should be based on clinical needs and findings [18]. The concept of "unmet need," or an opportunity for improvement, is first established. Design thinking process is then utilized to define the proper target. The next step is to collect high-quality data, enhanced by expert definitions and labeling. Data scientists and software engineers need to work closely with the medical team to develop suitable algorithms for clinical application. The next critical part is clinical deployment. Post-deployment monitoring of algorithm performance is just as important, necessitating the collection of more high quality data and system maintenance to ensure clinical flow. AI deployment needs to be treated as a continuous feedback loop whereby subsequent real data is collected and in turn is used to improve the next version of the model in a continuous cycle.

The future of AI in medicine is bright. The full potential of AI will be realized once it becomes a reliable clinical assistant to care providers. By helping us cope with information overload, AI endowed machines may allow our faculties of reflection, imagination, and compassion to come to the fore when caring for fellow humans in distress. During each stage of hospitalization, deep learning can augment the clinician's decisions, leading to improved therapeutic quality [13]. The evidence is pointing toward the ability to improve outcomes, although significant challenges remain and even the most optimistic AI supporters acknowledge that successful prevention of clinically significant adverse effects remains difficult.

In summary, AI cannot currently replace the role and importance of clinicians and radiologists nor in the foreseeable future. The outputs of AI programs or devices remain largely statistical predictions. However, the importance of AI and its promises necessitate familiarity with this technology and further work in increasing its integration into clinical workflow for the benefit of our patients.

REFERENCES

1. Lovejoy, C. A., Buch, V. & Maruthappu, M. Artificial intelligence in the intensive care unit. *Crit. Care* **23**, 7 (2019).

2. Gutierrez, G. Artificial intelligence in the intensive care unit. *Crit. Care* **24**, 101 (2020).

3. Duong, M. T. *et al.* Artificial intelligence for precision education in radiology. *Br. J. Radiol.* **92**, 20190389 (2019).

4. Schork, N. J. Artificial intelligence and personalized medicine. In *Precision Medicine in Cancer Therapy* (eds. Von Hoff, D. D. & Han, H.), 265–283 (Springer International Publishing, 2019).

5. Stead, W. W. Clinical implications and challenges of artificial intelligence and deep learning. *JAMA* **320**, 1107–1108 (2018).

6. Cheng, C.-T. *et al.* A scalable physician-level deep learning algorithm detects universal trauma on pelvic radiographs. *Nat. Commun.* **12**, 1066 (2021).

7. Wang, X. *et al.* ChestX-Ray8: Hospital-scale chest X-ray database and benchmarks on weakly-supervised classification and localization of common thorax diseases. In *2017 IEEE Conference on Computer Vision and Pattern Recognition (CVPR)* (IEEE, 2017). doi:10.1109/cvpr.2017.369.

8. Setio, A. A. A. *et al.* Validation, comparison, and combination of algorithms for automatic detection of pulmonary nodules in computed tomography images: The LUNA16 challenge. *Med. Image Anal.* **42**, 1–13 (2017).

9. Aggarwal, R. *et al.* Diagnostic accuracy of deep learning in medical imaging: a systematic review and meta-analysis. *NPJ Digit Med.* 4, 65 (2021).

10. Wang, F., Casalino, L. P. & Khullar, D. Deep learning in medicine—promise, progress, and challenges. *JAMA Intern. Med.* **179**, 293–294 (2019).

11. Jing, B., Xie, P. & Xing, E. On the automatic generation of medical imaging reports. *arXiv [cs. CL]* (2017).

12. Benjamens, S., Dhunnoo, P. & Meskó, B. The state of artificial intelligence-based FDA-approved medical devices and algorithms: an online database. *NPJ Digit Med.* 3, 118 (2020).

13. Tsega, S. & Cho, H. J. Prediction and prevention using deep learning. *JAMA Network Open.* **2**, e197447 (2019).

14. Keane, P. A. & Topol, E. J. With an eye to AI and autonomous diagnosis. *NPJ Digit Med.* **1**, 40 (2018).

15. Maassen, O. *et al.* Future medical artificial intelligence application requirements and expectations of physicians in German University Hospitals: Web-based survey. *J. Med. Internet Res.* **23**, e26646 (2021).

16. Oh, S. *et al.* Physician confidence in artificial intelligence: An online mobile survey. *J. Med. Internet Res.* **21**, e12422 (2019).

17. Machine Learning Development Lifecycle. *Keras to Kubernetes®*, 243–264 (John Wiley & Sons, Inc., 2019).

18. Yock, P. G. *et al. Biodesign: The Process of Innovating Medical Technologies.* (Cambridge University Press, 2015).

2 Technological View

Suely Fazio Ferraciolli, Edson Saito, Eduardo Farina, Léo Max Feuerschuette Neto, Osvaldo Landi Junior, and Felipe Campos Kitamura

INTRODUCTION

The human body is highly complex. A massive number of interactions take place every milli-second to maintain function. These interactions need to be well programmed and highly tuned. Furthermore, the human body depends on the strength of its physical parts, from cells to organs and systems. When these and other aspects are in balance, we can achieve optimal health.

Technological systems also depend on something physical (hardware) associated with specific orders and interactions occurring together (software) to function. For a long time, "health" and "technology" were distant subjects. However, it is a natural path for the two themes to continually exert more influence on each other. In fact, recent molecular biology models suggest that intracellular dynamics resemble very complex robots.

Despite relatively slow progress, medicine has been changing. Professionals and institutions worldwide contribute to a culture of integration of medicine with other areas. Multidisciplinary teams are increasingly common. Other ways of thinking bring innovative solutions, accelerating medicine's ability to improve people's health.

Many recent technologies are poised to widen the possibilities in medicine. CRISPR-Cas9 is a promising therapeutic tool for all diseases amenable to gene editing. In conjunction with the Human Genome Project, CRISPR may be one of the major game-changers in medicine.

Artificial intelligence (AI) is an area that has one of the most significant impacts on all aspects of human development. In medicine, it is no different, and AI has extraordinary capabilities that can significantly impact physician performance. Initially, the field was dominated by computer and engineering professionals, but more individuals from other fields are venturing in, including doctors. In the following sections, we will discuss the main technologies used in AI and the different data inputs, such as text, audio, image, and video, highlighting the ones applied to medicine and radiology.

TEXT AND AUDIO

One of the most promising software technologies in healthcare is artificial intelligence (AI), and its successful implementation will require the community's attention to:

1. Managing over-expectations of AI

2. Designing work routines for AI

3. Addressing opposing user needs of AI

4. Integrating domain expertise with AI [1]

Deep learning (DL), one field of AI, is well known for its use in medical imaging with high performance for tasks such as classification, detection, segmentation, and more recently, denoising. However, DL can be used for other kinds of data besides images, including text, one of the most common types of data in healthcare, for example in the form of radiology reports and clinical notes in the electronic health record.

Many significant advances in AI for text processing have been made in recent years, including in medical applications. The previous AI models focused on medical text were based on "bag-of-words" and "TF-IDF" [2], which, generally speaking, uses the frequency of words in a document to classify them into different categories. Eventually, the word2vec technique was developed, which maps texts into an N-dimensional space where words with similar semantic meaning are closer together in this space, which increased performance over prior methods. The next major development came with the development of attention-based mechanisms and transformer models, with encoding and decoding techniques such as BHERT, a deep neural sequence transduction model using a transformer for electronic health records, resulting in further improvements compared to previous techniques [3].

AI for audio has also improved over the years, though with fewer medical applications than image- and text-based AI. It was shown that speech recognition with eye-tracking can be helpful for accurately annotating a brain tumor data set [4]. Another paper discussing the applications of

DOI: 10.1201/9781003095279-3

voice-enabled AI introduces possible use cases such as patient guidance within hospitals or voice-enabled glasses that could allow a physician to record essential aspects of a patient's history or physical exam for later review [5].

In non-radiology scenarios, other demonstrated applications include COVID-19 detection using audio recordings of coughs and sneezes [6]. The audio-based data can be converted into an array that represents the signals of the voice, or it can be converted to Mel Spectrograms and use image techniques, which will be further discussed.

IMAGE AND VIDEO

In recent decades, there have been several technological advances in medical imaging and video, which have changed how physicians work and have contributed to improving healthcare delivery. The development of new medical imaging equipment has empowered early diagnosis and non-invasive procedures in medical fields such as vascular surgery, cardiology, oncology, and neurology [7–12].

In vascular surgery, progress in medical imaging acquisition techniques such as ultrasound, computed tomography, and magnetic resonance have fostered non-invasive diagnosis, reducing the need for angiography [7]. Cardiac magnetic resonance imaging now plays a crucial role in early heart failure investigation. Its uses include ruling out active ischemia, assessing myocardial viability before revascularization, and guiding appropriate medical therapy [8]. Multiparametric resonance imaging of the prostate has made great strides in reducing the number of unnecessary biopsies and the misdiagnosis of clinically significant cancer when combined with prostate-specific antigen (PSA) screening and targeted biopsy [9]. Mammography, which dates back to the early 1900s [10], is now vital to reducing the risk of death from breast cancer through screening programs [11].

Image-guided procedures have also been favored by the growth of medical imaging acquisition techniques, now becoming the standard of care in several areas of medicine, such as neurosurgery [12]. Advances in imaging and video technology have been essential to improving laparoscopic and minimally invasive techniques and has enabled significant advancements in robotic surgery over the past few decades [13].

Along with telecommunications technology growth, advances in imaging and video technology have also allowed medical centers to connect with distant regions to provide medical information and services by telemedicine [14]. Data generated by these new rising medical image acquisition techniques has transitioned to the digital format with the introduction of picture archiving and communication systems (PACS) and the Digital Imaging and Communications in Medicine (DICOM) standard in the early 1980s [15].

This explosion of medical imaging data, combined with the exponential growth of processing power over the last decades, has created the perfect environment for data science applications to thrive in healthcare [16–21]. Big data and analytics, for example, have shown many uses in healthcare, including averting fraudulent payments, reducing hospital readmission rates, and assisting healthcare professionals in data-driven decisions for disease prediction, diagnosis, and treatment [16–18].

Artificial intelligence is a field of data science that has made significant breakthroughs in all sorts of medical imaging modalities, especially radiographs, computed tomography, and magnetic resonance. Its use cases range from image quality improvements, structure segmentation, quantification and extraction of features, detection, and classification, to name a few [20].

Radiomics is another field of data processing that is focused on quantitative mapping, extraction, analysis, and modeling for medical images. It has shown promising results for screening and staging of various malignancies, identification of prognostic imaging biomarkers, and recurrence predictions [21].

As we move toward value-based healthcare models, we may see professionals leveraging a combination of different medical image acquisition modalities, video technology, and some of these data science techniques as essential tools for personalized precision medicine.

TECHNOLOGY IN MEDICINE

As cited above in text, audio, image, video, and gene editing tools the advancement of technology in healthcare shows how technology can impact all aspects of daily physician practice (as shown in Figure 2.2). Even in the general population, medical devices and wearables are seeing increasing usage. Wearables have been popularized by millennials, a generation which has demonstrated more interest in using technology to track health, fitness, and lifestyle [22]. The adoption of health

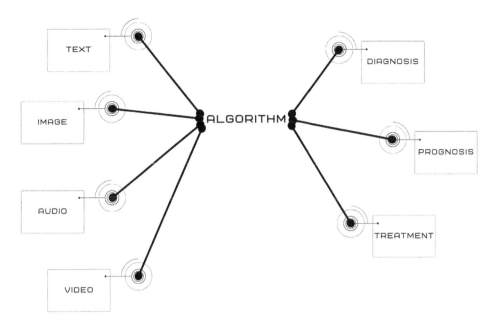

Figure 2.1 Main inputs and outputs for AI algorithms in medicine (illustration by Laura Polsin)

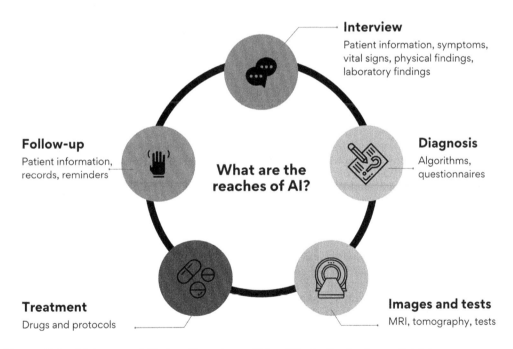

Figure 2.2 Main uses of AI algorithms in medicine (illustration by Laura Polsin)

data monitoring using wearables to evaluate and choose the best treatments that can lead to better medical outcomes is likely to increase as technology evolves and the quality and quantity of data collected increases. The velocity of wearables adoption by the population is driven by a customer-centric approach in the development of these gadgets [23]. The technologies involved in this development are further boosted by artificial intelligence (AI) [24]. Simultaneously, AI is increasingly being incorporated in various medical specialties, like dermatology, ophthalmology, pathology, gastroenterology, cardiology, and oncology.

In dermatology, algorithms for diagnosing melanoma using convolutional neural networks have been the object of research and studies. A meta-analysis of 70 studies showed computer-assisted diagnostic accuracy comparable to human specialists [25]. Future challenges related to equity, technical issues, model generalizability, model confidence calibration, and interpretability will need to be overcome in order to see more widespread adoption.

In ophthalmology, DL models have been used to assess the fundus with optical coherence tomography and visual field analysis, demonstrating promising results in diabetic retinopathy, retinopathy of prematurity, and macular edema of age-related macular degeneration. Challenges include technical issues (photo quality, Internet connection), model training with homogeneous datasets, little data on rare diseases, annotation variations of training datasets in diseases such as glaucoma and retinopathy of prematurity, and model generalizability. Issues related to interpretability are also relevant [26].

In pathology, the use of digitized whole slide images allows for creation of large databases where AI tools could be used for diagnostic support or generate new insights into disease biology [27, 28].

In gastroenterology, models that identify polyps or suspicious areas in colonoscopies have been proposed, with good metrics and some currently undergoing validation by the FDA [29].

In cardiology, the use of the Apple Watch to monitor cardiac rhythm is an FDA-cleared use case. When an abnormal rhythm is identified, the user is notified, and a recommendation for ECG is issued. If allowed, the ECG is recorded and can be sent to an emergency department or a referring physician [30, 31]. Despite other challenges to widespread adoption, wearables to monitor other biologic functions could follow the usability and customer-centric approach [23].

In oncology, despite the underwhelming initial efforts of IBM's Watson [32], the core principle of integrating clinical, genomic, laboratory, and imaging data to inform clinical decision-making in diagnosis and treatment of cancer remains valid. Several companies and startups have continued in its pursuit, and substantial research has been performed, especially in tumor imaging and pathology.

TECHNOLOGY IN RADIOLOGY

In radiology, there are many implementations of AI algorithms, not only related to imaging acquisition but also to image ordering, scheduling, protocoling, interpretation, report generation, and communication [33]. Potential applications include predicting missed appointments, allowing for more efficient scheduling, and clinical decision support for the referring physician at the time of ordering, to ensure that the proper exam is ordered.

Other potential uses are also apparent: preprocessing and reconstruction, denoising, and quality enhancement, contrast reduction, organ or lesion segmentation, lesion classification and monitoring, and prediction of prognosis and treatment response [34].

Tadavarthi et al. [35] analyzed 119 radiology AI algorithms available in the market, separating them into diagnostic (49%), quantitative (30%), repetitive (20%), and explorative (1%) tasks (with an online table available on: hitilab.org/pages/ai-companies). These tools would primarily help radiologists in segmentation and quantification tasks.

Mehrizi et al. [36] authored a technographic study about the applications of AI in diagnostic radiology, analyzing not only their primary function but also their leading technologies and implementations. They identified 269 applications offered by 99 companies. They have shown that these applications are mainly narrow in terms of tasks (1–2 tasks), modalities (single-modality), and anatomic regions (specific organs or lesion types). They mainly focus on supporting the perception and reasoning tasks of the radiologist. The AI functionalities related to perception and reasoning were segmentation (8%); quantification and extraction of features (28%); detecting and highlighting the suspicious areas (42%); comparison, cross-referencing, and longitudinal analysis (8%); diagnosis and classification of abnormalities (11%); prognosis (2%); patient profiling/synopsis and case prioritization (1%). Therefore, many improvements could be made, such as multimodality and multi-pathology AI algorithms, usage of clinical and genetic data which can help determine the prognosis of diseases and treatment response, and user-friendly interface which allows for real-time feedback.

There are many applications of radiology-based AI in oncology, especially utilizing radiomics and radiogenomics. Several review articles, such one by Bera et al. [37], explain the differences between hand-crafted radiomics and DL approaches, addressing the most common outputs, such as stratifying patients by disease severity and prognosis; predicting treatment response and benefit; identifying unfavorable treatment outcomes; distinguishing confounding responses (such as pseudo-progression from true disease progression), and non-invasively predicting salient molecular and genotypic traits. They also address the general framework of AI-enabled imaging biomarkers, discussing the technical underpinnings of commonly used methods and describing the AI tools used in complex decision-making tasks, summarizing some of the challenges and obstacles on clinical adoption of these approaches.

For most applications in radiology, the primary input data are images, sometimes with associated clinical or genetic data, usually utilizing deep learning algorithms, especially convolutional neural networks (CNN). Morid et al. [38] provide a meta-review of CNN approaches in radiology transfer learning studies, with detailed statistics on what architectures are most popular in different imaging modalities and anatomical sites. Khan et al. [39] provide a more thorough exploration of popular CNNs in computer vision, discussing a more extensive range of architectures and their contributions in greater technical detail.

Natural language processing (NLP) is also gaining more traction over the years. Pons et al. [40] wrote a systematic review of NLP in 2016 and found four major application categories for NLP in radiology: diagnostic surveillance, cohort building for epidemiological studies, quality assessment of radiological practice, and clinical support services. The AI algorithms described were mainly in a proof-of-concept stage at that time. Continued progress has been made, as noted by a 2020 review of NLP by Luo and Chong [41], who expanded upon the aforementioned four categories to include automatic protocoling; extraction of language features and coded ontology from charts; medical imaging appropriate use and clinical follow-up; and summarization of patient dossiers and conversational interfaces. Most applications use RNN-based models, such as the long short-term memory network (LSTM), which better handle sequence-based data, such as text, than other techniques. More recently, the majority of NLP models, such as ChatGPT, leverage the transformer architecture which leverages attention-based techniques.

CONCLUSION

By understanding the importance of thoroughly validating and subsequently integrating new technologies into medicine, clinicians can positively impact patient outcomes and experiences. In the years to come, radiologists and physicians in general will need to interpret and integrate the knowledge derived from these technological tools, explaining their results to patients, who will have wider access to medical technologies for personalized medicine [42]. New technologies alone will not result in substantial change unless the organizational culture incentivizes continued collaboration and innovation [43].

REFERENCES

1. May A, Sagodi A, Dremel C, van Giffen B. *Realizing Digital Innovation from Artificial Intelligence*. ICIS; 2020.

2. Soguero-Ruiz C, Hindberg K, Rojo-Alvarez JL, et al. Support vector feature selection for early detection of anastomosis leakage from bag-of-words in electronic health records. *IEEE J Biomed Health Inform*. 2016;20(5):1404–1415. doi: 10.1109/JBHI.2014.2361688

3. Li Y, Rao S, Solares JRA, Hassaine A, Ramakrishnan R, Canoy D, ... Salimi-Khorshidi G. BEHRT: Transformer for electronic health records. *Sci Rep*. 2020;10(1):1–12.

4. Stember JN, Celik H, Gutman D, Swinburne N, Young R, Eskreis-Winkler S, ... Bagci, U. Integrating eye tracking and speech recognition accurately annotates MR brain images for deep learning: Proof of principle. *Radiol Artif Intell*. 2020;3(1):e200047.

5. Isbitski D, Fishman EK, Rowe SP. Connecting with patients: The rapid rise of voice right now. J Am Coll Radiol. 2021;18(4):627–629.

6. Al-Dhief FT, Latiff NMAA, Malik NNNA, Salim NS, Baki MM, Albadr MAA, Mohammed MA. A survey of voice pathology surveillance systems based on Internet of Things and machine learning algorithms. *IEEE Access*. 2020;8:64514–64533.

7. Perry JT, Statler JD. Advances in vascular imaging. *Surg Clin N Am*. 2007;87(5):975–993.

8. Gonzalez JA, Kramer CM. Role of imaging techniques for diagnosis, prognosis and management of heart failure patients: Cardiac magnetic resonance. *Curr Heart Fail Rep.* 2015;12(4):276–283.

9. Ghai S, Haider MA. Multiparametric–MRI in diagnosis of prostate cancer. *Indian J Urol.* 2015;31:194–201.

10. Gold RH, Basset LW, Widoff BE. Highlights from the history of mammography. *RadioGraphics* 1990;10:1111–1131.

11. Pace LE, Keating NL. A systematic assessment of benefits and risks to guide breast cancer screening decisions. *JAMA* 2014;311(13):1327–1335.

12. Cleary K, Peters TM. Image-guided interventions: Technology review and clinical applications. *Annu Rev Biomed Eng* 2010;12:119–142.

13. Siddaiah-Subramanya M, Tiang KW, Nyandowe M. A new era of minimally invasive surgery: progress and development of major technical innovations in general surgery over the last decade. *Surg J.* 2017;3:e163–e166.

14. Zundel KM. Telemedicine: History, applications, and impact on librarianship. *Bull Med Libr Assoc* 1996;84(1):71–79.

15. Huang HK. Short history of PACS. Part I: USA. *Eur. J. Radiol.* 2011;78:163–176.

16. Islam S, Hasan M, Wang X, Germack HD, Noor-E-Alam. A systematic review on healthcare analytics: Application and theoretical perspective of data mining. *Healthcare* 2018;6:54.

17. Ghassemi M, Celi LA, Stone DJ. State of the art review - the data revolution in critical care. *Critical Care* 2015;19:118.

18. Tomar D, Agarwal S. A survey on data mining approaches for healthcare. *Int. J. Biosci. Biotechnol.* 2013;5:241–266.

19. Kansagra AP, et al. Big data and the future of radiology informatics. *Acad Radiol.* 2106;23:30–42.

20. Mehrizi MHR, Ooijen P, Homan M. Applications of artificial intelligence in diagnostic radiology - a technography study. *Eur Radiol.* 2021;31:1805–1811.

21. Lambin P, et al. Radiomics: The bridge between medical imaging and personalized medicine. *Nat Rev Clin Oncol.* 2017;14(12):749–762.

22. Medical Futurist. <https://medicalfuturist.com/millennials-and-their-views-about-living-a-healthy-way-of-life/>

23. Fast Company.<https://www.fastcompany.com/90688912/the-case-for-making-hearing-aids-and-insulin-monitors-as-sexy-as-an-apple-product>

24. Zhao, M., Wang, D. Li, J. Data management and visualization of wearable medical devices assisted by artificial intelligence. *Netw Model Anal Health Inform Bioinforma.* 2021;10:53. https://doi.org/10.1007/s13721-021-00328-0

25. Dick V, Sinz C, Mittlböck M, Kittler H, Tschandl P. Accuracy of Computer-Aided Diagnosis of Melanoma: A Meta-analysis. *JAMA Dermatol.* 2019 Nov 1;155(11):1291–1299. doi: 10.1001/jamadermatol.2019.1375. PMID: 31215969; PMCID:PMC6584889.

26. Ting DSW, Pasquale LR, Peng L, Campbell JP, Lee AY, Raman R, Tan GSW, Schmetterer L, Keane PA, Wong TY. Artificial intelligence and deep learning in ophthalmology. *Br J Ophthalmol*. 2019 Feb;103(2):167–175. doi: 10.1136/bjophthalmol-2018-313173.

27. Cui M, Zhang DY. Artificial intelligence and computational pathology. *Lab Invest*. 2021;101:412–422. https://doi.org/10.1038/s41374-020-00514-0

28. Benjamin Moxley-Wyles, Richard Colling, Clare verrill, artificial intelligence in pathology: An overview. *Diagn Histopathol*. 2020;26(11):513–520, ISSN1756-2317. https://doi.org/10.1016/j.mpdhp.2020.08.004

29. Kudo SE, Mori Y, Misawa M, Takeda K, Kudo T, Itoh H, Oda M, Mori K. Artificial intelligence and colonoscopy: Current status and future perspectives. *Dig Endosc*. 2019 Jul;31(4):363–371. doi: 10.1111/den.13340

30. The Verge. <https://www.theverge.com/2018/9/13/17855006/apple-watch-series-4-ekg-fda-approved-vs-cleared-meaning-safe>

31. Meskó B, Görög MA short guide for medical professionals in the era of artificial intelligence. *NPJ Digit. Med*. 2020;3:126. https://doi.org/10.1038/s41746-020-00333-z

32. IEEE SPECTRUM. <https://spectrum.ieee.org/how-ibm-watson-overpromised-and-underde-livered-on-ai-health-care>

33. Kapoor N, Lacson R, Khorasani R. Workflow applications of artificial intelligence in radiology and an overview of available tools. *J Am Coll Radiol*. 2020 Nov;17(11):1363–1370. doi: 10.1016/j.jacr.2020.08.016. PMID: 33153540.

34. Montagnon E, Cerny M, Cadrin-Chênevert A, Hamilton V, Derennes T, Ilinca A, Vandenbroucke-Menu F, Turcotte S, Kadoury S, Tang A. Deep learning workflow in radiology: A primer. *Insights Imaging*. 2020 Feb 10;11(1):22. doi: 10.1186/s13244-019-0832-5. PMID: 32040647; PMCID: PMC7010882.

35. Tadavarthi Y, Vey B, Krupinski E, Prater A, Gichoya J, Safdar N, Trivedi H. The state of radiology AI: Considerations for purchase decisions and current market offerings. *Radiol Artif Intell*. 2020 Nov 11;2(6):e200004. doi: 10.1148/ryai.2020200004. PMID: 33937846; PMCID: PMC8082344.

36. Rezazade Mehrizi MH, van Ooijen P, Homan M. Applications of artificial intelligence (AI) in diagnostic radiology: A technography study. *Eur Radiol*. 2021 Apr;31(4):1805–1811. doi: 10.1007/s00330-020-07230-9. Epub 2020 Sep 18. PMID: 32945967; PMCID: PMC7979626.

37. Bera K, Braman N, Gupta A, Velcheti V, Madabhushi A. Predicting cancer outcomes with radiomics and artificial intelligence in radiology. *Nat Rev Clin Oncol*. 2021 Oct 18. doi: 10.1038/s41571-021-00560-7. Epub ahead of print. PMID: 34663898.

38. Morid MA, Borjali A, Del Fiol G. A scoping review of transfer learning research on medical image analysis using ImageNet. *Comput Biol Med*. 2021;128:104115.

39. Khan A, Sohail A, Zahoora U, Qureshi AS. A survey of the recent architectures of deep convolutional neural networks. *Artif Intell Rev*. 2020;53:5455–5516.

40. Pons E, Braun LM, Hunink MG, Kors JA. Natural language processing in radiology: A systematic review. *Radiology*. 2016 May;279(2):329–343. doi: 10.1148/radiol.16142770. PMID: 27089187.

41. Luo JW, Chong JJR. Review of natural language processing in radiology. *Neuroimaging Clin N Am*. 2020 Nov;30(4):447–458. doi: 10.1016/j.nic.2020.08.001. PMID: 33038995.

42. Topol E. Deep medicine: How artificial intelligence can make healthcare human again. 2019:111–135.

43. Scalable data systems require creating a culture of continuous learning. *DataEBioMedicine*. 2021;74:103738, ISSN: 2352–3964. https://doi.org/10.1016/j.ebiom.2021.103738.

3 Societal View

Heena Shah, Ribhav Gupta, Rajiv Dharnipragada, and Ronit Gupta

THE ROLE OF TECHNOLOGY IN MEDICINE

Medicine has experienced large shifts in the last century that have been ushered in by the advent of technology. From the development of more efficacious medication to the invention of novel imaging machines, technology has integrated itself into every facet of medicine, and has promoted the advancement of healthcare delivery, diagnostic medicine, and the exchange of patient data.

Recent developments of medical technology have enabled the automation of tasks that were once manually performed by healthcare professionals. In doing so, technological innovations have allowed healthcare workers to concentrate their efforts on other responsibilities and decrease the likelihood of errors. This has ultimately resulted in improved health care delivery [1]. One example of this trend is in the implementation of automated infusion pumps. Intravenous (IV) infusions frequently contribute to medication errors due to their complex preparation and administration requirements [1]. Before the invention and implementation of IV smart pumps in hospitals, healthcare workers were required to perform complex calculations of the rates of infusion while simultaneously manipulating various units of calculations [2]. IV smart pumps now allow users to select desired medication from built-in drug libraries and to automatically calculate infusion rates based on pertinent patient input values, often enhanced by built-in dose error reduction systems that immediately alert the user if the calculated infusion rate exceeds the acceptable dosing limits [2]. IV smart pumps serve as one of many examples demonstrating how technology can be integrated into medical care to improve consistency and patient safety.

Diagnostic medicine is yet another facet of medical care that has benefited immensely from technological evolution. For instance, imaging technology, such as computerized tomography (CT), magnetic resonance imaging (MRI), and ultrasound, has led to the timely and relatively noninvasive detection, monitoring, and treatment of many diseases that lead to improved patient outcomes [3]. Additionally, diagnostic decision support systems (DDSS), which consist of software that matches patients' disease characteristics to a digital clinical knowledge base and makes diagnostic recommendations to providers, have demonstrated improved accuracy for certain use cases when validated with diagnostic testing [4]. A 2018 study comparing the diagnostic accuracy of 87 residents with and without DXplain, a DDSS tool, showed a significantly higher average score on diagnosis tests in the DXplain intervention group than the control group [5]. In this way, diagnostic technologies can serve as a conduit connecting isolated individuals or communities to healthcare networks to achieve greater consistency and quality in healthcare.

Technology has enabled the autonomous extraction, analysis, and exchange of health data, especially through the increasing use of electronic health records (EHR) and electronic case reporting (eCR) that have enhanced the quality and effectiveness of public health surveillance [6]. Such benefits are particularly important considerations for individuals of lower socioeconomic status, minorities, and rural communities who lack reliable access to healthcare and health information, and who often face disproportionately higher disease burdens [7]. For example, when it comes to infectious disease monitoring, a large body of literature has demonstrated the racial disparities in HIV outcomes for African Americans in the United States [8]. In response, policies such as the National AIDS Strategy have been developed and federal resources have been allocated to address such inequities [8]. This demonstrates one of many ways in which technology has enhanced the volume and scope of public health and quality improvement research.

DISRUPTIVE TECHNOLOGY AND PRECISION MEDICINE

Broadly speaking, developments in technology as applied to medicine have decreased the prevalence of medical errors, aided in diagnostics, and facilitated novel research inquiries. However, certain technological milestones have undeniably fostered relatively more significant medical progress and evolution. These notable developments range from establishing standardized hygiene protocols, to the invention of antibiotics and vaccines, to discovering the structure of deoxyribonucleic acid (DNA) [9]. More recently, advancements in genomics, biotechnology, wearable sensors, and artificial intelligence (AI) have allowed medicine to transition from trial-and-error-based to evidence-based approaches, and from finding generalized solutions for the greatest number of patients to individualized care through precision medicine [10]. Precision medicine

DOI: 10.1201/9781003095279-4

is defined as tailoring medical treatment to characteristics of individual patients and applying preventative or therapeutic interventions to those who will derive the greatest benefit and bear the least burdens from side effects or costs associated with such interventions [11]. The following precision medicine technologies have disrupted the state of medicine in the twenty-first century and will continue to transform healthcare delivery and accessibility [10].

GENOMICS AND BIOTECHNOLOGY

Genomics is a branch of molecular biology that initially aimed to characterize genomes or exomes through the use of first-generation sequencing methods such as Sanger sequencing [12]. However, in the last few decades, novel sequencing methods, such as Next Generation Sequencing (NGS), have enabled larger genomes to be more efficiently sequenced [13]. DNA sequencing complements genetic engineering methods such as clustered regularly interspaced short palindromic repeat (CRISPR)-associated 9 (Cas9) nuclease, zinc-finger nucleases (ZFNs), and transcription activator-like effector nucleases (TALENs) that allow modifications to be made to existing sequences [13, 14]. Genome sequencing and editing have enabled scientists to discern the significance of genetics in various diseases and develop gene therapies that target specific DNA sequences or proteins [13]. A key example of gene therapy that has seen remarkable success is the treatment of sickle cell disease (SCD), a recessive gene–linked condition resulting in defective β-globins which can induce the sickling of red blood cells in hypoxic environments, leading to the occlusion of blood vessels that produce immense pain and potential organ failure in affected individuals [14]. A 2015 study highlights the success of gene therapy in transferring a β-globin sequence that reversed sickling in a 13-year-old patient with severe SCD who was unresponsive to standard drug management therapies [15]. Not only were sickling-related crisis symptoms alleviated, as conventional pharmaceutical therapies can only hope to achieve, but rather the probability of future SCD-related events was greatly decreased, thus treating the patient's underlying pathology rather than the symptoms alone [15]. This breakthrough treatment of SCD exemplifies progress based on scientific inquiry and reinforces the promises of advancements in genomics and biotechnology.

WEARABLE DEVICES

Wearable devices are a rapidly expanding hub for technological innovation in medicine that has been utilized in the treatment and monitoring of a plethora of conditions ranging from neurological to cardiovascular [16]. These devices capture physiological, biochemical, and motion data to facilitate the diagnosis and continued surveillance of medical conditions and have extended the remote assistance of specialists to rural areas or low-income countries where physicians may be scarce [16]. Individuals living in such areas travel farther to see providers and have less access to specialists, often leading to worse clinical outcomes for common conditions such as diabetes and heart attacks [16]. In such circumstances, wearable sensors can use wireless communication to send collected data to a remote center for real-time monitoring and assessment. Some devices can also integrate alert systems that detect critical values and inform emergency services and patients' caregivers [16]. In addition to monitoring disease, wearable technology has been found to effectively encourage healthier habits. A 2016 study found that young adults spent significantly less time being sedentary, compared to baseline, when receiving feedback from their wearable device [17]. Overall, wearable medical technologies have a wide array of applications, ranging from expanding access to limited medical resources, to informing and encouraging individuals to pursue healthy behaviors [18].

ARTIFICIAL INTELLIGENCE

Precision medicine requires detailed collection and interpretation of patient's health and disease data to improve clinical decision-making [10]. Before the adoption of digital health, clinicians relied on personal learning and experience combined with individual problem-solving and often rudimentary clinical tools to make such decisions. AI, with access to big data and a broad vantage point, has led to the development of advanced tools for clinical decision-making for use by both physicians and patients [10]. A 2018 study found that the FareWell digital therapeutic, an interactive digital intervention method for diabetes treatment, significantly reduced hemoglobin A_{1c} (HbA$_{1c}$) levels and increased patients' confidence in managing their diabetes [19]. While technological advancements in genomics, biotechnology, and wearable sensors have greatly augmented the breadth of health data collection, without AI, it would be seemingly impossible for physicians to individually consolidate and analyze such large data sets. The integration of AI in healthcare has also enabled the efficient collection, extraction, and mining of health data. For example, Deep

Genomics has used AI to identify genomic mutations and their linkages to diseases using genetic data sets mapped against medical records [10]. These disruptive technologies are advancing into medicine a culture of prevention, personalization, and precision.

SOCIETAL LIMITATIONS FOR TECHNOLOGICAL INNOVATION IN MEDICINE

While there are many potential benefits to technological innovation, it is important to consider limitations and potential challenges in technological implementation within modern health systems. One limitation is the need for a workforce with specialized skill sets to promote a modernized vision of health care, along with recognizing that certain jobs will be transformed or even eliminated as a result [20]. For example, as AI in radiology continues to improve, radiological images may one day be autonomously and possibly even more accurately interpreted by AI enabled algorithms. Therefore, physicians will eventually need to adapt to computerized tools becoming standard clinical practice [20]. As a result, it will be important to determine responsibilities that will be delegated to humans and those that will be delegated to machines, as well as the best skills required to perform different tasks in the healthcare chain [20].

As data collection and exchange becomes more ubiquitous, the consideration of data ownership becomes increasingly relevant to the ethical advancement of healthcare. With patient data underpinning many technologies and medical advancements, it is paramount to recognize and respect that patients have a right to control the use of their personal data. Thus, the need for regulatory systems, protocols, and repercussions for misused data must be established in order to protect patients' autonomy over their personal health information [20].

As digital health by means of electronic medical records (EMRs) and public health databases saturate healthcare systems, such systems, particularly hospitals, become increasingly vulnerable to cyberattacks [20]. Increasing vulnerabilities arise from hospital organizations needing to manage growing numbers and the complexity of information technologies and connected medical devices, compounded by internal politics, state and federal regulatory pressures, and their patient-centered focus [21]. Cyberattacks may range from nuisance malware, to ransomware, to attacks on hospitals' connected electric grids, leading to degraded or complete cessation of hospital function [20]. These attacks can cause significant data loss, monetary theft, eroded patient trust, and, most importantly, jeopardize human life [22]. Therefore, it becomes especially important for healthcare organizations to invest financial capital and resources to protect their systems [21].

Lastly, as new innovations saturate the market, there has been a drive for hospitals and healthcare providers to continually explore and invest in new drugs and technology [23]. While the approval of newly developed drugs often requires them to be more efficacious or less expensive than currently utilized treatments, the standards for approving new medical technologies are generally much less stringent [24]. As a result, expensive new medical technologies are sometimes adopted without significant evidence supporting improved patient outcomes or cost-saving potential [24], leading to increased costs incurred by health organizations and patients, and contribute to the overutilization of expensive medical services [25].

PATIENTS' ATTITUDES ON TECHNOLOGICAL INNOVATION IN MEDICINE

The overuse of expensive diagnostic tests and therapeutic procedures, especially in high-income countries, has led to increased financial burdens on health systems and patients [25]. Moreover, incidental findings associated with over-testing can influence asymptomatic patients to seek unnecessary prophylactic care, including many therapies without evidence of improvements to long-term survival or morbidity [25]. While medical technology, by way of imaging, laboratory tests, and screenings can reveal asymptomatic diseased states or increased risk of diseases, such information when provided to patients can alter their behavior and cause heightened anxiety and vulnerability [26]. Furthermore, the increased use of digital technologies can place protected patient health information at risk of security breaches or misuse [27]. In spite of these considerations, patients generally favor the use of new technologies due to the general belief that new technologies would result in improved management of their health [27]. Furthermore, there has been a positive trend in patient willingness to share their data collected from these technologies for research studies, having recognized that their information could be useful in the advancement of medicine for themselves and society at large [28, 29]. Ultimately, the enormous amount of information made available to both healthcare providers and patients through disruptive medical technologies has generally empowered patients and led to shared decision-making and the democratization of care [30].

PHYSICIAN'S ATTITUDES ON TECHNOLOGICAL INNOVATION IN MEDICINE

Despite these valuable shifts in the culture of healthcare, there has been reluctance among physicians to adopt digital medicine. Physicians experience burnout for many reasons, such as from chaotic work environments, increased administrative tasks, and lack of autonomy. However, technology, specifically health information technology, can also play a major role in contributing to physician burnout. [31]. Physicians expressed discontent with navigating complicated EHR user interfaces and report insufficient time for documentation. These factors have contributed to less face-to-face time with patients and affect the physician–patient dynamic [31]. Furthermore, with the use of EHRs and a vast array of medical devices, physicians are expected to collect and process enormous amounts of clinical information rapidly, thus creating a hazard for information overload and error [32]. With increasing turnover rates of new innovations, it has been difficult to keep up—placing strains on already limited physician time, system resources, and department finances. The faster new technologies are pushed out to the work environment, the more incompatibilities emerge between old and new systems, and the gap between those who have immediate access to new technologies and those who do not can be exacerbated [33].

CONCLUSIONS AND FUTURE CONSIDERATIONS

Technology has improved healthcare delivery and outcomes by enhancing physicians' ability to make more accurate diagnoses and treatment decisions. Patients benefit from more informed, tailored, and potentially more efficacious treatment options through precision medicine. However, rapid technological advancement must be tempered by recognition of its limits and potential risks for society, in order to anticipate and address secondary effects such as cybersecurity, physician burn-out, and rising healthcare costs. Such foresight will enable the development of regulations to protect patient interests and to facilitate further technological innovation.

REFERENCES

1. Sutherland A et al. European Journal of Hospital Pharmacy (2020). DOI: 10.1136/ejhpharm-2018-001624.

2. Giuliano KK. Biomedical Instrumentation & Technology (2015). DOI: 10.2345/0899-8205-49.s4.13.

3. Frija G et al. eClinical Medicine (2021). DOI: 10.1016/j.eclinm.2021.101034.

4. Sutton RT et al. NPJ Digital Medicine (2020). DOI: 10.1038/s41746-020-0221-y.

5. Martinez-Franco AI et al. Diagnosis (2018). DOI: 10.1515/dx-2017-0045.

6. Guthrie S et al. Annual Review of Public Health (2015). DOI: 10.1146/annurev-publhealth-031914-122747.

7. Crilly J et al. American Journal of Public Health (2011). DOI: 10.2105/AJPH.2010.300003.

8. Nunn A et al. AIDS and Behavior (2019). DOI: 10.1007/s10461-019-02631-4.

9. Boudoulas KD et al. Hellenic Journal of Cardiology (2017). DOI: 10.1016/j.hjc.2017.05.001.

10. Mesko B. Expert Review of Precision Medicine and Drug Development (2017). DOI: 10.1080/23808993.2017.1380516.

11. Ginsburg GS et al. Health Affairs (2018). DOI: 10.1377/hlthaff.2017.1624.

12. Li H et al. Signal Transduction and Targeted Therapy (2020). DOI: 10.1038/s41392-019-0089-y.

13. Giani AM et al. Computational and Structural Biotechnology Journal (2019). DOI: 10.1016/j.csbj.2019.11.002.

14. Doudna JA. Nature (2020). DOI: 10.1038/s41586-020-1978-5.

15. Cavazzana M et al. Blood (2015). DOI: 10.1182/blood.V126.23.202.202.

16. Patel S et al. Journal of NeuroEngineering and Rehabilitation (2012). DOI: 10.1186/1743-0003-9-21.

17. Ellingson LD et al. Translational Journal of the American College of Sports Medicine (2016). DOI: 10.1249/TJX.0000000000000001.

18. Lu L et al. JMIR mHealth and uHealth (2020). DOI: 10.2196/18907.

19. Berman MA et al. JMIR Diabetes (2018). DOI: 10.2196/diabetes.9591.

20. Dzau VJ et al. Science Translational Medicine (2018). DOI: 10.1126/scitranslmed.aau4778.

21. Jalali MS et al. JMIR Publications (2018). DOI: 10.2196/10059.

22. Mohan DN et al. International Journal of Research in Engineering, Science and Management (2020). ISSN (Online): 2581-5792.

23. Sorenson C et al. ClinicoEconomics and Outcomes Research (2013). DOI: 10.1001/virtual mentor.2014.16.2.pfor1-1402.

24. Burke LA et al. American Medical Association Journal of Ethics (2014). DOI: 10.1001/virtual mentor.2014.16.2.pfor1-1402.

25. Brownlee S et al. Lancet (2017). DOI: 10.1016/S0140-6736(16)32585-5.

26. Hofmann B et al. Life Sciences, Society, and Policy (2018). DOI: 10.1186/s40504-018-0069-y.

27. Boeldt DL. JMIR Publications (2015). DOI:10.2196/jmir.4456.

28. Petersen C. Yearbook of Medical Informatics (2018). DOI:10.1055/s-0038-1641193.

29. Page SA et al. BMC Med Ethics (2016). DOI:10.1186/s12910-016-0130-4.

30. Mesko B et al. JMIR Publications (2019). DOI: 10.2196/12490.

31. Gardner RL et al. Journal of the American Medical Informatics Association (2019). DOI:10.1093/jamia/ocy145.

32. Feblowitz JC et al. Journal of Biomedical Informatics (2011). DOI: 10.1016/j.jbi.2011.03.008.

33. Thimbleby H. Journal of Public Health Research (2013). DOI:10.4081/jphr.2013.e28.

4 Financial View

Charlene Liew Jin Yee

There is no shortage of innovative platform technologies which have expanded their utility into medicine, to name a few: augmented reality, 3D printing, robotics, genome editing, and artificial intelligence (AI). Some have a greater potential to cause disruption than others. In the last decade and in the decade to come, the innovations which have demonstrated the greatest potential to disrupt medicine can be broadly described as automation technologies, robotics, and data science including artificial intelligence. Much has been discussed about the societal and ethical impact of disruptive innovation in medicine, and yet—some of the most powerful forces which will drive the adoption of these technologies arise from financial factors and cost leadership that disruptive innovation strategies bring to the healthcare sector.

Financial considerations impact disruptive innovation in several ways:

1. Competitive advantage to the adopters of innovative technology

2. Financial cost of adopting innovative technology

3. Impact on labor and the future of work

4. Competitive Advantage to the Adopters of Innovative Technology

A common observation amongst industry players has been that the market growth of robotics and AI in medicine has been slower than anticipated, but a growing momentum toward adoption of the technology in tandem with progress in scientific research are positive signs of an accelerating growth trend.

Only a small percentage of medical occupations can be fully automated by adopting technologies such as robotics and AI, but some work activities of almost all medical occupations could be automated. In 2017, a study by McKinsey found that 50% of the activities that people are paid to do in the global economy have the potential to be automated. In about 60% of occupations, including those in healthcare, at least 30% of activities could technically be automated. The study also found that technically automatable activities would affect the equivalent of 1.2 billion employees and $14.6 trillion in wages worldwide [1].)

Achieving the full potential of disruptive technology will take several more years as technology adopters and decision-makers study the impact of using these technologies in the real world. Greater confidence in robotics and AI needs to be built beforehand, and financial executives will expect concrete data on the value and the return on investment for disruptive technologies. Despite the cautious approach toward disruptive innovation in executive boardrooms, there is a consensus amongst key stakeholders that disruptive technology is strategically important to all institutions. The competitive advantage of adopting disruptive technology can be seen to outweigh the costs to the company—hence the rhetoric becomes "can the institution afford not to adopt disruptive innovation," rather than the reverse.

One of the competitive strategies to drive any industry forward is cost leadership. Take AI in radiology for example: the integration of machine learning in imaging diagnosis has the potential to cut costs for patients and insurance companies by half [2] and it may cost as little as $1000 USD to install machine learning–enabled technology capable of processing 260 million images per day [3]. To put this number into perspective, that is more than the total number of all MRI and CT scans performed in the USA in a day. A thousand dollars is approximately the current cost to payer for a single MRI study.

Institutions utilizing AI to diagnose disease may apply this technology as a business differentiation strategy if patients (customers) perceive this as having value. Apart from creating value by increasing diagnostic accuracy, hybrid human–AI intelligence may increase patient access to imaging in remote areas by allowing non-experts to arrive at a provisional diagnosis. Furthermore, healthcare access for non-urgent conditions may be improved by provision of around-the-clock services for routine studies, increasing consumer convenience—a lucrative business model in a highly competitive industry.

From a government-level perspective, disruptive technology in medicine can help to close a workforce growth gap resulting from declining growth rates of working-age populations. Peak employment will occur in most countries within 50 years due to declining birthrates and an aging workforce in advanced and some emerging economies [4, 5]. A case in point is the current

DOI: 10.1201/9781003095279-5

shortage of radiologists in the United Kingdom, which was widely reported beginning from 2016. The Royal College of Radiologists' clinical radiology UK workforce census 2016 report found that 97% of radiology departments in the UK said that they had been unable to meet their diagnostic reporting requirements in 2016 and found that 22% of the consultant radiologist workforce were predicted to retire in the next five years. To overcome the shortage, the UK National Health Service (NHS) paid nearly £88m ($116m) in 2016 for backlogs of radiology examinations to be reported, 92% of radiology departments paid radiologists to work overtime, 78% outsourced reporting to independent teleradiology companies, and 52% employed ad hoc locums [6, 7].

For policymakers worldwide, the greying workforce poses a worrying challenge since approximately half of the sources of GDP growth from the past half century (employment growth) will disappear as their populations age. Disruptive technology would be able to mitigate some of the effects of these demographic trends: McKinsey Global Institute predicted that the productivity boost from automation could potentially add between 0.8% and 1.4% of global GDP annually, and that by 2065, automation could potentially add productivity growth in the largest economies in the world the equivalent of an additional 1.1 billion to 2.3 billion full-time workers. Therefore, governments worldwide have a strong incentive to encourage the implementation of these technologies to support future economic targets. Policymakers will need to start a conversation with stakeholders to anticipate the types of jobs that could be created [8]. In a similar vein, healthcare industry stakeholders should capitalize on this unique opportunity to initiate dialogues with their medical workforce about the type of work that needs doing, along with the societal impact on patients.

Healthcare workers will have to learn how to integrate their workflows with automation and disruptive technology, potentially freeing up time to improve upon and leverage on traditionally "high touch" patient interaction skills which sets medicine aside from other industries. For their managers, there will be greater opportunity for them to spend more time on coaching and developing human potential.

It is possible to examine the economic impact upon different groups of healthcare workers according to their level of specialization in certain skills:

1. High-skilled healthcare workers who work closely with technology will likely be in strong demand. They will be able to leverage new opportunities for independent work if there are further healthcare business shifts toward work being outsourced by companies. Telemedicine adoption will accelerate this trend and radiologists may fall into this category.

2. Middle-skilled workers have the highest technical potential for automation (predictable, repetitive physical activities, data collection and analysis). This group of workers will require upskilling to prepare for shifts in the nature of work toward complementing disruptive technologies. Laboratory technologists and research scientists are examples of these workers.

3. Low-skilled workers working with disruptive technology—in particular, repetitive physical work such as porterage—may experience the most disruption if technologies such as self-driving robots displace their work entirely.

In summary, disruptive technologies such as AI, robotics, and automation could potentially unleash a renaissance in the development of intrinsic human skills that machines struggle to replicate: creativity, social and emotional capabilities [9].

FINANCIAL COST OF ADOPTING INNOVATIVE TECHNOLOGY

Since the adoption of disruptive technologies often requires enormous investment and capital spending by healthcare institutions, the cost of developing and deploying technological solutions invariably affects the pace of adoption. Hardware and software solutions are highly designed and application-specific, such as surgical robots with robotic arms requiring surgeon-level dexterity. Expensive sensors are required for perceptive tasks, such as cameras and laser imaging, detection, and ranging (LIDAR) for visual tasks, and high-fidelity microphones for voice recognition. Costs can be prohibitive for institutions investing in large-volume data storage and cloud solutions.

Increasingly, cost escalation in medicine is a topic which policymakers grapple with, and top-of-mind awareness (TOMA) is the role that medical technology plays. It has been estimated by health care economists that 40–50% of annual cost increases can be attributed to new technologies or the intensified use of old ones [10, 11]. Ergo, in the eyes of the policymaker, decisions regarding the adoption of new technologies become the key consideration when it comes to keeping costs down.

The predicament is that medical technology has become a fundamentally valued feature of modern medicine in all jurisdictions. In the words of Daniel Callahan, cofounder of The Hastings Center:

Patients expect it, doctors are primarily trained to use it, the medical industries make billions of dollars selling it, and the media loves to write about it. The economic and social incentives to develop and diffuse it are powerful, and the disincentives so far weak and almost helpless. Cutting the use of technology will seem wrong—even immoral—to many. [10]

It seems inevitable that economists will eventually concede that disruptive technology is central to modern medicine and unavoidable. They would not be wrong, seeing as how mRNA vaccines for COVID-19, telemedicine, and older historical innovations such as antibiotics, anesthesia, cardiac electrophysiology, and cancer therapy owe their existence to disruptive innovations of the day.

Nonetheless, several European healthcare systems have been lauded for their ability to rise to the challenge of sustainable healthcare spending and cost control by eliminating wasteful and ineffective healthcare. These healthcare systems achieve this by utilizing tools such as price controls, negotiated physician fees, hospital budgets with limits on expenditures, and stringent policies on the adoption and diffusion of new technologies. The figures vary, but these economies have been able to contain annual cost increase to within the 3–4% range, with better health outcomes than other high-income countries such as the US [12, 13].

More recently, and expanding the analogy of the competitive advantage afforded by innovative technology, data analytics, business intelligence, and medical informatics offer an attractive solution to cost-control by reducing inefficiencies in healthcare delivery, along with developing preventive medicine, population screening, personalized patient-directed healthcare, and clinical decision support tools (CDST).

The Hastings Center, which was instrumental in establishing the field of bioethics in 1969, poses several rhetorical questions that should help policymakers design their overall technology adoption strategy [10]:

1. Should death be seen as the greatest evil that medicine should seek to combat, or would a good quality of life within a finite life span be a better goal?

2. Do the elderly need better access to intensive care units and more high-tech medicine to extend their lives, or better long-term and home care and improved economic and social support?

3. Does it make any sense that the healthier we get, the more we spend on healthcare, not less?

In a more local context, the deployment of disruptive technologies requires significant upfront capital expenditure (CaPex) spending for each institution, resulting in high initial costs compared to the wages of a hospital's workforce. Notably however, both hardware and software costs decline over time, eventually increasing the attractiveness and commercial viability of these solutions [14].

The pace of disruptive technological adoption in medicine may be influenced by local labor market dynamics in different countries. The quality and availability of the workforce, as well as the demand and costs of human labor, determine how soon certain tasks will be automated. Labor market dynamics in different countries are therefore affected by evolving demographics (e.g., aging of the workforce, migration) as well as different wage rates. For example, job automation is more likely to be adopted sooner in countries with high wages, such as North America and Western Europe, than in developing countries with lower wages.

Beyond labor cost savings, developed economies may also place a larger emphasis on performance gains such as increased patient throughput and revenue, and, importantly, improved safety and quality, which sometimes exceed the benefits of labor cost savings.

IMPACT OF DISRUPTIVE TECHNOLOGY ON LABOR AND THE FUTURE OF WORK

Medical doctors are familiar with the use of case studies as an instructional tool to enable them to quickly gain insight into novel scenarios. Similarly, let us stretch the analogy to radiology as a case study for how disruptive technology may impact medicine.

Radiologists are not newcomers to the field of artificial intelligence, pioneering work in medical imaging perception in the 1980s [15]. To most outsiders, radiology would appear ripe for disruption by imaging AI. In the past 5–10 years, there has been substantial progress in deep learning methods of image detection and classification [16]. Current artificial neural networks have accuracy rates that surpass those of radiologists in nodule detection, pathologists in detecting lymph node metastasis from breast cancer [17] and are likely as accurate as ophthalmologists for detecting vision-threatening diabetic retinopathy [18].

In the 1970s, Michael Porter described his classic generic strategies for competitive advantage, which would go on to transform industrial strategy [19]. The three generic strategies determining competitive advantage are cost reduction, differentiation, and focus. The differentiation and focus advantage possessed by the specialty is a formidable barrier to entry for competition from AI technology—likely accounting for the pivot of the AI industry toward augmenting radiologists' work rather than the piecemeal replacement of radiological tasks. Diagnostic work forms a large proportion of the tasks that radiologists perform, but a radiologist's greatest strength lies, counterintuitively, not in the ability to accurately detect and classify disease but in the ability to tap into collective and individual experience to make clinical judgments based on data. It is close to impossible to replicate the complexity and expansiveness of human perception in training datasets, and narrow AI algorithms have not been able to reproduce the intuition of human experience.

The strategic goal of the AI industry has therefore moved toward one that aims to create an AI–human intelligence dyad and an ecosystem to sustain and expand the field of radiology and, collectively, all of the medical specialties into a much broader discipline that fuses all available data, including those derived from an individual's proteomics, metabolomics, microbiome, genomics, informatics metadata, electronic health records, and biometrics, to arrive at a timely diagnosis (preferably before clinical manifestation).

Apart from the obvious new roles and jobs these combinatorial "medical information specialist" fields will create, such as imaging biomarker-metageneticists and the like, artificial intelligence and other disruptive technologies, such as robotics, have the potential to create a completely new classification of healthcare jobs. In an article by Wilson et al., a new type of jobholder, termed "explainers," could emerge, whose job it will be to bridge the gap between technologists and decision-makers. Explainers in medicine could, for example, help elucidate the inner workings of an AI system to help convince executives who are uneasy with the opaqueness of AI algorithms [20]. This may in turn be enforced by governments, for example, the European Union's General Data Protection Regulation (GDPR), has created a "right to explanation," allowing consumers to question and challenge decisions produced purely on an algorithmic basis. Another example could be "algorithm auditors" who would be in charge of holding any algorithm accountable for its results. Auditors may be tasked to examine errors and decisions made by AI systems when these result in unintended adverse outcomes.

When the economist David Autor studied the future of work, he found that over the centuries and through four industrial revolutions, technological innovations that automate tasks have had a consistent effect: an increase in the numbers of jobs, and an evolution toward more cognitively demanding jobs. An example he cites are automated teller machines (ATMs), which had two countervailing effects on bank teller employment. As teller tasks were automated, the number of tellers per branch fell. But it had become cheaper to open new branches, so the number of bank branches increased by about 40%. The net result was more branches and more tellers. However, he found that the nature of bank tellers' work had evolved. "As their routine, cash-handling tasks receded, they became less like checkout clerks and more like salespeople, forging relationships with customers, solving problems and introducing them to new products like credit cards, loans and investments" [21].

Ultimately, using the time freed by increased productivity, clinicians can revert to being humanistic practitioners of the "art of medicine," by increasing face-to-face interactions, doing things "for" rather than "to" patients. In the end, if we contemplate the value of achieving this vision, no price tag can be too high.

REFERENCES

1. Manyijka J, Chui M, Bughin J, George K, Willmott P, Dewhurst M. A Future that works: Automation, employment, and productivity. *McKinsey Glob Inst.* 2017 Jan;60:1–28.

2. Frost & Sullivan. *Cognitive Computing and Artificial Intelligence Systems in Healthcare.* Cognitive Computing and Artificial Intelligence Systems in Healthcare; 2015 [cited 2017 Nov 18]. Available from: http://www.frost.com/sublib/display-report.do?id=NFFE-01-00-00-00&bdata =aHR0cDovL2NvcnBjb20uZnJvc3QuY29tL2UvZjI%2FZWxxRm9ybU5hbWU9QnV5bm93X0JsaW5kRm9ybSZlbHFTaXRlSQ9MTU0NCZDX0VtYWlsQWRkcmVzcz0mQ2FtcGFpZ25JR D1OQV9QUl9LQmVsY2hlcl9ORRkZFXzE4RGVjMTUmZWxxPTAwM

3. Molteni M. *If You Look at X-Rays or Moles for a Living, AI Is Coming for Your Job.* Wired; 2017 [cited 2017 Nov 18]. Available from: https://www.wired.com/2017/01/look-x-rays-moles-living-ai-coming-job/

4. Manyika J, Woetzel J, Dobbs R. *Global Growth: Can Productivity Save the Day in an Aging World?* McKinsey Global Institute; 2015.

5. US Bureau of labor statistics job openings and labor turnover survey database. 2017 [cited 2016 Apr 16]. Available from: https://www.bls.gov/jlt/data.htm

6. The Royal College of Radiologists. Clinical radiology UK workforce census 2016 report [BFCR(17)6]. *WwwRcrAcUk.* 2016 Oct:1–54. Available from: https://www.rcr.ac.uk/system/files/publication/field_publication_files/cr_workforce_census_2016_report_0.pdf

7. Rimmer A. Radiologist shortage leaves patient care at risk, warns royal college. *BMJ Br Med J.* 2017;359.

8. O'Reilly T. Don't replace people. Augment them. XRDS crossroads. *ACM Mag Students.* 2016;23(2):18–21.

9. Autor DH. Why are there still so many jobs? The history and future of workplace automation. *J Econ Perspect.* 2015;29(3):2015–3. Available from: https://economics.mit.edu/files/11563.

10. Callahan D. Health care costs and medical technology. In: Hastings Center (ed.) *From birth to death bench to clinic: The hastings center bioethics briefing book for journalists, policymakers, and campaigns.* Hastings Center, New York, 2008:79–82.

11. Keehan S, Sisko A, Truffer C, Smith S, Cowan C, Poisal J, et al. Health spending projections through 2017: The baby-boom generation is coming to medicare: Accelerating growth in Medicare spending by the end of the projection period is the first sign of the coming demographic shift. *Health Aff.* 2008;27(Suppl1):w145–55.

12. Borgonovi E, Adinolfi P, Palumbo R, Piscopo G. Framing the shades of sustainability in health care: Pitfalls and perspectives from Western EU countries. *Sustain.* 2018;10(12):1–20.

13. Simonet D. Healthcare reforms and cost reduction strategies in Europe: The cases of Germany, UK, Switzerland, Italy and France. *Int J Health Care Qual Assur.* 2010;23(5):470–488.

14. Gwynne P. Predicting the progress of technology. *Res Technol Manag.* 2013;56(4):2.

15. Krupinski EA. The future of image perception in radiology: Synergy between humans and computers. *Acad Radiol.* 2003 Jan 1;10(1):1–3 [cited 2017 Nov 11]. Available from: http://www.ncbi.nlm.nih.gov/pubmed/12529022

16. Liew CJY. Medicine and artificial intelligence: a strategy for the future, employing Porter's classic framework. *Singapore Med J.* 2020;61(8):447.

17. Ehteshami Bejnordi B, Veta M, Johannes van Diest P, van Ginneken B, Karssemeijer N, Litjens G, et al. Diagnostic assessment of deep learning algorithms for detection of lymph node metastases in women with breast cancer. *JAMA.* 2017 Dec 12;318(22):2199 [cited 2018 Mar 8]. Available from: http://jama.jamanetwork.com/article.aspx?doi=10.1001/jama.2017.14585

18. Ting DSW, Cheung CYL, Lim G, Tan GSW, Quang ND, Gan A, et al. Development and validation of a deep learning system for diabetic retinopathy and related eye diseases using retinal images from multiethnic populations with diabetes. *JAMA - J Am Med Assoc.* 2017;318(22):2211–23.

19. Porter M. How competitive forces shape strategy. *Strateg Plan Readings*. 1979 Mar–Apr:102–17. Available from: http://faculty.bcitbusiness.ca/KevinW/4800/porter79.pdf

20. Wilson HJ, Daugherty P, Bianzino N. The jobs that artificial intelligence will create. *MIT Sloan Manag Rev*. 2017;58(4):14.

21. Autor, D. Will automation take away all our jobs?. *Ideas.Ted.Com*. 2017 [cited 2017 Nov 6]. Available from: https://ideas.ted.com/will-automation-take-away-all-our-jobs/

PART II
RADIOLOGY'S ROLE IN MEDICINE

5 Clinical View

Christian Bluethgen

Radiology is a unique medical specialty that forms an integral part of all levels of healthcare [1]. Its primary aim is to provide useful information gathered from medical imaging to answer questions arising from clinical problems. To this end, radiology handles the acquisition, utilization, and dissemination of information, the punctuality of information delivery, and the application of appropriate levels of specialization to interpretation [2]. Radiology tailors this information to cater to referring physicians (e.g., from emergency departments, wards, and primary care) and other intermediate actors, such as insurance companies and forensic institutions.

Throughout its history and starting with its inception, radiology was largely driven by ground-breaking technological discoveries, developing an understanding of these technologies, translating and implementing them in a clinically useful way. Radiology maintains close relationships with not only its clinical partners but is also translating clinical problems by engaging in research at the intersection of image capturing technology, engineering, and information technology.

To be able to understand how radiology as a specialty can benefit from recent technological developments in artificial intelligence, this chapter seeks to give a high-level overview of the role of radiology in medicine from a clinical point of view, by analyzing the specialty's components, its various tasks and functions.

THE RADIOLOGIST'S TOOLBOX

Understanding the role of radiology requires familiarity with its technological toolbox built upon different complex physical principles. In 1885, German physicist Wilhelm Conrad Röntgen experimented with an electrical discharge tube shielded by a cardboard and noticed fluorescence of a barium platinocyanide screen on his desk a few feet away [3]. Röntgen discovered that the rays he suspected of causing this phenomenon (originally labeled "X-Strahlen" for their unknown nature) could pass through the human body and, to his amazement, he was able to display the bones of a hand clearly (Figure 5.1).

Because of its immediately apparent tremendous potential for medicine, it took unusually little time from this discovery to the introduction of X-rays in clinical practice: Charles Thurstan Holland was one of the first physicians to offer radiology as a service as early as 1896 and is referenced as a pioneer in establishing it as a medical specialty in its own right [4].

Radiographs are widely available, entail only short acquisition times, and come with comparably little costs and radiation exposure. Although other modalities often provide more detailed information, radiographs remain a method of choice in the first-line assessment of many pathologies, e.g., for detection of fractures or pneumonia. The discovery of X-rays also enabled the development of a variety of derived techniques: mammography uses low-energy X-rays to examine the human breast for both screening and diagnosis of lesions. Fluoroscopy allows for a real-time assessment of moving structures like the gastrointestinal tract, often enhanced by the use of radio-opaque contrast agents. In a similar fashion, angiography is used to dynamically visualize the lumen of blood vessels after injecting contrast agents.

Computed tomography (CT) uses one or more X-ray tubes and detectors revolving around a patient to acquire attenuation profiles used to create cross-sectional axial images of a patient. The acquired data can be reconstructed in other anatomical (i.e., sagittal, coronal) or custom planes, or rendered as a volume. Short acquisition times and a high spatial resolution make CT an ideal tool in settings that mandate an urgent and detailed assessment of the patient's condition. CT comes at the cost of exposing the patient to considerably higher amounts of radiation compared to radiographs, depending on selected protocol and examined body region.

Magnetic resonance imaging (MRI) for medical imaging is based on the use of a strong external magnetic field and radio frequency pulses, using the nucleus of hydrogen atoms abundantly found in water and fat to produce detailed images. This modality features superior soft-tissue contrast and an absence of ionizing radiation, but comes with comparably long acquisition times and a possibly discomforting examination conditions in a long, narrow tube.

Diagnostic medical ultrasonography makes use of ultrasound waves that propagate through soft tissues at varying speeds and are reflected on interfaces of tissues with differing acoustical impedances. The information from the returning ultrasound waves can be visualized in real

DOI: 10.1201/9781003095279-7

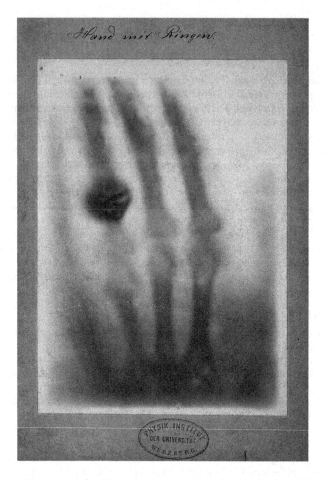

Hand mit Ringen.

Figure 5.1 Wilheim Roentgen's first human x-ray image, of his wife Anna Bertha's hand

time and allows the creation of anatomical maps (anatomical ultrasound) or the gaining of insight about characteristics like blood velocity and the stiffness of tissue (functional ultrasound).

Many radiology departments include physicians that specialize in nuclear medicine, a field that uses a variety of radioactive tracers (e.g., F-18 labeled deoxyglucose) to derive images with techniques like scintigraphy (e.g., of the lung or the thyroid gland), single photon emission computed tomography (SPECT), and positron emission tomography (PET). These methods are able to assess bodily functions, e.g., metabolic activity and treatment response. Radiopharmaceuticals applied by nuclear medicine physicians can also be used to treat diseases.

The theoretical and applied mastery of these tools provides the foundation for addressing the clinical questions radiologists face on a daily basis.

A RADIOLOGIST'S WORKFLOW
Example From the Emergency Department

A patient presents to the emergency department with symptoms of acute abdominal pain. The patient is triaged to be evaluated by an emergency physician. The emergency physician obtains the patient's history, examines her or him, and orders blood tests. The gathered information leads to the formulation of an initial working diagnosis, or, more likely, a list of possible diagnoses (differential diagnosis). Clinical findings are summarized (e.g., "Unclear abdominal pain for 6 hours, focus right lower quadrant, elevated leukocyte count"), a question is formulated (e.g., "Appendicitis? Signs of intestinal inflammation? Cholecystitis? Other pathology?"), and an ultrasound scan is requested. The radiologist confirms the appropriateness and performs the procedure (in some countries, trained radiology technicians may perform the ultrasound). After the exam is completed, the images are uploaded to a picture archiving and communication system (PACS). The

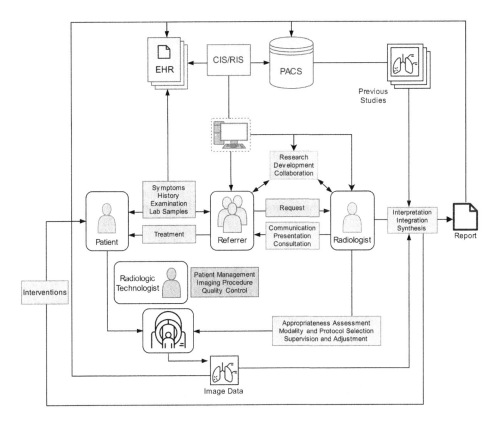

Figure 5.2 Radiology workflow

radiologist informs the clinician of the result of the study (e.g., "Enlarged appendix. No perifocal free fluid. Findings consistent with appendicitis") in a timely manner. The clinician can use this information to narrow down the differential diagnosis, ideally confirm or exclude the working diagnosis, and choose the appropriate management strategy. The radiologist writes a report with a more detailed assessment of all imaged structures, which is added via the radiology information system (RIS) to the electronic health record (EHR) of the patient and can be accessed by other healthcare professionals through the clinical information system (CIS).

There are many variations to this example of a basic workflow in radiology (Figure 5.2).

Patient Setting

Imaging studies are not only performed on patients who present with acute symptoms. Sometimes, studies are ordered for patients without symptoms but with predisposing recent history (e.g., elderly patient who fell) to rule out injuries and other pathologies. In in-patient care, imaging may be ordered to investigate abnormal laboratory results (e.g., to exclude post-renal causes of elevated levels of creatinine) or symptoms developed in the hospital setting (e.g., exclusion of deep vein thrombosis for patients with long resting hours). There are also patients with a long history of a chronic disease that requires follow-up studies (e.g., cancer patients). Fast and secure Internet connections enable the analysis of imaging performed at remote sites (teleradiology), resulting in increased, around-the-clock availability of specialized radiologic assessment, at the cost of reduced direct contact with limited supervision and intervention possibilities. Individuals undergoing an imaging study might also be asymptomatic but may be part of a risk group and be eligible for secondary prevention (screening). Lastly, radiologic imaging may be performed in forensic, post-mortem settings.

Appropriateness Assessment

Not every requested imaging study is indicated, necessary, or appropriate. Indications for imaging studies need to be critically evaluated, as imaging is costly in terms of money, equipment utilization, and the time of patient, doctors, and technical staff.

Some medical questions may be answered by multiple imaging modalities with comparable information gain. Most of the time, however, one or two modalities are favored over others for answering a specific question; in other cases, the medical question or the circumstances of patient or examination settings may limit the number of applicable modalities. Additionally, in the case of radiographs and derived modalities like CT, imaging studies are accompanied by exposure to ionizing radiation.

Radiologists ensure the appropriate use of equipment and imaging procedures and apply radiation protection principles [5]. While it might be reassuring to have a CT scan rule out acute injuries for liability reasons, appropriateness must be factored into the decision-making process. This is complicated by the fact that referring parties requesting imaging studies often impose the associated costs without incurring them directly, and thus might be less aware of the efficient and economic use of imaging resources [1]. Good cooperation and communication between referring clinician and radiologists are pivotal for the assessment of appropriateness [6]. For this reason, guidelines for medical imaging like the American College of Radiology's appropriateness criteria [7] or the European Society of Radiology's iGuide [8] have been established and should be consulted.

Selection of Imaging Modality and Protocol

In the abovementioned example, the question and the subsequently selected imaging approach was broad, to assess a range of possible diagnoses. Other questions might require more specific imaging protocols. The radiologist must carefully select imaging modality and the respective settings, e.g., the type of contrast agent (e.g., extracellular vs. hepatospecific gadolinium-based contrast agents in MRI studies of the liver), contrast agent application parameters (e.g., total amount, injection rate, bolus composition), whether to add oral or rectal contrast, acquisition timing(s), special sequences in MRI or special phases in CT imaging, scan ranges, field of view, scanning parameters (e.g., tube voltage, pitch), and patient positioning.

Although it is not always feasible in the setting of high-throughput practices, image quality for critical exams should be evaluated right after acquisition and the study repeated if the images are deemed undiagnostic. This is usually done by the radiologic technician performing the study (e.g., to assess motion artifacts). In high-stake situations, however, the radiologist should also supervise the examination while the patient is still in the scanner. For example, a radiologist should check whether the pulmonary arteries are thoroughly contrasted to be able to exclude peripheral pulmonary embolisms or repeat the study if necessary.

Certain imaging studies may require modification of the study protocol and scan parameters during the acquisition. Especially in the case of MRI, sequences that are not included in the standard protocols may become interesting after acquiring the first few images and can be added on the fly. Another prime example would be cardiac MRI, where the radiologist often has to identify the optimal inversion time to ensure myocardial "nulling," to maximize the image intensity difference between infarcted and normal myocardium [9]. Techniques to reduce this time-consuming step are an area of active research [10].

Radiologists continuously optimize protocols to reduce scan times, increase image quality, and minimize the risk through ionizing radiation.

Image Interpretation

The interpretation of medical images is a challenging task heavily reliant on experience, which alone warrants radiology to be a specialty of its own right. Prospective radiologists have to undergo years of professional training. In the United States, for instance, this entails completing at least 13 years of medical training, including medical school, a four-year residency, and often an additional one- or two-year fellowship of subspecialized training like diagnostic neuroradiology, pediatric radiology, or interventional radiology. Similar training times of five to six years after completion of medical school are required in most other countries. Radiologists are certified by national organizations like the American Board of Radiology and have detailed requirements for continuing medical education throughout their practicing years.

The basic steps of image interpretation are image perception, hypothesis generation, data interpretation and verification, and the generation of a conclusion, all of which are susceptible to cognitive biases [11]. First of all, image interpretation requires profound knowledge of the relevant anatomy, both as described and illustrated in anatomy textbooks such as *Gray's Anatomy* [12] or Netter's *Atlas of Human Anatomy* [13] but also taking into account a wide range of anatomic variants, which may or may not become symptomatic and may or may not mimic pathologies [14]. Real imaging findings must be distinguished from a large variety of possible imaging artifacts that can mask pathologies or create false-positives, necessitating familiarity with the underlying imaging technique [15–18]. Sometimes, artifacts may also be useful as surrogates for underlying conditions, e.g., for the evaluation of pulmonary edema through the use of B-line artifacts in bedside ultrasound [19, 20].

Image interpretation requires extensive knowledge of both general and specific pathology and corresponding imaging signs. In a healthcare environment that extensively relies on imaging, highly specialized clinicians will often rely on their own interpretation of imaging studies, as they are familiar with the specific anatomy and pathology information they need for optimal treatment planning. However, diseases are often not limited to one specific system. Imaging might reveal unexpected conditions ranging from trivialities to management-altering findings (e.g., by requiring follow-ups) and to potentially life-threatening pathologies that require immediate attention. On the other hand, a radiologist might be essential to mitigate over-treatment by reducing false-positive findings [21].

With the ongoing development of new therapies and new diagnostic tools, radiologists need to subscribe to a career-long endeavor of self-education. Nonetheless, it is nowadays considered virtually impossible to acquire expert levels of competence in all areas of radiology [21, 22]. Radiologists who underwent the general training are therefore often subspecializing in areas like thoracic, cardiovascular, or oncologic imaging. Subspecialized radiologists can form teams to provide comprehensive radiological service or can join secondary or tertiary referral centers working primarily in a subspecialty [21].

The interpretation process requires the skill of integrating imaging findings with the information gathered from the clinical assessment, from the patient's history and from the laboratory test results. As imaging is not only used to make a first diagnosis, but also to monitor treatment and to predict outcomes, image findings have to be compared with previous studies of the same patient from the same or different modalities, put into the context of received treatments and possible effects thereof, and other conditions the patient might suffer from. If the patient received previous imaging, a trajectory (i.e., categorizing the findings as "new," "better," "stable," or "worse") and the magnitude of change for each finding has to be determined.

The initial interpretation task is finished when all images in all acquired series have been analyzed, medically relevant conditions have been identified or excluded, and, ideally, the information relevant to answering the clinical question has been extracted. Sometimes, additional information becomes available to the involved parties after the imaging procedure, so that subsequent secondary interpretation is not uncommon.

Communication

In the clinical setting, information gathered from imaging is only valuable if it affects a decision and is available to the decision-maker at the time of the decision [6]. Therefore, after interpreting the acquired image data with respect to other available information, the resulting findings need to be reported to the decision-maker in a timely manner. Important findings (e.g., vascular occlusion, intracranial hemorrhage, pneumothorax) warranting immediate attention and change of management must be communicated without delay, while other findings (e.g., an incidental pulmonary nodule) may be reserved for the final report.

The cost associated with getting the relevant information out of a larger body of data is not always readily apparent to all those involved in a multi-party environment. A CT scan typically involves hundreds of images that need to be carefully analyzed—a visual task that requires focus and time. To deliver information valuable to the referring physician, the radiologist must understand and appreciate the clinical problem. This makes it easier to filter and structure the relevant information in a way that benefits the receiving party, although it might come with increased workload for the sending party.

The radiology report represents the amalgamation of available information and the radiologist's compiled insight into a patient's condition and is the most common and complete way of dissipating information distilled by radiologic assessment [23]. The report is targeted at the referring physician, but its audience includes many other parties, such as healthcare providers in other specialties, other radiologists, medical support personnel, and interested patients and relatives.

Reports usually include a findings section, representing a list of all objective observations that can be made from the study at hand, ideally reflecting the approach that has been pursued while reading the study. The collection of findings may be structured according to templates provided by the department or an organization (i.e., "structured reporting"), which provides standardization, adds a level of quality control, and may facilitate automatic extraction; on the other hand, this approach may reduce flexibility compared to free-text reporting, e.g., by artificially fragmenting related conditions as they appear in multicompartmental diseases [23]. This section may also include findings of a clearly benign nature (e.g., simple cortical renal cysts) that are omitted in the impression section for lack of clinical relevance.

The impression section is the product of the radiologist's thought process with respect to all available information and represents a compact synthesis of the clinically relevant, ideally actionable interpretations in the form of a (differential) diagnosis and/or recommendations. Sometimes, despite a well-defined clinical question, a definitive answer remains impossible, e.g., due to a lack of specific imaging findings, insufficient quality of the study, or inherent limitations of the modality. Therefore, statements in the impression section should be accompanied by expressions modulating the degree of confidence conveyed in the interpretation (e.g., "consistent with," "possible").

Consultative Services

Sometimes, not all relevant information is sufficiently conveyed in the written and finalized radiology report. Reasons could be an omission of clinically relevant findings, possible findings in the images spotted by the clinician but left uncommented by the radiologist, uncertainty about localization or meaning of findings described in the report itself, or because additional information became available after reporting the study. The face-to-face communication and exchange of information in the reading room for the discussion of complicated cases is an important part of adding value to the healthcare process through radiology. Additional contribution happens in the setting of multidisciplinary conferences like oncological tumor boards, during which radiologic assessment can be supplemented by additional information from different clinical parties and jointly discussed. This adds to the integrative process of selecting the optimal treatment strategy for each patient. As the former executive director of the International Society of Radiography Otha Linton put it:

> The obligation or responsibility or opportunity of a radiologist to go beyond a terse dictated report is what allows the specialty to be more than a technical service. […] If a radiologist provides nothing more than an observation of abnormal densities, radiology will be minimized or eliminated. [24]

BEYOND THE DIAGNOSTIC PROCESS

This chapter focused primarily on the role of radiology in diagnostic imaging. However, image interpretation and other roles directly associated with the diagnostic process are only a subset of the functions the radiology department plays in healthcare.

Importantly, the sophisticated imaging techniques developed over the course of the development have led to the creation of interventional radiology, a subspecialty that uses the available imaging modalities to guide minimally invasive procedures. These procedures are either diagnostic (e.g., biopsies, angiographies), therapeutical (e.g., trans-catheter chemo-embolization procedures to occlude vessels feeding tumors or stent angioplasty), or both. The image-guided, minimally invasive approach minimizes risk of adverse events compared to open-access techniques, while at the same time allowing for high levels of precision and control by ensuring optimal placement of the therapeutic devices or agents. Interventional radiology thereby often paves the way, establishing new technologies that are later performed routinely and not only by interventional radiologists [25].

As imaging is a key component of medical diagnostics in nearly all specialties, radiology departments take important leading roles in the organization of how the imaging data is transmitted, organized, distributed, and accessed in a hospital. Therefore, the department works in close collaboration with IT professionals and companies offering technological solutions for picture archiving communication systems (PACS) and radiology information systems (RIS) and the integration of this information in the clinical information systems (CIS) of hospitals.

As any other specialty in academic institutions, radiology is intimately involved in research. This is particularly supported by the nature of radiologists' workflow, which encompasses daily data collection, data analysis, and data labeling. This especially enables the conduction of high-volume retrospective studies. Research may focus on technological development, e.g., how to optimize images and scan conditions while reducing radiation dose [26], where collaborations with engineering and computer science are frequent, or on clinical topics, e.g., to develop new, imaging-based biomarkers for interstitial lung disease [27]. Radiologic consultation can also significantly improve the quality of scientific studies and case reports that involve imaging studies [28].

RADIOLOGY IS A KEY PLAYER IN MODERN MEDICINE

Radiology is a key player in modern medicine with contact points to nearly all medical specialties. Its services and functions are vital to the goal of personalized precision medicine by engaging in the screening, diagnosis, and monitoring of disease and the application and monitoring of therapy. Beyond the diagnostic process, radiology offers a broad range of consultative services and is deeply engaged in research at the intersection of image capturing technology, engineering, computer science, and information technology.

REFERENCES

1. Brady A, Brink J, Slavotinek J. Radiology and Value-Based Health Care. *JAMA*. 2020 Oct 6;324(13):1286–7.

2. Brady AP, Bello JA, Derchi LE, Fuchsjäger M, Goergen S, Krestin GP, et al. Radiology in the Era of Value-Based Healthcare: A Multi-Society Expert Statement From the ACR, CAR, ESR, IS3R, RANZCR, and RSNA. *Can Assoc Radiol J*. 2021 May 1;72(2):208–14.

3. Thomas AMK, Banerjee AK. *The History of Radiology*. Oxford: Oxford University Press; 2013. 239 p.

4. Cope R. Radiologic History Exhibit. Charles Thurstan Holland, 1863-1941. *Radiogr Rev Publ Radiol Soc N Am Inc*. 1995 Mar;15(2):481–8.

5. Krestin GP. Maintaining Identity in a Changing Environment: The Professional and Organizational Future of Radiology. *Radiology*. 2009;250(3):612–7.

6. Larson DB, Langlotz CP. The Role of Radiology in the Diagnostic Process: Information, Communication, and Teamwork. *Am J Roentgenol*. 2017 Nov 1;209(5):992–1000.

7. American College of Radiology. ACR Appropriateness Criteria® [Internet]. [cited 2021 Jul 1]. Available from: https://acsearch.acr.org/list

8. ESR iGuide | Imaging Referral Guidelines published by the European Society of Radiology [Internet]. [cited 2021 Jul 9]. Available from: https://www.esriguide.org/

9. Kim RJ, Shah DJ, Judd RM. How We Perform Delayed Enhancement Imaging. *J Cardiovasc Magn Reson*. 2003;5(3):505–14.

10. Polacin M, Gastl M, Kapos I, Eberhard M, Weber L, Gotschy A, et al. Novel Magnetic Resonance Late Gadolinium Enhancement With Fixed Short Inversion Time in Ischemic Myocardial Scars. *Invest Radiol*. 2020 Jul;55(7):445–50.

11. Busby LP, Courtier JL, Glastonbury CM. Bias in Radiology: The How and Why of Misses and Misinterpretations. *RadioGraphics*. 2018 Jan 1;38(1):236–47.

12. Gray H. *Gray's Anatomy: Classic Illustrated Edition*. 15th ed. New York: Lewes: Barnes & Noble Inc; 2013.

13. Netter F. *Atlas of Human Anatomy*. 7th edition. Philadelphia, PA: Elsevier; 2018. 640 p.

14. Anderson MW, Keats TE. *Atlas of Normal Roentgen Variants That May Simulate Disease: Expert Consult*. 9th edition. Philadelphia, PA: Saunders; 2012. 816 p.

15. Triche BL, Nelson JT, McGill NS, Porter KK, Sanyal R, Tessler FN, et al. Recognizing and Minimizing Artifacts at CT, MRI, US, and Molecular Imaging. *RadioGraphics*. 2019 Jul 1;39(4):1017–8.

16. Feldman MK, Katyal S, Blackwood MS. US Artifacts. *RadioGraphics*. 2009 Jul 1;29(4):1179–89.

17. Barrett JF, Keat N. Artifacts in CT: Recognition and Avoidance. *RadioGraphics*. 2004 Nov 1;24(6):1679–91.

18. Zhuo J, Gullapalli RP. MR Artifacts, Safety, and Quality Control. *RadioGraphics*. 2006 Jan 1;26(1):275–97.

19. Dietrich CF, Mathis G, Blaivas M, Volpicelli G, Seibel A, Atkinson NS, et al. Lung Artefacts and Their Use. *Med Ultrason*. 2016 Dec 5;18(4):488–99.

20. Blüthgen C, Sanabria S, Frauenfelder T, Klingmüller V, Rominger M. Economical Sponge Phantom for Teaching, Understanding, and Researching A- and B-Line Reverberation Artifacts in Lung Ultrasound. *J Ultrasound Med*. 2017;36(10):2133–42.

21. European Society of Radiology 2009. The Future Role of Radiology in Healthcare. *Insights Imaging*. 2010 Jan 1;1(1):2–11.

22. Dunnick NR, Applegate K, Arenson R, Levin D. Training for the Future of Radiology: A Report of the 2005 Intersociety Conference. *J Am Coll Radiol*. 2006 May 1;3(5):319–24.

23. Hartung MP, Bickle IC, Gaillard F, Kanne JP. How to Create a Great Radiology Report. *RadioGraphics*. 2020 Oct 1;40(6):1658–70.

24. Linton O. Consulting. *Acad Radiol*. 2004 May 1;11(5):602.

25. Cornelis F-H, Solomon S-B. Image Guidance in Interventional Radiology: Back to the Future? *Diagn Interv Imaging*. 2020 Jul;101(7–8):429–30.

26. Alkadhi H, Euler A. The Future of Computed Tomography: Personalized, Functional, and Precise. *Invest Radiol*. 2020 Sep;55(9):545–55.

27. Martini K, Baessler B, Bogowicz M, Blüthgen C, Mannil M, Tanadini-Lang S, et al. Applicability of Radiomics in Interstitial Lung Disease Associated with Systemic Sclerosis: Proof of Concept. *Eur Radiol*. 2021;31(4):1987–98.

28. Bosmans JM, Luyckx E, Broeckx BJ, Ceyssens S, Parizel PM, Snoeckx A. Radiologists as Co-Authors in Case Reports: Does Their Involvement Make a Difference? *Acta Radiol*. 2020 Mar;61(3):338–43.

6 Technological View

Abhishta Bhandari

Artificial intelligence (AI) has the potential to greatly transform radiology, and medicine at large; however, it currently remains a field that has yet to be translated fully into the clinical realm. Radiologists will play a key role in the optimal translation and adoption of AI into healthcare. Here, we describe specific clinical tasks AI could perform, the relevance of these task, and how AI can ultimately be integrated into clinical decision-making.

As described previously, tasks that an AI algorithm can perform include classification (for example, classifying an unknown lesion into a particular tumor subtype), detection (for example, localizing pulmonary nodules or intracranial hemorrhage), and segmentation (for example, delineating brain tumors in MRI).

For many of these tasks, AI can make the mundane more bearable in clinical practice. For example, automatic pulmonary nodule detection can save the radiologist from combing through hundreds of slices on a chest CT for small nodules. For other tasks, there is potential to augment the radiologists' clinical decision-making or provide added clinical value to the referrer, such as radiogenomics. For example, isocitrate dehydrogenase mutation status in brain tumors has implications for treatment decisions and prognostication. Other clinically relevant tasks that have been defined in the literature include differentiating true disease progression from pseudo-progression in various cancers. There will be important implications for these algorithms in multi-disciplinary meetings with other specialists and stakeholders (Figure 6.1).

A major current barrier to AI implementation is trust, or rather lack thereof, in the predictions, made more difficult by the black-box nature of many of these algorithms. The radiologist needs to be aware of the technical limitations of algorithms and how these affect the final prediction. Just as blood tests have imperfect sensitivity and specificity, so do AI algorithms. Guidelines have been released by various American, European, and Australian radiological societies, for example, and, interestingly, Microsoft has also recently released a guideline that may be useful in a hospital setting.

The other issue is trust in radiologists if they adopt these algorithms in routine clinical practice. How clinicians from other specialties will react if these algorithms are adopted in clinical practice is unknown. Also, how this affect patient care, the legality of using "subpar" algorithms, and whether clinicians will trust the judgment of a machine. If it were to be adopted, the initial stages will need radiologists to become advocates of its use to combat questions regarding its validity.

Radiologists are likely to be data integrators as well as clinicians. This is inherent in the output of the algorithms that aim to solve specific tasks. Outputs include an AUC and diagnostic test accuracy statistics (such as sensitivity, specificity, negative predictive value, positive predictive value, and accuracy). The data output of the algorithms will be integrated with the signs the radiologist sees within the image. For example, take brain tumors: there exists the T2 mismatch sign which is highly specific for genetic status of IDH positive, 1p19q-codel glioma (100%) with a moderate sensitivity (42%), therefore this sign is good at "ruling in" tumors. This may be combined with a radiomic/deep learning-based algorithm determining genetic status. So, the radiologists will have the visual feature on his screen, with a set of diagnostic statistics on the other screen and use this in combination with clinical acumen to render a final genetic diagnosis. Other roles of AI algorithms would be in patient counseling given that algorithms can prognosticate brain tumors. For example, the BraTS challenge has looked at predicting survival of brain tumors through their collated brain MRI database.

In the future patients may request to use an algorithm to predict the nature of their tumor. The advantage of AI algorithms is that in tumors they can provide a "virtual biopsy" which negates the need for invasive biopsy which comes with its own risks such as bleeding, infection, tumor seeding, etc. Thus, patients in the future may request to use the tumor images to predict survival, grade, and genetic status of their tumor during their treatment process. Another advantage of AI algorithms is that they can be used in resource-poor settings, where biopsy is not feasible and there is a movement to make the code of these algorithms free to use, within the medical science literature.

There are, however, disadvantages to the use of these algorithms. There is the thought that AI algorithms will make radiologists faster and more efficient at their job. However, does this translate to a better working environment for the radiologist? It is likely more scans will need to be read

DOI: 10.1201/9781003095279-8

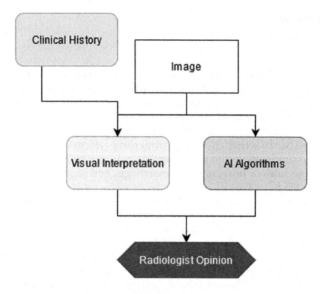

Figure 6.1 Radiologist decision-making algorithm

at a more rapid rate given that practice administration will expect a return on investment to justify the use of the software. For example, if an algorithm is used to automate protocolling of scans, it is likely that the protocolling radiologist/trainee is to be assigned to read more scans throughout the day. Another case is that if reading X-rays are automated with possible pathology already highlighted by the algorithm, it is likely that more X-rays are expected to be read. This may have benefits since many X-rays are read by the ordering clinicians (for example, orthopedic doctors read their own X-rays before even considering what is going on in a patient, emergency department (ED) physicians read their own chest X-rays during resuscitation) given there may be delays in reporting. This means radiologists are freed up to do other tasks.

   Given the implications of AI in radiological decision-making and the vast potential for change in clinical practice, the radiologist role will also be to facilitate the uptake of the technology into clinical practice. For example, radiologists may have an essential role in task definition, labeling of ground truths and steering the research into practical clinical applications. The combination of human and machine may drive more accurate decision-making and provide more objective clinical opinions and make the radiology profession more advanced in this age of rapidly developing healthcare.

BIBLIOGRAPHY

Amershi S, Weld D, Vorvoreanu M, Fourney A, Nushi B, Collisson P, et al., editors. Guidelines for Human-AI Interaction. Proceedings of the 2019 Chi Conference on Human Factors in Computing Systems; 2019.

Baid U, Ghodasara S, Bilello M, Mohan S, Calabrese E, Colak E, et al. The RSNA-ASNR-MICCAI BraTS 2021 Benchmark on Brain Tumor Segmentation and Radiogenomic Classification. 2021. http://arxiv.org/abs/2107.02314.

Bakas S, Akbari H, Sotiras A, Bilello M, Rozycki M, Kirby JS, et al. Advancing The Cancer Genome Atlas glioma MRI Collections with Expert Segmentation Labels and Radiomic Features. *Scientific Data*. 2017;4:170117.

Bhandari A, Purchuri SN, Sharma C, Ibrahim M, Prior M. Knowledge and Attitudes Towards Artificial Intelligence in Imaging: A Look at the Quantitative Survey Literature. *Clinical Imaging*. 2021;80:413–419.

Bhandari AP, Liong R, Koppen J, Murthy SV, Lasocki A. Noninvasive Determination of IDH and 1p19q Status of Lower-grade Gliomas Using MRI Radiomics: A Systematic Review. *AJNR Am J Neuroradiol*. 2020;42(1):94–101.

Lasocki A, Rosenthal MA, Roberts-Thomson SJ, Neal A, Drummond KJ. Neuro-Oncology and Radiogenomics: Time to Integrate? *Am J Neuroradiol*. 2020;41(11):1982–1988.

Menze BH, Jakab A, Bauer S, Kalpathy-Cramer J, Farahani K, Kirby J, et al. The Multimodal Brain Tumor Image Segmentation Benchmark (BRATS). *IEEE Trans Med Imaging*. 2015;34(10):1993–2024.

Mongan J, Moy L, Charles E. Kahn J. Checklist for Artificial Intelligence in Medical Imaging (CLAIM): A Guide for Authors and Reviewers. *Radiol Artif Intell*. 2020;2(2):e200029.

Qian X, Tan H, Zhang J, Zhao W, Chan MD, Zhou X. Stratification of Pseudoprogression and True Progression of Glioblastoma Multiform Based on Longitudinal Diffusion Tensor Imaging without Segmentation. *Med Phys*. 2016;43(11):5889.

Soffer S, Ben-Cohen A, Shimon O, Amitai MM, Greenspan H, Klang E. Convolutional Neural Networks for Radiologic Images: A Radiologist's Guide. *Radiology*. 2019;290(3):590–606.

7 Societal View

Krithika Rangarajan and Devasenathipathy Kandasamy

INTRODUCTION

Radiology is a fairly young specialty; much of the work radiologists perform today has evolved in just the last 50 years. Much of what is done in modern radiology has evolved from tasks which were not possible to do prior to the evolution of specific imaging modalities. As a result, radiologists have had to carve out their niche from roles distinct from other traditional medical specialties and have shuttled between the extremes of being seen as a paraclinical assistant on one hand and a highly specialized clinician taking important decisions in diagnosis and treatment on the other.

The history of radiology as a specialty has been punctuated at regular intervals by groundbreaking technological advances, which have increased the complexity of the work performed. As a result, the roles played by radiologists have also been altered with each technological advance. From the era of interpreting radiographs and double contrast barium exams only to rule out gross pathologies, radiological exams are now an important part of patient work-up at every stage. The unequivocal rise of radiology, however, has not been paralleled by an equally unequivocal rise in the importance of the radiologist, as several times in the past traditional clinical specialties have attempted to incorporate medical imaging within their own domain. While this has been rendered difficult due to the constant evolution of complexity in performing and interpreting these images, the fact that traditionally radiology is not a patient-facing field makes the work of the radiologist largely invisible to the patient.

While the past was replete with advancements in hardware and methods of medical imaging, which often added complexity to medical imaging, software advances which are aimed at making image interpretation more objective and easier will drive future advancements. So, would such improvement in the ease of image interpretation make the radiologist obsolete? Would software advances now bring imaging interpretation closer to incorporation into respective clinical subspecialties? Much of all this depends not only on the added value the radiologist can bring to accurate image interpretation, but also on patient perception of the role of the radiologist.

PATIENT PERCEPTION OF THE RADIOLOGIST

Traditionally, a radiologist communicates with the referring physician as well as with the patient by means of a radiology report. Radiology teaching has also revolved around making the report complete, objective, legally proven, and capable of answering questions which have a direct bearing on patient management. However, radiologists and referring physicians alike look at the radiology report as being aimed at the treating physician, enabling them to take a decision, rather than helping the patient understand what their condition might be. As a result, in most medical practices across the world, no direct communication exists between patients and radiologists during most diagnostic investigations.

It is therefore not surprising that patients are often unaware that radiologists are trained physicians, and that image interpretation is performed by the radiologist, not the referring clinician. In a survey performed among patients attending a breast clinic, an overwhelming 76% of patients thought radiologists were technical staff, and 40% did not believe radiologists had a significant role to play in their management [1]. A campaign carried out by the American College of Radiology, called the "Face of Radiology," also showed that the general public was largely unaware that radiologists are trained physicians [2]. On the other hand, in surveys pertaining to patient satisfaction with radiology services, patients have expressed a desire to better understand their diagnostic procedures. In a survey conducted in 40 countries, 36% of patients expressed their dissatisfaction at the non-availability of radiologists for consultation at the time of their scan [3]. Studies have indicated that patients often have a poor understanding of possible risks involved during radiological scans such as from radiation exposure and potential contrast reactions, and this lack of communication leads to dissatisfaction with healthcare [4]. More recent studies show that 85% of patients wish to meet the radiologists interpreting their scans, and up to 60% of them are also willing to pay an extra fee for the consultation [5].

This lack of patient awareness is not restricted to individuals but is also propagated by popular media. News channels often report on radiology-related health problems of famous personalities,

DOI: 10.1201/9781003095279-9

but only call upon non-radiologist clinicians for comment [6]. Radiologists are often non-existent in most popular television shows, or depicted in roles played by technologists, and, even worse, in some cases even depicted as specialists who are unhelpful to patients, unwilling to be available for emergencies [7].

The lack of awareness of the role of radiologists is further carried forward into the realm of law. For example, the PC-PNDT law in India, which aims to curb sex determination of the fetus in order to reduce the female feticide that is unfortunately prevalent in the country, equates radiologists with any registered medical practitioner. A provision of the law (though later stayed by the court) disallowed radiologists from visiting multiple imaging centers in order to curb the practice of prenatal sex determination, while placing no such restriction upon gynecologists and obstetricians who actually perform medical abortions[8].

RADIOLOGISTS PERCEPTION OF THEIR ROLE IN HEALTHCARE

Radiologists are largely content with their practice, in that radiology continues to rank top among the most sought-after and highest-paid specialties in medicine [9]. Surveys showed that despite long working hours, radiologists tend to enjoy being at the forefront of technological innovation. This satisfaction with current practice patterns perhaps explains why many radiologists do not find it necessary to fill existing gaps in patient perception by interacting directly with the patient. In one survey it was found that while 76% of patients wanted to meet a radiologist immediately after a diagnostic investigation, only about 24% actually did [10]. Many senior radiologists perceive themselves as providing services exclusively to the referring clinician, rather than to the patient.

This attitude, however, has undergone significant change in the past decade. A drive toward "patient-centric" radiology is emerging to the forefront [11]. Surveys have been conducted and numerous articles have been written emphasizing the need for radiologists to play more active roles in not only accurately interpreting scans but also in patient education, patient selection for specific imaging modalities, and conveying the results of these investigations to patients [12]. As medicine is increasingly focused on patient satisfaction, there is an increasing need to assess radiology services not just in terms of interpretative accuracy, but also in terms of patient satisfaction. Patient-centric radiology lies at the forefront of ensuring that patient services make for a good end-to-end experience [13].

The evolving role of increased radiologist–patient interaction, however, does not garner universal acceptance among radiologists. The spectrum of beliefs among radiologists ranges from those who believe one must never meet the patient, to those who believe one occasionally conveys findings to patients, depending on the findings and patient preferences, to some radiologists who believe every report must be conveyed to patients [14]. In addition to variability in radiologists' perceptions, operational difficulties also limit the execution of patient-facing radiology. For instance, how is the time "penalty" involved in explaining findings to patients in a busy radiology department recognized and compensated? Does the improvement in patient satisfaction justify lower radiologist productivity as measured by reporting metrics? How are radiologist consultations reimbursed? Does the consultation and patient interaction equate to no added monetary benefit for the radiologist? Would insurance companies find such additional compensation acceptable? Finally, not all patients' questions are answerable by a radiologist. So, what should a radiologist do if the patient wants to know about prognosis and treatment options? Or if a clinical question cannot be adequately answered by imaging alone? While most radiologists acknowledge the need to find solutions for these problems, many believe patient-centric radiology is a step in the right direction.

EVOLUTION OF THE ROLE OF THE RADIOLOGIST: Imaging 3.0

The birth of radiology enabled medical practitioners to look inside the human body and avoid invasive procedures. The earliest form of radiology that evolved soon after the discovery of X-rays was restricted to ruling out gross pathologies. Much of the interpretation was left to the imagination of the radiologist, and anything more extensive than basic evaluation can be fraught with errors. This was the era where radiologists worked in dark rooms illuminated only by view-boxes, where radiographs may take hours to process before it could be interpreted. Radiologists were only visited by other doctors for occasional discussions, and radiology was seen as a paraclinical field, and indeed many debated the need for a specialized medical practitioner to take up this field.

This restricted role of radiology, however, changed with technological advances. With CT scans, accuracy of interpretation improved significantly, and much more could be visualized to reach a specific diagnosis, rather than just exclude pathologies. The advent of magnetic resonance imaging (MRI) also enabled new dimensions in accurate diagnosis. It was popularly believed that the advent of MRI would make image interpretation very easy, and that the clinician would be able to interpret the images with such little doubt that radiologists would become superfluous. But as imaging technology continued to evolve, imaging protocols became complex, with multiphase studies, different types of oral and intravenous contrast agents, and multiple different imaging sequences and planning cross-sectional imaging for a patient by itself proliferated into a complex proposition. With improvement in acquisition times, more studies could now be scheduled, and each study generated hundreds and thousands of images to interpret. With improvements in accuracy, indications for imaging also expanded quickly, and imaging became a frequently indispensable part of the patient's work-up for most clinical specialties. The radiologist, however, went from interpreting a few images in an hour, to up to thousands of images per patient. The increase in workload meant the radiologist was essentially tied to viewing and interpreting images with little time for anything else.

This invisibility of the radiologist was exacerbated by improvement in telecommunications, where reports could be relayed to patients and physicians without need for interaction with the radiology department. While teleradiology and telecommunication vastly helped in reducing patient visits, and in making imaging more easily accessible, it moved the radiologist even further away from the patient.

The American College of Radiology began dissemination of a strategic initiative called "Imaging 3.0." Through the initiative they recognize and propagate the evolution of radiology as a field, from the time of discovery of X-rays by Roentgen in 1895 (Imaging 1.0), on to the times of rapid proliferation in complex imaging modalities, including CT, MRI, Ultrasound, and PET (Imaging 2.0), to the present day (Imaging 3.0). The initiative emphasizes bringing stakeholders together to ensure that the imaging experience is optimal for the patient [15].

With this evolution of radiology, there is continued emphasis on imaging, and indeed radiologists continuing to play hugely important roles in patient management. Much of this currently happens behind the scenes. Further technological developments would be driven by whether radiologists as a community choose to harness the power of technology to further increase throughput of patients, or whether radiologists choose to focus on patient-centered changes in practice workflow. Radiologists are gatekeepers to imaging facilities, and whether they would play a more active role in ensuring radiation safety, public health, and quality improvement would now be determined by the practice patterns they choose [16]. Figure 7.1 shows how the work of the radiologist has changed from the past to the present, and how it may change in future.

MASTER DIAGNOSTICIAN OR SIGNATORY AUTHORITY? WHAT DOES THE FUTURE LOOK LIKE? OPINION

Radiology today stands at a cusp. An era of rapid technological advances made imaging complex and enabled patient screening, complex diagnosis using imaging, and even imaging-directed therapy. Technological developments in the future, however, promise to simplify and aid interpretation, make it more objective, as well as promise to deliver more information out of images.

A mix of initial heightened expectations and fear led many to believe that radiology as a field would be absorbed into clinical subspecialties, facilitated by increasing availability of imaging decision support provided by advances in information technology (IT) and artificial intelligence (AI). It was famously said by Geoffery Hinton, a pioneer in AI, in 2016 that "it is obvious that we should stop training radiologists now." Subsequent measured analysis highlighted several important practical considerations, such as who takes the blame when the software goes wrong? However, the question then arises as to whether radiologists would then be reduced to signatory authorities who would in reality remain only to take the blame in case of misdiagnosis.

It is, however, widely recognized now that the task of a radiologist goes much beyond what an AI algorithm can currently perform. AI provides a narrow intelligence specific to a task it is trained to perform, unlike a radiologist who can adapt to looking to answer questions unique to a particular clinical scenario. In addition, AI, radiomics and similar advances in information technology are likely to bring in an entirely new dimension in image analysis. Much like new modalities for imaging, they would work within constraints. It would then be essential for radiologists to know what may influence the decisions of such systems, similar to the requirement to understand imaging artifacts from MRI and CT scanners. Reports would then become more complicated,

Figure 7.1 The past, present, and future of the radiologist. The role of the radiologist has evolved from ruling out major pathologies on conventional 2D imaging to interpreting complex cross-sectional imaging. However, this evolution has largely taken place behind the scenes, and patients continue to have limited interaction with radiologists. While technological advances promise to make image interpretation more objective and faster, time would determine if these advances would make the radiologist more visible to the patient, or would they be instead be largely be used to increase radiologist output. Created with BioRender.com

including such possibilities as response to therapy and presence of genetic mutations, adding to complexity and additional dimensions to the existing job of image interpretation.

Indeed, information technology is capable of providing some solutions that radiologists seek in order to facilitate meeting patients [17]. For example, techniques such as natural language processing (NLP) and large language models can help generate a report in lay language that the patient can understand. Object detection networks trained for medical imaging questions can potentially reduce the time spent by the radiologist in detecting findings, thus freeing up time for patient interaction. Auto-generated editable reports can potentially reduce time spent in typing/dictating reports, and thus facilitate making the report immediately available after an imaging study. IT may thus provide opportunities to radiologists to realize patient-centric radiology and enable a shift in perception among patients of radiologists as being paraclinical image interpreters to being physicians who hold the key to their diagnosis. Figure 7.2 shows how the workflow of such a radiology department may be structured in the future. The direction of evolution of this change, however, depends on a tug-of-war between patient and clinician expectations on one hand and financial pressures on radiologists to perform more imaging studies on the other. Which direction these forces take radiologists in will be determined by where radiologists stand in this tug-of-war and how they choose to utilize the saved time and improved diagnostic confidence that increased use of computer-assisted tools promise to deliver.

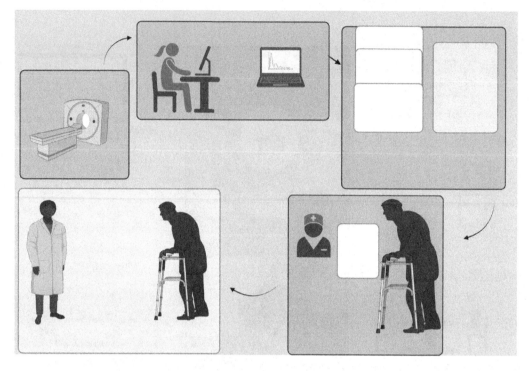

Figure 7.2 Potential workflow of radiology departments. Advances in information technology and artificial intelligence promise to make image interpretation faster, and automate some processes. Harnessing these capabilities can potentially enable a work-flow where the radiologist becomes more visible and enable patients to better understand their imaging studies. Created with BioRender.com

REFERENCES

1. O'Mahony N, McCarthy E, McDermott R, O'Keeffe S. Who's the doctor? Patients' perceptions of the role of the breast radiologist: a lesson for all radiologists. *Br J Radiol.* 2012 Dec;85(1020):e1184–9.

2. Neiman HL. Face of Radiology campaign. *AcadRadiol.* 2009 May;16(5):517–20.

3. Fuchsjäger M, Derchi L, Hamm B, Brady AP, Catalano C, Rockall A, et al. Patient survey of value in relation to radiology: results from a survey of the European Society of Radiology (ESR) value-based radiology subcommittee. *Insights Imaging.* 2021 Jan 7;12(1):6.

4. Baumann BM, Chen EH, Mills AM, Glaspey L, Thompson NM, Jones MK, et al. Patient perceptions of computed tomographic imaging and their understanding of radiation risk and exposure. *Ann Emerg Med.* 2011 Jul 1;58(1):1–7.e2.

5. Domina JG, Bhatti ZS, Brown RKJ, Kazerooni EA, Kasotakis MJ, Khalatbari S. Journal CLUB: Patient perception of radiology and radiologists: A survey analysis of academic and community institutions. *Am J Roentgenol.* 2016 Aug 4;207(4):811–9.

6. Mamlouk MD. Using the media to improve the public's perception of radiologists. *J Am Coll Radiol.* 2014 Jan 1;11(1):6.

7. Gunderman RB, Mortell KE. Radiologists on television. *J Am Coll Radiol JACR.* 2009 Mar;6(3):144–6.

8. Onkar P, Mitra K. Important points in the PC-PNDT Act. *Indian J Radiol Imaging*. 2012;22(2):141–3.

9. Survey: Radiologists are Happy at Work [Internet]. Everything rad. 2014 [cited 2021 Aug 4]. Available from: https://www.carestream.com/blog/2014/04/16/survey-radiologists-are-happy-at-work/

10. Radiologist, Meet Patient [Internet]. Diagnostic imaging. [cited 2021 Aug 4]. Available from: https://www.diagnosticimaging.com/view/radiologist-meet-patient

11. Itri JN. Patient-centered Radiology. *RadioGraphics*. 2015 Oct 1;35(6):1835–46.

12. Gunn AJ, Mangano MD, Choy G, Sahani DV. Rethinking the role of the radiologist: Enhancing visibility through both traditional and nontraditional reporting practices. *RadioGraphics*. 2015 Mar 1;35(2):416–23.

13. Towbin AJ. Customer service in radiology: Satisfying your patients and referrers. *RadioGraphics*. 2018 Oct 1;38(6):1872–87.

14. Smith JN, Gunderman RB. Should we inform patients of radiology results? *Radiology*. 2010 May 1;255(2):317–21.

15. Ellenbogen PH. Imaging 3.0: what is it? *J Am Coll Radiol JACR*. 2013 Apr;10(4):229.

16. Knechtges PM, Carlos RC. The evolving role of the radiologist within the health care system. *J Am Coll Radiol JACR*. 2007 Sep;4(9):626–35.

17. Rubin DL. Informatics methods to enable patient-centered radiology. *AcadRadiol*. 2009 May 1;16(5):524–34.

8 Financial View

Youngmin Chu

Radiology serves patients in a unique way. Radiology is an essential part of patient care, tightly integrated to all aspects of medicine including screening, diagnosis, treatment, and surveillance. Despite being an integral part of the entire spectrum of patient care, personal contacts with patients are limited when compared to other medical specialties in most of the diagnostic radiology services. Furthermore, imaging studies are ordered by clinical colleagues, where radiologists play little-to-no role in an initial ordering process. Radiologists are also tasked with managing technically sophisticated equipment as well as minimizing potential harm that may be caused from imaging.

When it comes to financial aspects of medicine, Centers for Medicare & Medicaid Services (CMS) rules play an important role as most payment plans take some variation of CMS rules. Value-based care has been progressively emphasized in the last few decades in medicine. For example, financial aspects of medicine developed many strategies in an attempt to capture the value of patient care rather than a number of services. Currently, value-based care is implemented as part of a quality payment program including the Merit-based Incentive System (MIPS), incentivizing practices following quality measures that have plans to be updated in 2024 by integrating patient experiences as well [1, 2].

The ultimate goal of CMS is population-based care where a single lump sum of funding is given to a group of healthcare professionals who provide value-based health care. This concept raises concerns for radiology because having a finite amount of funding for the population will decrease the volume of imaging because of associated costs, thereby reducing care for the population. Furthermore, the recent 2021 CMS final physician fee schedule (PFS) is projected to include a significant reduction of imaging reimbursement [2]. To survive this turmoil, multiple strategies have been proposed with promising results. For example, one institution hired a group of additional staff to assist imaging authorization to prevent write-offs, which not only saved millions of dollars for the institution by decreasing authorization write-offs, but it also increased ordering office satisfaction as well as decreased incidences of authorization clarification between physicians [3].

Value can be expressed as quality over cost where quality further divides patient outcome and patient experience [4]. In radiology, most of the work product includes radiology reports and a radiology report value that can be expressed as follows: {(Accurate + Useful) × (Clear + Concise + Timely)} / Cost [5]. Each component of this equation can provide an opportunity of improvement, thereby increasing the value of radiology.

Accuracy may also be improved by peer consultation, continued education of radiologists, as well as proofreading reports to decrease errors [5]. The usefulness of radiology reports increases when the report provides an evidence-base for the next steps [5]. Concise reports can also increase efficiency by the use of systematically organized templates and limiting radiologic signs [5]. Improving clarity of reports may be achieved by using well-defined lexicons such as the Breast Imaging Data and Reporting System or unambiguous, widely understood terminology [5]. To promote timely reporting, radiologists can utilize standardized reports, thereby allowing succinct communication of normal and abnormal findings [5].

Additional propositions can be made in light of the recent 2021 CMS final PFS: facilitating appropriate imaging, providing guides for follow-up or surveillance imaging, optimizing screening programs, generating billing friendly reports, and reducing repeat imaging.

Facilitating appropriate imaging has been a long-standing challenge for radiologists as the ordering providers are usually non-radiologists who may not be able to choose the best possible imaging for a given patient. Inappropriate imaging not only increases the cost for patient care, but it also increases unnecessary radiation exposure and patient inconvenience. The financial aspect of this challenge has escalated such that CMS reimbursement will no longer be available in the near future unless clinicians provide documentation of an appropriate use criteria (AUC) consultation before ordering the imaging [6]. Evidence-based resources have been generated including ACR® appropriateness criteria [7], and more recently, R-SCAN® where ordering providers can take case-based online lessons and provide relevant patient education information [8].

In addition to the initial diagnostic imaging, follow-up and surveillance imaging can potentially add to the cost whilst adding little-to-no value to the outcome. For example, incidental findings without specific recommendations for follow-up guidance in radiology reports can generate

DOI: 10.1201/9781003095279-10

unnecessary imaging. Numerous evidence-based algorithms are also available in literature including white papers from ACR®, which radiologists can include in their reports, guiding ordering physicians on how to request additional workups [9]. Similarly, ill-timed or poor preparation for surveillance imaging may decrease value by negatively affecting patient experience as well as adding to the total cost.

"Prevention is better than a cure." This phrase captures the clinical importance of preventative medicine, especially in a population-based health care era. By the same token, image screening can assist with early detection of diseases, thereby leading to less invasive and less costly treatments. The cost effectiveness of imaging screening programs has been postulated [10, 11]. Refining patient selection criteria, as well as developing more efficient ways of utilizing image screening tools, can maximize the benefit of screening that can also be addressed by radiologists.

Despite prolonged education requirements to become a medical practitioner, little is known about billing. Most of the professional component reimbursements in radiology come from specific information or diagnosis in the radiology report. Thus, it is essential to include key information for billing to ensure proper coding and reimbursement [12]. One of the ways to alleviate this process is to use standardized templates with necessary information sections for billing, some of which have been shown to be successfully increase compliance [13, 14].

The need for higher-level care, seeking second opinions, or regional centralized care are but a few examples that may involve patient transport from one institution to another. Repeat imaging can be ordered when the initial imaging is not readily available at the transfer center, benefiting little to the patient's outcome yet increasing total cost. Enabling availability of inter-facility imaging is a difficult task but can increase the value of care in the long term.

Radiology has always been adapting to new challenges such as technological advances, development in new imaging modalities, and creating quality measures for our practice. Providing an accurate, clear, and concise radiology report with evidence-based recommendations in a timely manner will increase the quality of our reports. The cost of radiology as a whole can be decreased by facilitating appropriate imaging, providing evidence-based guides for follow-up and surveillance imaging, optimizing screening programs, and reducing repeat imaging. As such, the financial scope of providing radiology services continues to be in flux as new laws, technologies, and cost-saving practices are developed.

REFERENCES

1. Golding LP. The future of MIPS 2021 quality payment program update [web streaming video]. Online: Radiology Society of North America; 2020 December 5 [cited 2020 February 18]. Available from: rsna2020.rsna.org

2. American College of Radiology. ACR® detailed MPFS final rule summary covers payment policies. [Internet]. Reston Virginia: ACR®; 2020 December 9 [Updated 2020 December 9; Cited 2021 February 18]. Available from: https://www.acr.org/Advocacy-and-Economics/Advocacy-News/Advocacy-News-Issues/In-the-Dec-12-2020-Issue/ACR-Detailed-MPFS-Final-Rule-Summary-Covers-Payment-Policies

3. Brink J. Financial outlook of a large academic department [web streaming video]. Online: Radiological Society of North America; 2020 December 5 [cited 2020 February 18]. Available from: rsna2020.rsna.org

4. Chen M. Getting paid for population health in radiology [web streaming video]. Online: Radiology Society of North America; 2020 December 5 [cited 2020 February 18]. Available from: rsna2020.rsna.org

5. Eberhardt SC, Heilbrun ME. Radiology report value equation. *Radiographics*. 2018 Oct;38(6) 1888–96.

6. Centers for Medicare & Medicaid Services. Appropriate Use Criteria Program. [Internet]. Baltimore Maryland: CMS; 2020 August 12 [Updated 2020 August 12; Cited 2021 Feb 21]. Available from: https://www.cms.gov/Medicare/Quality-Initiatives-Patient-Assessment-Instruments/Appropriate-Use-Criteria-Program

7. American College of Radiology. ACR® Appropriateness criteria® overview [Internet]. Reston Virginia: ACR®; 2021 January 28 [Updated 2021 January 28; Cited 2021 February 18]. Available from: https://www.acr.org/Clinical-Resources/ACR-Appropriateness-Criteria

8. American College of Radiology. R-SCAN® Radiology support, communication and alignment network [Internet]. Reston Virginia: ACR®; 2020 November 25 [Updated 2020 November 25; Cited 2021 February 18]. Available from: https://www.acr.org/Practice-Management-Quality-Informatics/rscan

9. American College of Radiology. Incidental Findings. [Internet]. Reston Virginia: ACR®; 2019 July 2 [Updated 2019 July 2; Cited 2021 February 21]. Available from: https://www.acr.org/Clinical-Resources/Incidental-Findings

10. Mohan G, Chattopadhyay S. Cost-effectiveness of leveraging social determinants of health to improve breast, cervical, and colorectal cancer screening: A systemic review. *JAMA Oncol.* 2020 Sep 1;6(9):1434–44.

11. Raymakers AJ, Mayo J, Lam S, et al. Strategies using low-dose computed tomography: A systematic review. *Appl Health Econ Health Policy.* 2016 Aug;14(4): 409–18.

12. Thorwarth Jr WT. Get paid for what you do: Dictation patterns and impact on billing accuracy. *J Am Coll Radiol.* 2005 Aug;2(8):665–9.

13. Kohli M, Schonlau D. Radiology quality measure compliance reporting: An automated approach. *J Digit Imaging.* 2016 Jun:29(3): 297–300.

14. Wang KC, Patel JB, Vyas B, et al. Use of radiology procedure codes in health care: The need for standardization and structure. *Radiographics.* 2017 Jul–Aug;37(4)1099–110.

PART III

WHAT IS AI?

9 Clinical View

Christian Federau

Artificial intelligence can be understood as computer methods that perform tasks normally requiring human intelligence [1]. Machine learning is one type of artificial intelligence wherein computer algorithms are trained to perform certain tasks from existing data without being explicitly programmed to perform those tasks. Deep learning is a form of machine learning using computer models composed of multiple processing learning to represent data with multiple levels of abstraction [2].

Deep learning discovers structure in large data sets by computing an internal representation of the data in each layer from the representation in the previous layer of some network [2]. Deep learning develops by using the backpropagation algorithm to modify the parameters used to perform this computation, by minimizing a loss function that is set for a given purpose. The deeper layers of the networks typically contain a more conceptual representation of the information contained in the data. The use of convolutional layers in deep learning networks dramatically improved results in several fields, particularly in image processing, such as in semantic segmentation tasks of radiological images. Deep learning methods have also been decisively driven by the increasing computational power of graphics processing units, which were originally developed for video gaming, and the increasing availability of very large datasets made possible by the increasing digitalization of data throughout the world.

A further advantage of deep learning–based algorithms is their simplicity of use. For decades, conventional machine learning methods have tried, with limited success, to extract features and information using domain expertise and cumbersome engineering. Deep learning networks can extract those features with a straightforward design: all it requires is a well-labeled database, an adequate network, and a loss function designed for the expected outcome.

Deep learning methods can be classified into *supervised methods*, where some "ground truth" has been defined and is used to train the algorithms, and *unsupervised methods*, where the computer itself determines some internal structure of the data. Deep learning methods can be used to develop *predictive models* or *generative models.* Predictive models can be used, for example, to classify data in some specific diseases entity, or to segment structures on images, such as stroke volumes in diffusion-weighted images [3]. Generative models can be used, for example, to generate new images from noise [4], or to modify images in a specific way, for example, adding an image of a stroke in a normal diffusion-weight magnetic resonance image [5]. Variational autoencoders [6] and generative adversarial networks [7] are two dominant approaches among the multiple types of generative models that have been proposed and extensively studied in recent years. Both approaches have their own strengths and weaknesses. Variational autoencoders are known to generate blurry images, but might be easier to train, and generate more sample diversity than generative adversarial networks. Generative adversarial networks are known to be able to generate sharper images but might lack sample diversity and can be more difficult to train.

Currently, the most common roles for deep learning networks in the clinical setting are in tasks involving clinical decision support and medical imaging analysis [8]. Clinical decision support tools can help physicians make decisions about diagnosis, treatments, or medications by providing fast analysis of a large amount of information. In medical imaging, deep learning networks are being used on X-rays, computed tomography, magnetic resonance, or ultrasound images. Some examples of deep learning network–enabled imaging tasks include detecting lesions that a radiologist might miss, such as lung lesions visualized by computer tomography, fractures on skeletal X-rays, or multiple sclerosis lesions on magnetic resonance brain imaging.

Many interesting applications of AI have also been developed outside of the medical imaging field. One productive application of AI involves the recognition of relatively complex, but common, patterns in datasets that are too large to be correctly evaluated by a human being. One example is the detection of atrial fibrillation from ECG [9]. A particularly interesting application is in the permanent ambulatory monitoring of physiological parameters using wearable devices, allowing for the detection of abnormal patterns such as cardiac arrhythmia [10] or generalized epilepsy [11].

For medical image semantic segmentation tasks, the current state-of-the-art is the so-called U-Net [12]. Best in class results were obtained with U-Net architectures in a broad range of pathologies and image types [13], such as retinal diseases from optical coherence tomography [14], or

DOI: 10.1201/9781003095279-12

stroke segmentation from diffusion-weighted magnetic resonance images [3]. Attempts to improve results using architectural changes from the U-Net have so far failed to prove superior [13].

Despite these recent success, fully automated end-user applications remain limited in the daily clinic, and this is unlikely to radically change in the near future. First, deep learning methods require large datasets to train adequately, and access to such large datasets is often restricted for various reasons. Further, many medical conditions are constantly evolving, which renders the constitution of a representative dataset challenging. For example, new pathologies appear; disease incidences evolve with changes in collective population habits, environmental, and other external factors; and with the introduction of new treatment methods. Image quality generally improves through constant technical developments. Finally, while the attitude of the patients and the general public toward applications of artificial intelligence in medicine is currently positive, many, understandably, have reservations and favor application of AI with human supervision [15].

In many fields of machine learning applications, such as in autonomous driving, full replacement of the human is the key compelling goal, as such an accomplishment would help solve many problems in transportation, potentially leading to significant reductions in the number of accidents and allowing for cars to be utilized by several users during any given day, facilitating overall reductions in the number of automobiles and freeing up valuable spaces currently used for parking in cities. In contrast, full replacement of human activities in healthcare is currently undesirable, for various reasons, but in particular given the critical role of the doctor–patient relationship [16].

Nevertheless, there are numerous ways deep learning can, and will, positively impact the practice of medicine. Deep learning methods could generate great value in clinical workflow by preparing information for physicians, who would remain in control but are less encumbered by information overload. Artificial intelligence for the foreseeable future would support existing physicians and the methods currently in place, rather than replacing physicians [17]. Machine learning could also significantly impact medicine by speeding up the pace and the quality of research. Translation of these new methods to the clinics will remain nevertheless challenging and demonstrating that AI provides tangible direct benefits to patients in clinical care will be necessary [18].

REFERENCES

1. Zaharchuk G, Gong E, Wintermark M, Rubin D, Langlotz CP. Deep learning in neuroradiology. *AJNR Am J Neuroradiol*. 2018;39:1776–1784.

2. LeCun Y, Bengio Y, Hinton G. Deep learning. *Nature*. 2015;521:436–444.

3. Federau C, Christensen S, Scherrer N, et al. Improved segmentation and detection sensitivity of diffusion-weighted stroke lesions with synthetically enhanced deep learning. *Radiol Artif Intell*. 2020;2:e190217.

4. Hirte AU, Platscher M, Joyce T, Heit JJ, Tranvinh E, Federau C. Realistic generation of diffusion-weighted magnetic resonance brain images with deep generative models. *Magn Reson Imaging*. 2021;81:60–66.

5. Platscher M, Zopes J, Federau C. Image translation for medical image generation -- ischemic stroke lesions. arXiv:201002745 [cs, eess] [online serial]. Epub 2020 October 5. http://arxiv.org/abs/2010.02745. Accessed October 24, 2021.

6. Kingma DP, Welling M. An Introduction to variational autoencoders. *FNT in Machine Learning*. 2019;12:307–392.

7. Goodfellow IJ, Pouget-Abadie J, Mirza M, et al. Generative adversarial networks. arXiv:14062661 [cs, stat] [online serial]. Epub 2014 June 10. http://arxiv.org/abs/1406.2661. Accessed September 12, 2018.

8. Benjamens S, Dhunnoo P, Meskó B. The state of artificial intelligence-based FDA-approved medical devices and algorithms: An online database. *NPJ Digit Med*. 2020;3:118.

9. Halcox JPJ, Wareham K, Cardew A, et al. Assessment of remote heart rhythm sampling using the alivecor heart monitor to screen for atrial fibrillation: The REHEARSE-AF study. *Circulation*. 2017;136:1784–1794.

10. Turakhia MP, Desai M, Hedlin H, et al. Rationale and design of a large-scale, app-based study to identify cardiac arrhythmias using a smartwatch: The apple heart study. *Am Heart J*. 2019;207:66–75.

11. Regalia G, Onorati F, Lai M, Caborni C, Picard RW. Multimodal wrist-worn devices for seizure detection and advancing research: Focus on the Empatica wristbands. *Epilepsy Res*. 2019;153:79–82.

12. Ronneberger O, Fischer P, Brox T. U-Net: Convolutional networks for biomedical image segmentation. arXiv:150504597 [cs] [online serial]. Epub 2015 May 18. http://arxiv.org/abs/1505.04597. Accessed September 12, 2018.

13. Isensee F, Jaeger PF, Kohl SAA, Petersen J, Maier-Hein KH. nnU-Net: A self-configuring method for deep learning-based biomedical image segmentation. *Nat Med*. 2021;18:203–211.

14. De Fauw J, Ledsam JR, Romera-Paredes B, et al. Clinically applicable deep learning for diagnosis and referral in retinal disease. *Nat Med*. 2018;24:1342–1350.

15. Young AT, Amara D, Bhattacharya A, Wei ML. Patient and general public attitudes towards clinical artificial intelligence: A mixed methods systematic review. *The Lancet Digital Health*. 2021;3:e599–e611.

16. Norden JG, Shah NR. What AI in health care can learn from the long road to autonomous vehicles. *Epub*. 2022;3(2):8.

17. Hainc N, Federau C, Stieltjes B, Blatow M, Bink A, Stippich C. The bright, artificial intelligence-augmented future of neuroimaging reading. *Front Neurol*. 2017;8:489.

18. Keane PA, Topol EJ. With an eye to AI and autonomous diagnosis. *NPJ Digital Med*. 2018;1:40, s41746-018-0048-y.

10 Technological View

Bilwaj Gaonkar and Luke Macyszyn

Artificial intelligence (AI) is a group of technologies which aims to mimic the human capacity for intelligence in machines. Since we do not understand on a mechanistic level how a collection of neurons becomes intelligent, mimicking such intelligence *in silico* is a challenging endeavor. Thus, a neurobiological understanding of natural intelligence and the development of artificial intelligence have been pursued as tandem goals by the scientific community. While progress has been made toward both goals, much remains to be understood about human intelligence, and engineered intelligence continues to underperform compared to its natural counterpart.

However, what has emerged from the vast multi-decade effort to develop artificial intelligence is an array of technologies that fail to achieve the generalized capacity of human intelligence and yet can mimic and even surpass human intelligence on narrow well-defined tasks [1, 2]. When computer scientists refer to AI, they usually refer to this narrow subset of technologies. Within this narrow but useful definition of AI lies a development of AI technology that begins in the 1960s and continues to this day. Here, we briefly review the history of this field in the context of major philosophical and technological developments.

The development of AI algorithms has explored two distinct paradigms. The first of these is *symbolic AI*, the dominant space of AI exploration in the twentieth century [3, 4]. The second of these was *machine learning* (ML), the dominant space of AI exploration in the twenty-first century. Somewhere in between these two competing paradigms lies the *sub-symbolic AI*, which we describe as well. Next, we describe each of these paradigms in detail.

SYMBOLIC AI

Explicitly encoding various elements of human knowledge into a computer program is the basis of symbolic AI methodologies. One of the simplest symbolic AI systems is the semantic-net which allows for programmatic usage of related concepts—wherein the relationships between concepts are explicitly outlined by human raters. Several explicit rules-based AI systems have been developed since the 1950s. This approach to AI has had several early successes. Symbolic AI algorithms may be divided into two broad classes—rules-based and knowledge-based (Figure 10.1).

Rules-Based Symbolic AI

This type of AI aimed to develop a set of pre-defined rules which turn data into information. Rules may be complex as in the case of cognitively motivated or logical programming-based AI or relatively simple which is typical of knowledge-based systems.

Cognitive Modeling-Based Rules

Some of the first attempts at rules-based AI attempted to generate rules for AI programs by computationally modeling the human cognitive process. This branch of rules-based AI involved studying human cognition using experimental psychology and then explicitly modeling abstracted motifs of human behavioral patterns *in silico*. Since human behavior is typically diverse and multi-layered, such systems naturally incorporate multiple modules that might interact with one another in a pre-specified manner. For example, a particularly successful culmination of the rules-based AI approach is the SOAR project [5]—which continues to this day. This project was used to model human-like behavior of virtual agents in puzzles, video games, and even military applications. Another example of cognitively modeled AI is work based on script theory. Script theory posits that human behavior can be structured into relatively few scripts of sequences of modular actions [6]—and that such scripts are the basis of intelligence.

Non-Cognitive Rules-Based AI

Some rules-based AI theorists abandoned the study of human cognition and reasoned that AI systems could be built based on formal logic that is independent of the human cognitive process. The Prolog [7] (logic programming) language was a culmination of this approach to AI. Prolog remains a powerful logic programming language that is used to this day in various commercial and industrial applications. Less formal versions of logic-based AI systems were the so called "scruffy" systems [4] which departed from formal logic but were based on pre-specified ad hoc rules.

DOI: 10.1201/9781003095279-13

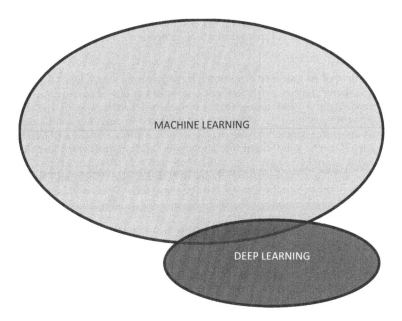

Figure 10.1 A visual representation of AI and its subfields

Expert Systems

The culmination of symbolic AI was the development of expert systems [8–10]. These contained a knowledge base and an inference engine. The knowledge base was effectively an expert-curated database relevant to the problem at hand and the inference engine was a set of human coded rules used to drive software systems using the knowledge base. They also typically contained a knowledge acquisition mechanism and a user interface mechanism. The expert curated database was always extremely expensive to create and maintain. Furthermore, to make a knowledge base truly comprehensive requires an enormous amount of effort and large amounts of computer memory. As advances in silicon manufacturing unfolded in the 1970s and 1980s, computer memory started increasing exponentially, enabling the development of knowledge-based systems. Large well-maintained ontologies of "concepts" are a product of this approach—and several large-scale ontologies underlie expert systems today. Multi-decade projects aiming to collect, annotate, and curate linguistic knowledge have spawned several knowledge-based AI systems. The CYC project [11, 12] is one of the largest databases of such concepts that has been in continuous development since the 1980s.

Abandonment

Despite its several successes, the relative level of interest in symbolic AI has reduced over time. There were several philosophical reasons for this failure [13, 14]. Symbolic AI relies on computer programs starting off with vast curated databases. It is extremely difficult to continually build and maintain vast databases—and hire experts to keep curating such databases. Secondly, inference that depends on pre-written sets of rules is inevitably limited—and it is infeasible to create a comprehensive inference engine. As a concrete example, consider the fact that human language is continually evolving and concepts that are considered closely related in one decade may not be considered as such in another decade. Thus, no single rule or set of rules can adequately capture the essence of intelligence. It is paramount that machine intelligence has the capacity to evolve alongside concepts. In the 1990s, as computing became ubiquitous and increasingly powerful, the non-symbolic form of AI was shown to be capable of "automatically" learning rules of inference from statistics underlying the data itself. This form of inference became known as statistical learning or machine learning. Machine learning eventually superseded symbolic AI and became the dominant paradigm for automating various tasks in computer vision and natural language processing. Likewise, in the last decade, a specific form of machine learning called deep learning has dominated the field of machine learning itself. With these developments, symbolic AI was more or less abandoned by most of the AI community. Yet, some researchers contend that symbolic AI will form an important component of future systems that compete with human intelligence.

MACHINE LEARNING

Since the 1990s, the field of AI has been dominated by algorithms which learn directly from data. Machine learning systems identify patterns that exist within data and can use such patterns to make useful inferences [15]. An example of this is a machine learning system that can "learn" to identify what constitutes a "face" in a large set of annotated image data and use this "learned pattern" to identify faces in a scene that was not a part of the annotated data [16]. In broad strokes, machine learning itself can be subdivided into supervised and unsupervised learning as well as a set of algorithms which cannot be cleanly classified into either of these paradigms, termed semi-supervised learning [17]. Next, we present a history of technologies which have had a high level of success under each of these paradigms. We present these techniques with specific focus on the radiological sciences.

Supervised Machine Learning

Supervised machine learning involves "learning" from data for which human-generated labels are present. A vast amount of work done toward automating specific radiological processes falls under this category [18–20]. A concrete example of a supervised learning task would be learning to identify and delineate spinal canals or neural foramina in MR images, given a large database of MR images with the boundaries of these structures annotated by human operators [21, 22]. In the radiological sciences, supervised algorithms are used to learn how to accurately delineate organ boundaries [19]; differentiate organs with pathology from those without [23, 24]; identify, delineate, and measure lesions and tumors [25, 26]; and diagnose, quantify, and even manage various pathologies [27, 28], among a host of other applications. Datasets used to train these algorithms usually consist of thousands of raw or preprocessed MR images (inputs) and associated delineations (outputs). Large databases geared toward the identification of a specific disease, or the segmentation of a specific organ, have been the workhorses for applying supervised ML in radiological applications [29, 30]. Despite the vast variation in available algorithms, most supervised ML algorithms typically have a set of trainable components called "parameters" and tunable components called "hyperparameters." The differentiation between parameters and hyper-parameters is that parameters will typically change based on the labeled data being used for training and hyper-parameters are usually selected by the modeler and largely fixed during the training process. "Training" an algorithm on a dataset is essentially an optimization problem whereby parameters are adjusted to specific values based on the statistics of the training data. While conventional wisdom requires hyperparameters to be fixed during training, it is easy to find algorithms which tune hyper-parameters in a semi-automated or fully automated fashion. Conventionally, the set of parameter values that are fixed using the training data are called "the learned (or trained) model." The model aims to ingest the human annotated data and distill the essence of the mapping between inputs and outputs using various methodologies. For most supervised learning tasks, determining optimal hyperparameters is as much an art as a science and is precisely where expert data scientists differ from novice operators, though recent advances in automatic hyperparameter optimization have significantly narrowed the gap. Supervised learning has also been the dominant paradigm of machine learning research in image analysis, addressing tasks involving image classification, image segmentation, and image generation. Several classes of algorithms have been used to address each of these tasks; to enumerate them all is beyond the scope of this chapter. In what follows, we aim to describe three broad classes of supervised ML algorithms without delving into specific application areas. We believe these to be especially important in radiological image analysis. These are deep learning, decision tree–based algorithms and kernel machines. We describe these algorithmic frameworks next.

Deep Learning Algorithms

This class of supervised learning algorithms has been the dominant form of image analysis technologies since 2015 [31]. Deep learning–based algorithms have dominated the majority of machine learning benchmarks in the last five years [32–34]. It involves the use of giant networks which consist of artificial neurons. Each "neuron" is effectively a simple mathematical function with a few adjustable numeric parameters. In deep learning, neurons are organized into layers. The outputs of neurons from one layer are fed as inputs to the next layer and so on [35]. The networks are said to be "deep" because there are more than a few "layers of neurons"—and training is implemented using backpropagation [36] or some variant of it. Note that backpropagation adjusts numeric parameters associated with neurons in a layer-wise fashion beginning from the output layer all the way back to the input layer. Modern deep neural networks tend to have up to 100+ "layers" [37, 38] and differ from traditional neural networks (pre-1990) in a few key aspects. These aspects are the mathematical functions used by their neurons, network architecture, and specifics of the optimization algorithm for training. Enormous historical effort has been involved in carefully choosing each of these aspects. This effort, along with the development of accessible and well-maintained software packages such as TensorFlow (https://www.tensorflow.org/) and PyTorch (https://pytorch.org/), have contributed to the tremendous surge in activity around deep learning in radiology in the last decade. Several deep learning algorithms have pushed the enve-lope in anatomical segmentation [37–40] and disease classification [41, 42]. A key appeal of deep learning in radiology has been the ease of transferring knowledge from other-domains—resulting in a reduced need for *de novo* training data. For example, the initial layers of a deep neural network trained on natural color images extract a series of imaging features which can be put to work for finding and delineating tumors in brain scans [43–45]. This type of transfer learning with deep networks is frequently done in medical image analysis, reducing the need to collect, archive, and segment anatomy *de novo*. Given its effectiveness and ease of use, deep learning is likely to be the dominant form of AI that finds its way into the radiology clinic in the near future.

A second set of supervised algorithms which have enjoyed enormous success in the AI in medi-cine space are *decision tree*–based algorithms. Several protocols that are commonplace in medical practice can be thought of as decision trees—and, as such, decision trees are highly amenable to automating such processes. Decision trees can also be "learnt" from data (e.g., ID3/C4.5 [46] algo-rithm). Learnt decision trees can be combined in various ways including *random forests* [47], *gradi-ent boosted trees* [48], and bagging, among many others. These can be used with either pre-designed features or "learnt" features for image analysis as well as non-image analysis tasks. Typical use for image analysis involves the extraction of fixed features such as histogram of oriented gradients (HOG [49]) or scale invariant feature transform (SIFT [50]) features from images followed by the application of a random forest or gradient boosting for inferring boundaries of an organ or pres-ence of a disease from images. While this type of approach is still competitive with deep learn-ing when dataset sizes are small and transfer learning is not possible, it is not justified in most cases where generating additional data is possible. The cases where decision tree ensembles truly shine is when heterogeneous datasets need to be ingested to make a single decision. For example, when categorical data, genetic data, text data, imaging data, and time series data are all required to be combined to render a treatment decision, random forests are a convenient solution. This is because they are able to handle multiple types of data more easily than other algorithms, which may require complex transformations of the input data. Furthermore, training these algorithms requires 10x less time than deep learning, which makes them ideal for prototyping and for algo-rithm development efforts in settings where computing power is limited. That said, these algo-rithms are also very scalable and are able to take advantage of increased computing power when available to create more powerful models.

Kernel Machines

Machine learning between 1995 and 2013 was dominated by *kernel machines* [51]. An archetype of this class of algorithms is the "support vector machine" [52, 53] which was popular throughout the early 2000s. Kernel machines typically employ the so-called "kernel trick" wherein data is projected from a low-dimensional space to a "high or even an infinite"–dimensional space. The hope is that the relationships relevant to decision-making that are not obvious in low-dimensional space will become obvious in high-dimensional space thus allowing for "learning" using simple linear statistical models. Kernel machines, especially support vector machines, continue to be used to this day given their ease of interpretation and ability to analyze data that are not easily interpre-table to human beings. However, their popularity for image analysis has waned given the enor-mous improvements ushered in by deep learning. Kernel machines can also be highly effective in low-data environments as compared to deep learning. However, deep learning supersedes kernel machine performance once a large amount of labelled data becomes available.

Other Supervised Methods

Apart from the classes of supervised methods described above, several techniques have been described in literature including discriminant analysis, statistical regression, and shallow neural networks—all of which have niche application areas but have not been of particularly high relevance to machine learning in radiology.

Unsupervised, Weakly Supervised, and Semi-Supervised Machine Learning

Unsupervised learning refers to a class of machine learning algorithms that "learn" from data without any prior human expert input. For example, the distribution of intensity values may differ between grey matter and white matter in the brain on an MRI. This difference can be utilized by a clustering algorithm to differentiate the two tissue types on an image [54]. While supervised learning currently dominates radiological image analysis, unsupervised algorithms have been used extensively in the past and have a role to play in the future. For instance, unsupervised algorithms can be used in conjunction with minimal user input to ease the process of manual data labelling [55]. Weak supervision is the formal machine learning paradigm that implements this idea of using minimal user input to "train" machine learning algorithms. Given that data generation remains an expensive affair, weakly supervised methods have been gaining popularity in radiological image analysis [56–59]. It is likely that these techniques will come to dominate AI in radiology in the future.

REFERENCES

1. Liu S, Deng W. Very deep convolutional neural network based image classification using small training sample size. In: *Proceedings - 3rd IAPR Asian Conference on Pattern Recognition, ACPR 2015*; 2016. doi:10.1109/ACPR.2015.7486599

2. Kaiming H, Xiangyu Z, Shaoqing R, Jian S. Delving deep into rectifiers: surpassing human-level performance on imagenet classification kaiming. *Biochem Biophys Res Commun.* 2018;498(1):254–261.

3. Haugeland J. *Artificial Intelligence: The Very Idea*; 1985. The MIT Press.

4. Minsky ML. Logical versus analogical or symbolic versus connectionist or neat versus scruffy. *AI Mag.* 1991;12(2):34–34. doi:10.1609/AIMAG.V12I2.894

5. Soar Tutorial 9.6.0 - Soar Cognitive Architecture. Accessed January 3, 2022. https://soar.eecs.umich.edu/articles/downloads/soar-suite/228-soar-tutorial-9-6-0

6. Schank RC. Dynamic memory : A theory of reminding and learning in computers and people. Published online 1982:234.

7. Clocksin WF, Mellish CS. Programming in prolog. *Program Prolog.* Published online 2003. doi:10.1007/978-3-642-55481-0

8. Jackson P. Introduction to expert systems. Published online January 1, 1986.

9. Henrion M, Breese JS, Horvitz EJ. Decision Analysis and Expert Systems. *AI Mag.* 1991;12(4):64–64. doi:10.1609/AIMAG.V12I4.919

10. Liebowitz J. Worldwide perspectives and trends in expert systems: An analysis based on the three world congresses on expert systems. *AI Mag.* 1997;18(2):115–115. doi:10.1609/AIMAG.V18I2.1297

11. Lenat DB, Prakash M, Shepherd M. CYC: Using common sense knowledge to overcome brittleness and knowledge acquisition bottlenecks. *AI Mag.* 1985;6(4):65–65. doi:10.1609/AIMAG.V6I4.510

12. Cyc | The next generation of enterprise AI. Accessed January 3, 2022. https://cyc.com/

13. Dreyfus HL. What computers still can't do: A critique of artificial reason. Published online 1992:354.

14. McDermott D. A critique of pure reason1. *Comput Intell.* 1987;3(1):151–160. doi:10.1111/J.1467-8640.1987.TB00183.X

15. Bishop CM. *Pattern Recognition and Machine Learning.* Vol 4.; 2006. doi:10.1117/1.2819119

16. Ortiz EG, Becker BC. Face recognition for web-scale datasets. *Comput Vis Image Underst.* 2014;118:153–170. doi:10.1016/j.cviu.2013.09.004

17. Goldberg X. Introduction to semi-supervised learning. *Synth Lect Artif Intell Mach Learn.* Published online 2009. doi:10.2200/S00196ED1V01Y200906AIM006

18. Shen D, Wu G, Suk H II. Deep learning in medical image analysis. *Annu Rev Biomed Eng.* Published online 2017. doi:10.1146/annurev-bioeng-071516-044442

19. Mary SP, Ankayarkanni, Nandini U, Sathyabama, Aravindhan S. A survey on image segmentation using deep learning. In: *Journal of Physics: Conference Series*; 2020. doi:10.1088/1742-6596/1712/1/012016

20. Litjens G, Kooi T, Bejnordi BE, et al. A survey on deep learning in medical image analysis. *Med Image Anal.* Published online 2017. doi:10.1016/j.media.2017.07.005

21. Gaonkar B, Villaroman D, Beckett J, et al. Quantitative analysis of spinal canal areas in the lumbar spine: an imaging informatics and machine learning study. *Am J Neuroradiol.* Published online 2019. doi:10.3174/ajnr.a6174

22. Gaonkar B, Beckett J, Villaroman MHSD, et al. Quantitative analysis of neural foramina in the lumbar spine: An imaging informatics and machine learning study. *Radiol Artif Intell.* 2019;1(2). doi:10.1148/ryai.2019180037

23. Varol E, Gaonkar B, Erus G, Schultz R, Davatzikos C. Feature ranking based nested support vector machine ensemble for medical image classification. In: *Proceedings - International Symposium on Biomedical Imaging*; 2012. doi:10.1109/ISBI.2012.6235505

24. Gaonkar B, Davatzikos C. Deriving statistical significance maps for SVM based image classification and group comparisons. *Med Image Comput Comput Assist Interv.* 2012;15(Pt 1):723–730.

25. Porz N, Habegger S, Meier R, et al. Fully automated enhanced tumor compartmentalization: Man vs. Machine reloaded. *PLoS One.* 2016;11(11):e0165302. doi:10.1371/journal.pone.0165302

26. Gaonkar B, Macyszyn L, Bilello M, et al. Automated tumor volumetry using computer-aided image segmentation. *Acad Radiol.* 2015;22(5):653-661. doi:10.1016/j.acra.2015.01.005

27. Gaonkar B, Pohl K, Davatzikos C. *Pattern Based Norphometry.* Vol. 6892. LNCS.; 2011. doi:10.1007/978-3-642-23629-7_56

28. Wang Y, Song Y, Rajagopalan P, et al. Surface-based TBM boosts power to detect disease effects on the brain: An N=804 ADNI study. *Neuroimage.* Published online 2011. doi:10.1016/j.neuroimage.2011.03.040

29. Jones-Davis DM, Buckholtz N. The impact of the Alzheimer's disease neuroimaging initiative 2: What role do public-private partnerships have in pushing the boundaries of clinical and basic science research on Alzheimer's disease? *Alzheimers Dement.* 2015;11(7):860-864. doi:10.1016/j.jalz.2015.05.006

30. SpineWeb : Main / HomePage: Browse. Accessed August 17, 2017. http://spineweb.digitalimaginggroup.ca/

31. Chan HP, Samala RK, Hadjiiski LM, Zhou C. Deep learning in medical image analysis. In: *Advances in Experimental Medicine and Biology*; 2020. doi:10.1007/978-3-030-33128-3_1

32. Szegedy C, Liu W, Jia Y, et al. Going deeper with convolutions. In: *Proceedings of the IEEE Computer Society Conference on Computer Vision and Pattern Recognition*. Vol. 07, 12 June; 2015:1–9. doi:10.1109/CVPR.2015.7298594

33. Purushotham S, Meng C, Che Z, Liu Y. Benchmarking deep learning models on large health-care datasets. *J Biomed Inform*. 2018;83. doi:10.1016/j.jbi.2018.04.007

34. Hesamian MH, Jia W, He X, Kennedy P. Deep learning techniques for medical image segmentation: Achievements and challenges. *J Digit Imaging*. 2019;32(4). doi:10.1007/s10278-019-00227-x

35. LeCun Y, Bengio Y, Hinton G. Deep learning. *Nature*. 2015;521(7553):436–444.

36. Schmidhuber J. Deep Learning in neural networks: An overview. *Neural Networks*. 2015;61. doi:10.1016/j.neunet.2014.09.003

37. Chen L-C, Zhu Y, Papandreou G, Schroff F, Aug C V, Adam H. deeplabv3+: Encoder-decoder with atrous separable convolution for semantic image segmentation. *Proc Eur Conf Comput Vis*. Published online 2018.

38. He K, Gkioxari G, Dollár P, Girshick R. Mask r-cnn. In: *Proceedings of the IEEE International Conference on Computer Vision*, 2017: 2961–2969.

39. Ronneberger O, Fischer P, Brox T. U-Net: Convolutional networks for biomedical image segmentation. In: *Medical Image Computing and Computer-Assisted Intervention MICCAI 2015*. Springer; 2015:234–241.

40. Milletari F, Navab N, Ahmadi SA. V-Net: Fully convolutional neural networks for volumetric medical image segmentation. In: *Proceedings - 2016 4th International Conference on 3D Vision, 3DV 2016*; 2016. doi:10.1109/3DV.2016.79

41. Ghesu FC, Georgescu B, Gibson E, et al. Quantifying and leveraging classification uncer-tainty for chest radiograph assessment. In: *Lecture Notes in Computer Science (Including Subseries Lecture Notes in Artificial Intelligence and Lecture Notes in Bioinformatics)*; 2019. doi:10.1007/978-3-030-32226-7_75

42. Araujo T, Aresta G, Castro E, et al. Classification of breast cancer histology images using convolutional neural networks. *PLoS One*. Published online 2017. doi:10.1371/journal.pone.0177544

43. Dheeraj D, Prasantha HS. Dr-unet: A hybrid model for classification of Gl ioma using transfer learning on MR images. *Int J Eng Trends Technol*. 2021;69(10). doi:10.14445/22315381/IJETT-V69I10P218

44. Tjahyaningtijas HPA, Rumala DJ, Angkoso CV, et al. Brain tumor classification in MRI images using En-CNN. *Int J Intell Eng Syst*. 2021;14(4). doi:10.22266/ijies2021.0831.38

45. Kaur A, Kaur L, Singh A. GA-UNet: UNet-based framework for segmentation of 2D and 3D medical images applicable on heterogeneous datasets. *Neural Comput Appl*. 2021;33(21). doi:10.1007/s00521-021-06134-z

46. Hssina B, Merbouha A, Ezzikouri H, Erritali M. A comparative study of decision tree ID3 and C4.5. *Int J Adv Comput Sci Appl*. 2014;4(2). doi:10.14569/specialissue.2014.040203

47. Breiman L. Random forests. *Mach Learn*. Published online 2001. doi:10.1023/A:1010933404324

48. Natekin A, Knoll A. Gradient boosting machines, a tutorial. *Front Neurorobot.* 2013;7(Dec). doi:10.3389/fnbot.2013.00021

49. Dalal N, Triggs B. Histograms of oriented gradients for human detection. In: *Proceedings - 2005 IEEE Computer Society Conference on Computer Vision and Pattern Recognition, CVPR 2005.* Vol I. IEEE; 2005:886–893. doi:10.1109/CVPR.2005.177

50. Lindeberg T. Scale invariant feature transform. *Scholarpedia.* 2012;7(5). doi:10.4249/scholarpedia.10491

51. Müller KR, Mika S, Rätsch G, Tsuda K, Schölkopf B. An introduction to kernel-based learning algorithms. *IEEE Trans Neural Networks.* 2001;12(2). doi:10.1109/72.914517

52. Burges CJC. A tutorial on support vector machines for pattern recognition. *Data Min Knowl Discov.* 1998;2(2). doi:10.1023/A:1009715923555

53. Gaonkar B, Davatzikos C. Analytic estimation of statistical significance maps for support vector machine based multi-variate image analysis and classification. *Neuroimage.* 2013;78:270-283. doi:10.1016/j.neuroimage.2013.03.066

54. Mehidi I, Belkhiat DEC, Jabri D. A fast k-means clustering algorithm for separation of brain tissues in MRI. In: *Proceedings of the 2020 International Conference on Mathematics and Information Technology, ICMIT 2020;* 2020. doi:10.1109/ICMIT47780.2020.9046971

55. Yushkevich PA, Piven J, Hazlett HC, et al. User-guided 3D active contour segmentation of anatomical structures: Significantly improved efficiency and reliability. *Neuroimage.* 2006;31(3):1116–1128.

56. Zhou Z, Sodha V, Rahman Siddiquee MM, et al. Models genesis: generic autodidactic models for 3d medical image analysis. In: *Lecture Notes in Computer Science (Including Subseries Lecture Notes in Artificial Intelligence and Lecture Notes in Bioinformatics);* 2019. doi:10.1007/978-3-030-32251-9_42

57. Roth HR, Yang D, Xu Z, Wang X, Xu D. Going to extremes: Weakly supervised medical image segmentation. *Mach Learn Knowl Extr.* 2021;3(2). doi:10.3390/make3020026

58. Souly N, Spampinato C, Shah M. Semi supervised semantic segmentation using generative adversarial network. In: *Proceedings of the IEEE International Conference on Computer Vision;* 2017. doi:10.1109/ICCV.2017.606

59. Kervadec H, Dolz J, Tang M, Granger E, Boykov Y, Ben Ayed I. Constrained-CNN losses for weakly supervised segmentation. *Med Image Anal.* 2019;54. doi:10.1016/j.media.2019.02.009

11 Societal View

Amy Patel

RADIOLOGISTS

There are many radiologists that possess steadfast optimism and hope about the future of radiology and the implications of AI in the overall landscape when it comes to improving practice efficiency and diagnostic accuracy [1].

However, the fear of AI is not unfounded amongst radiologists and even trainees. Studies that demonstrate an AI machine learning algorithm outperforming a radiologist are now being demonstrated regardless of variables that may be perceived by some as having flawed methodology and conclusions, such as a study conducted where an AI deep learning–based system was trained using mammograms from women in the UK and the United States and then applied retrospectively to test sets, resulting in reductions of false-positive and false-negative detection of biopsy-proven breast cancers [4]. Of course, broad applicability across all patient types is still not established, and human monitoring of the AI system would need to be in place to ensure performance did not deteriorate over time. Nevertheless, studies such as this incite fear in some in the house of radiology that perhaps we could one day be replaced. When it comes to trainees, a survey was conducted whose results demonstrated that they

> were more likely to express doubts on whether they would have pursued diagnostic radiology as a career had they known of the potential impact artificial intelligence is predicted to have on the specialty, $P = 0.0254$ and were also more likely to plan to learn about the topic, $P = 0.0401$.

Also in this study, it was revealed that the attending radiologists lacked exposure to the latest medical literature on AI which may have contributed to the trainees' sentiments [5]. In another study performed in Germany, from a total of 452 participants, 90% had some sort of familiarity with AI, "but only 24% reported good or expert knowledge." However, most felt that the utilization of AI in healthcare was "positive" or "very positive (53.18%)" [3]. In all, the truth of the matter is that radiology itself is a more complex subspecialty that has nuances requiring the human element such as patient interaction when performing procedures or delivering bad news, but the performance aspect does alarm some all the same. As Pakdemirli poignantly states, "wherever you see yourself in the spectrum, nobody can stop the advancement, innovations, and implantation of AI in radiology" [2].

REFERRING PROVIDERS

There is a current shortage of literature regarding referring providers/non-radiologists' perception of AI in radiology, particularly its utilization in diagnostic medical imaging reports. However, in one study, referring providers, regardless of career level or subspecialty, felt significantly more apprehensive relying on an AI-issued report rather than one issued by a radiologist (p <0.001). However, when it came to an "AI-hybrid model of care" they felt "equally comfortable." When it came to medicolegal risk, the majority felt that this should fall upon radiologists (54.5%), hospitals, and providers of health service (65.9%) [6]. When it comes to healthcare and AI in general, physicians, regardless of subspecialty, agree that patients must have mutual cooperation and participation when it comes to the utilization of AI in their healthcare plan. "Patient empowerment" will be crucial as AI is increasingly adopted in the day-to-day diagnosis and treatment of patients [3].

PATIENTS

However, if one thinks of society as the consumer, in our case, patients, the concept is still very foreign to them and they perhaps have a broad picture of what this could mean but are not entirely aware of its implications, with a lack of knowledge and understanding. So far, the reaction is overall positive and there is belief that the benefits exceed the risks, but there are concerns of patients feeling that AI will replace doctors or at the minimum, decrease interaction with them. Thus, feelings of the absence of the human factor due to AI are of concern to patients. Also, patients currently have limited knowledge on the subject matter [7]. For example, in my practice, we are utilizing an AI breast ultrasound tool. When discussing with patients about this tool and

DOI: 10.1201/9781003095279-14

how we use it as a second opinion consult in coming to a management conclusion, it often is met with awe and excitement, but little is actually known about how it works and what it does until we discuss these elements with the patient, and even then, depending on the patient's health literacy level, there is still some element of confusion or lack of comprehension. In another study, it was noted that the general public and patients do not understand how to approach errors potentially generated by AI systems [8]. Additionally, there is ongoing concern from some patients that they will not be able to refuse an AI application used in their treatment plan, which may result in insufficient insurance coverage and rising healthcare costs [3].

Additionally, concerns of patient privacy are also at the forefront for all factions, including patients, in regard to who is using, storing, and overseeing the data [9]. In order for AI to be reproducible across all practice types and populations, this will require big datasets and so this is an ongoing concern and topic in the field of AI and amongst radiologists. Furthermore, some pose the question whether complete deidentification without the ability to reidentify the data is a feasible occurrence in AI? This could particularly be the case in a scenario of high-resolution 3D imaging and facial recognition, where one could potentially match images generated from an MRI or CT to an individual's photographs [10].

OTHER HEALTHCARE STAKEHOLDERS

The majority of other stakeholders' perceptions, including computer scientists and the general public, is that AI will improve the field of radiology and not replace radiologists. Unfortunately, a level of understanding of AI for those who are non-computer scientists is lacking. However, there is a thirst for education on the subject, and collaboration between computer scientist AI experts and radiologists can be mutually beneficial in enhancing patient care [8]. When it comes to the general public, the perception is not necessarily positive across the board. Discussions are ongoing regarding AI's ethical implications, including "loss of control" and unintended consequences by non-critical utilization of AI [3].

CONCLUSION

Overall, stakeholders in AI, including radiologists, trainees, referring providers, and patients, feel that AI can improve the field of radiology and will not replace radiologists [8]. However, demystification of AI, including more robust knowledge and collaboration with those who are experts in the AI space could improve one's knowledge base, regardless of faction. At a minimum, collaboration with radiologists and trainees is paramount to establish a baseline understanding from which these groups can then, in turn, educate referring providers and patients. The future of radiology will undoubtedly include AI and its various forms, and the sooner one can learn about and eventually adopt AI applications in practices and institutions, the more prepared one will be when AI becomes the rule and not the exception, elevating the way in which we practice.

REFERENCES

1. Bryson J. The future of AI's impact on society. *MIT Technol Rev* (2019). https://www.technologyreview.com/2019/12/18/102365/the-future-of-ais-impact-on-society/ Accessed 7/08/21

2. Pakdemirli E. Perception of artificial intelligence among radiologists. *Acta Radiol Open* 8, no. 9 (2019). 31632696

3. Fritsch S, Blankenheim A, Wahl A, et al. Attitudes and perception of artificial intelligence in healthcare: A cross-sectional survey among patients. *Digit Health* 8, (2022). 35983102

4. Killock, D. AI outperforms radiologists in mammographic screening. *Nat Rev Clin Oncol* 17, no. 3 (2020). 31965085

5. Collado-Mesa F, Alvarez E, Arheart K. The role of artificial intelligence in diagnostic radiology: A survey at a single radiology residency training program. *J Am Coll Radiol* 15, no. 12 (2018). 29477289

6. Lim S, Phan T, Law M, et al. Non-radiologist perception of the use of artificial intelligence (AI) in diagnostic medical imaging reports. *J Med Imaging Radiat Oncol* 66, no. 8 (2022). 35191186

7. Gao S, He L, Chen Y, et al. Public perception of artificial intelligence in medical care: Content analysis of social media. *J Med Internet Res* 22, no. 7 (2020). 32673231

8. Yang L, Ene I, Belaghi R, et al. Stakeholders' perspectives on the future of artificial intelligence in radiology: A scoping review. *Eur Radiol* 32, no. 3: 1477–1495, (2022)

9. Pisano, E. AI Shows Promise for breast cancer screening. *Nature* (2020). https://www.nature.com/articles/d41586-019-03822-8 Accessed 7/12/21

10. Lotan, E. Tschider C, Sodickson D, et al. Medical Imaging and privacy in the era of artificial intelligence: Myth, fallacy, and the future. *J Am Coll Radiol* 17, no. 9 (2020). 32360449

12 Financial View

Christian Park

GENERAL VIEW

Over the last decade, AI technologies have attracted substantial investment and interest across various sectors, a trend that continues to accelerate. In 2022, global private investment in AI was $91.9 billion, despite a 26.7% decrease from the previous year [1]. However, this is still a significant increase from the amount a decade ago, marking an 18-fold rise since 2013 [2]. The United States has been a leader in this space, with $47.4 billion invested in 2022, which was roughly 3.5 times the amount invested in China, the next highest country [3].

In 2023, worldwide spending on AI-centric systems is forecast to reach $154 billion, indicating a robust 26.9% increase over the previous year [4]. This spending trend is expected to maintain a compound annual growth rate (CAGR) of 27.0% from 2022 to 2026, with spending expected to surpass $300 billion by 2026 [5]. The integration of AI across a wide range of products underscores the technology's pervasive impact, with only one of the 36 AI use cases identified by IDC showing a CAGR of less than 24% over the forecast period [6].

The financial impact of AI is most evident in three key segments: automation, insights, and augmentation. These segments cover a broad spectrum of applications, from augmenting human abilities to automating repetitive tasks and providing personalized recommendations. The sectors that see the largest AI investments include banking and retail, with professional services, and discrete and process manufacturing also showing significant spending. Geographically, the United States accounts for more than 50% of all AI spending worldwide, with Western Europe following at more than 20% [7].

The AI technology landscape has been dynamic, with advances in generative AI, language processing, and ethical considerations gaining prominence. The AI Index Report of 2023 highlights the importance of balancing model performance with ethical issues, and the growing public consciousness about AI's capabilities and its potential misuse [1].

As AI continues to evolve, it is crucial to consider not only the financial benefits and investment trends but also the ethical implications and the need for responsible deployment of AI technologies.

Automation

The AI-driven automation market is on a significant growth trajectory, with various estimates projecting its size and scope. The hyper-automation market, which includes AI, is expected to surpass $44 billion USD by 2035, growing at a CAGR of about 11% during the forecast period from 2023 to 2035 [8]. North America is anticipated to dominate this market, owing to technological advancements in the region and a strong presence of leading technology companies.

Within the realm of AI-driven robotic automation, Amazon's commitment is notably profound, with over $100 billion invested in research and development [9]. The acquisition of Kiva Systems for $775 million and subsequent implementation of advanced packing, shipping, and fulfillment protocols have led to a reduction in operating expenses by approximately one fifth, or roughly $22 million per fulfillment center, plus an additional $800 million due to increased short-term implementation efficiency [10, 11].

For supply chain optimization, UPS has reported a savings of $50 million from mile reduction in daily driver routes using their ORION system AI tool [12]. Similarly, Lineage Logistics harnesses AI to optimize food freshness and prevent spoilage, thereby increasing efficiency by 20% [13–17]. AI's role in enhancing safety, asset maintenance, supply line prediction, energy efficiency, and freight management is becoming increasingly pivotal for value creation within supply chains.

The artificial intelligence market, in general, is also on a steep incline, with projections indicating an increase from $515.31 billion USD in 2023 to $2,025.12 billion USD by 2030, exhibiting a CAGR of 21.6% [18]. This growth is fueled by the increasing availability of data and the development of high-performance computing infrastructures, which are integral to AI's functionality.

The largest market share by end-use in the hyper-automation market by 2035 is expected to be held by healthcare, driven by the sector's vast patient treatment requirements [19–22]. Artificial intelligence (AI) technology, particularly deep learning, is leading AI market segments due to its capability to manage complex data-driven applications, such as content and speech recognition.

Deep learning captured around 36.4% of the global revenue in 2022 and is expected to remain a significant area of investment, aiding in overcoming challenges associated with high volumes of data.

Augmentation

AI-driven augmentation in various industries, particularly outside of healthcare, is anticipated to continue its trajectory of adding significant value. The AI augmentation market was projected to create $2.9 trillion in global value in 2021, with the financial services and retail sectors poised to reap the most benefits [23–25]. This growth is fueled by the increasing integration of AI into various business functions, such as service operations, risk management, and supply-chain management, with service operations expected to dominate the sector due to AI's ability to solve problems more quickly and enhance customer service.

In the financial sector, AI has revolutionized equities markets with algorithmic trading systems that offer considerable performance improvements over traditional trading methods. Despite the reliability of these systems, the COVID-19 crisis highlighted the importance of human oversight in algorithmic trading, as black swan events can expose certain limitations of AI.

The fintech sector is expected to see significant growth, with AI augmentation projected to grow at a compound annual growth rate of 23.17% through 2026, increasing the market from $7.91 billion in 2020 to $26.67 billion by 2026 [26–29]. AI in fintech enables more rigorous adherence to ever-changing rules and regulations, enhances fraud detection, and optimizes risk analysis for loans and credit.

Generative AI, which refers to AI that can generate new content and ideas based on training data, could further revolutionize the automation landscape. It has the potential to drive innovation in product and service development, marketing, and customer engagement by creating new and personalized experiences [24]. The growth of generative AI is particularly significant, with companies increasingly investing in and developing AI tools to cater to changing user requirements.

Globally, North America has been a dominant force in the AI market due to the presence of key industry players and favorable government initiatives that encourage the adoption of AI technologies [24]. The surge in adopting cloud computing technologies amid the pandemic has also contributed to the rise in AI deployments [30]. Asia Pacific is expected to experience the highest growth rate in the coming years, bolstered by significant investments and a growing number of startups focusing on AI to improve operational efficiency.

Key industry players are enriching their partner ecosystems to develop new opportunities, generate revenue, and expand their presence. Microsoft, IBM, Google, NVIDIA, and Intel are among the companies leading the way in AI developments through strategic alliances, product launches, and partnerships.

Artificial Intelligence-Driven Insights

AI-driven analytics are revolutionizing the way businesses harness big data to understand trends, patterns, and make predictions about future events. The global artificial intelligence market size, which includes AI-driven analytics, was valued at $428 billion USD in 2022 and is projected to grow from $515.31 billion USD in 2023 to $2,025.12 billion USD by 2030, exhibiting a CAGR of 21.6% [31–34]. The North American region, particularly the United States, has been leading in AI adoption due to a strong ecosystem of academia, industry, and government collaborations [35]. However, the Asia Pacific region is anticipated to grow at the highest CAGR in the upcoming years, driven by significant investments and a growing number of startups focusing on improving operational efficiency through AI [36].

In the analytics segment, tools like IBM's Watson have showcased their versatility and value across various industries [37, 38]. Watson's scalability and ability to be customized to specific business operations using existing data have proven economically beneficial. Independent studies have shown that Watson can generate millions in added value, with a return on investment over several years that is substantial [39–42]. These AI analytics tools help predict machinery breakdowns, determine insurance claim eligibility, and improve performance, leading to cost savings and revenue generation.

AI analytics is a part of a broader AI market, including various technologies such as machine learning, deep learning, NLP (natural language processing), and computer vision [43–46]. These technologies are integrated into multiple business functions, including finance, security, human resources, law, marketing and sales, operations, and supply chain management.

The key market players in the AI space are focusing on expanding their partner ecosystems to remain competitive. These companies, which include Google, Microsoft, IBM, NVIDIA, Intel, and others, have been adopting strategies such as new product launches, mergers and acquisitions, and partnerships to maintain their market dominance. For example, recent developments have seen Google launch new AI-powered solutions for the life sciences industry [47–49], and HPE enhance its Ezmeral Software platform to expand data and analytics capabilities for machine learning and AI initiatives.

The introduction of generative AI is also making waves, with companies investing in tools that can generate content and ideas based on training data, leading to innovations in product development, marketing, and customer engagement. This burgeoning field is part of the ongoing evolution of AI, with organizations looking to leverage these advanced capabilities to drive growth and efficiency.

The integration of AI-driven analytics is thus playing a pivotal role in transforming industries by enabling them to make more informed decisions based on data-driven insights, ultimately driving profitability and growth.

Conclusion

The landscape of AI is a tapestry of rapid technological advancement, ethical considerations, and transformative potential. It is a realm where the financial benefits are as tangible as the ethical dilemmas are complex. As this technology matures, the conversation around it deepens, moving beyond mere capabilities to include a conscientious deployment and the safeguarding of societal norms.

THE FINANCIAL IMPACT OF AI IN HEALTHCARE

Artificial intelligence (AI) in the healthcare industry has seen a meteoric rise in investment and application. The AI healthcare market, valued at $14,6 billion USD in 2023, is on an unprecedented growth path, with projections reaching $102.7 billion USD by 2028, and a compound annual growth rate (CAGR) of 47.6% [50].

Capital Influx and Innovation

The healthcare AI market is experiencing dramatic expansion, with software solutions constituting a significant portion of the market at a revenue share of 40.5%. This segment is poised for robust growth from 2024 to 2030 due to rising applications across various healthcare domains [51].

Investment is flowing into AI healthcare startups, indicating a fertile ground for innovation and growth. Strategic alliances, such as the partnership between Aidoc and Imbio to enhance medical image analysis, demonstrate the sector's trend toward leveraging AI for improved patient care and operational efficiency [52].

AI technologies are streamlining healthcare operations, significantly reducing costs associated with administrative tasks, and improving resource allocation. The automation of processes and intelligent systems are enabling healthcare facilities to operate more effectively [52].

AI is facilitating a broader range of healthcare services, including remote patient monitoring, which extends the reach of healthcare providers and introduces new revenue streams by tapping into previously inaccessible markets [52].

AI's application extends beyond direct patient care to encompass operational aspects of healthcare, from administrative workflows to patient management systems, enhancing the overall efficiency and financial performance of healthcare providers [52].

Regional Dynamics and Market Leadership Dominance of North America

With its developed healthcare IT infrastructure and a prominent presence of key market players, North America holds the largest revenue share. The Asia Pacific region, however, is projected to witness the fastest growth, signaling a shift towards a more globalized AI healthcare innovation landscape [53].

Top companies like Amazon Web Services, Microsoft, and NVIDIA are at the forefront, driving the market through strategic collaborations and acquisitions, such as Microsoft's acquisition of Nuance Communications, which is expected to enhance their AI offerings in healthcare [53].

Conclusion

AI's role in medicine and healthcare is reshaping the industry, promising not only cost reductions and operational improvements but also spawning new business models and revenue channels.

The projected market growth underlines the industry's confidence in AI's transformative potential for global healthcare [53].

As AI becomes deeply integrated into medical practices, it has the potential to redefine patient care, streamline healthcare operations, and create new paradigms in the economics of health services. The healthcare sector is on the cusp of an AI-driven transformation that promises enhanced efficiency and patient outcomes [53].

THE FINANCIAL IMPACT OF AI IN RADIOLOGY

Introduction

The advent of artificial intelligence (AI) in radiology marks a pivotal shift in the medical imaging domain [54]. This technology's integration is not just a leap in diagnostic and operational capabilities but also a significant economic driver within the healthcare industry [55]. The financial ramifications are extensive, impacting everything from cost efficiencies to the creation of novel revenue avenues [56]. This chapter delves deeper into the various economic aspects of AI's role in radiology, analyzing its implications on cost structures, operational efficiencies, revenue models, and overall healthcare economics.

Initial Costs and Funding

While there are many promising applications of AI, implementing AI in radiology departments necessitates a substantial initial investment [57]. This includes not only the procurement of sophisticated AI software but also the necessary hardware to support these advanced systems. The costs extend to integrating these technologies into existing radiological workflows and ensuring compatibility with current medical imaging equipment [58]. Additionally, significant financial resources are allocated to training medical personnel to proficiently use these AI tools. Training encompasses understanding the nuances of AI outputs and effectively integrating AI-driven insights into clinical decision-making processes [59].

The initial costs for implementing AI in radiology are offset by potential gains. While specific dollar amounts for these initial investments vary widely based on the scale and scope of implementation, the broader market trends provide insight into the overall financial landscape. For instance, the AI medical imaging market, which includes radiology, was valued at $1.9 billion in 2022 and is projected to reach $29.8 billion by 2032, growing at a CAGR of 32.1% [60]. However, these initial costs are typically offset by the long-term financial and operational benefits. AI's introduction in radiology promises a return on investment through enhanced diagnostic capabilities, reduced operational redundancies, and improved patient throughput [61].

Reducing Diagnostic Errors

One of the most significant financial impacts of AI in radiology is its capacity to substantially reduce diagnostic errors [62, 63]. AI algorithms, particularly in the interpretation of complex imaging, enhance accuracy, thereby reducing the incidence of misdiagnoses [64]. This accuracy is crucial, as diagnostic errors can lead to unnecessary and costly medical procedures, extended hospital stays, and increased patient morbidity, all of which impose substantial costs on healthcare systems [65, 66].

Streamlining Radiology Workflow

AI algorithms can quickly analyze medical images, providing radiologists with pre-analyzed data, which streamlines the decision-making process [67]. This capability is particularly beneficial in managing large volumes of imaging studies, as it prioritizes cases based on severity and urgency [68, 69]. Such streamlining leads to a more efficient allocation of radiologist time and resources, enhancing the overall productivity of radiology departments [70, 71]. The efficiency gains translate into financial savings, as faster turnaround times for imaging studies can increase the number of patients served and reduce the need for additional staffing [72].

Advanced Diagnostic Services

The integration of AI in radiology enables the provision of advanced diagnostic services [73]. These services, being more precise and often more comprehensive than traditional methods, can attract a higher level of reimbursement from health insurance providers [73]. The enhanced diagnostic capabilities offered by AI can position healthcare providers as leaders in medical imaging, potentially attracting a larger patient base seeking high-quality care [73]. This not only increases the revenue for healthcare providers but also enhances their reputation in the healthcare market [73].

The expansion of AI in radiology is not just about cost savings; it is also opening new revenue channels [74]. For example, RadNet Inc. reported a 109% increase in revenues year-over-year from its AI initiatives, though it is important to note that the business line was still operating at a loss in 2023 [74]. This growth is a positive indicator of the potential revenue generation from AI-enhanced radiology services [74].

AI significantly contributes to the expansion of tele-radiology services [75, 78]. By enabling remote interpretation of medical images with high accuracy, AI facilitates radiological consultations across geographical boundaries [76]. This expansion is particularly beneficial in providing access to high-quality radiological services in remote or underserved areas, where such expertise might be scarce [77]. The financial implications are twofold: it opens new markets for radiological services and reduces the costs associated with establishing physical radiology departments in these areas [78].

Cost Reductions Across the Healthcare System

The accurate and efficient diagnostic capabilities of AI in radiology have broader implications for healthcare economics [79, 80]. Early and precise diagnosis facilitated by AI can prevent the progression of diseases to more severe stages, which typically require more intensive and expensive treatments [81]. This early intervention capability of AI-equipped radiology not only improves patient outcomes but also significantly reduces the long-term costs associated with chronic disease management and acute care [82].

Direct and Indirect Costs

The implementation of AI involves various costs, including software purchases (which can range from $10,000 to over $100,000), hardware upgrades, and training. However, indirect benefits, such as increased throughput (potentially increasing revenue by 10–20%) and reduced operational costs (savings can range from 5–15% of operational budgets), can offset these expenses over time [83, 84].

Increasing Revenue through Enhanced Efficiency

In a fee-for-service model, AI has the potential to increase revenue or reduce costs significantly [85, 86]. For example, by shortening the time patients spend in the emergency department by even 10–30 minutes, hospitals can see cost savings ranging from a few hundred to several thousand dollars per patient [87, 88].

By increasing the number of studies a radiologist can interpret by 20–30%, AI can translate into higher revenue streams for radiology departments. This could mean an additional $100,000 to $300,000 per radiologist annually, depending on the volume and reimbursement rates [89, 90].

Enhancing Diagnostic Accuracy and Patient Care

AI applications like automated rib fracture detection have increased detection rates by up to 300% in specific cases. On average, enhanced detection rates for various pathologies range from 10–20%. This not only improves patient care but also has the potential to increase revenue through additional procedures and therapies, which could translate to an additional $50,000 to $200,000 per year for a busy radiology practice [91].

By reducing errors, AI can save a facility from the costs associated with misdiagnosis, which can range from legal fees to additional treatments, potentially saving hundreds of thousands to millions of dollars in the long term [92].

AI can increase radiologist efficiency by 30–50% in reporting tasks, helping to address the shortage of radiologists. This increased efficiency means that a radiology practice might handle a larger volume of cases without compromising quality, potentially increasing annual revenue by 10–20% per radiologist [89].

Large Language Models

In the rapidly evolving field of radiology, the introduction of large language models (LLMs) and generative AI is marking a significant turn toward efficiency and accuracy, leading to both financial gains and challenges in implementation [93].

LLMs such as Radiology-Llama2, developed with instruction tuning, are now being tailored specifically for radiological applications. They are trained on domain-specific datasets to understand and generate radiological reports with high accuracy, surpassing other models in performance metrics [94]. Radiology-Llama2, for example, demonstrated superior performance across various datasets, highlighting its potential to support radiologists in clinical settings by generating coherent and clinically useful impressions from radiological findings [95]. This kind of advancement promises not only improved patient care but also potential cost savings by reducing the time radiologists spend interpreting scans [96].

Generative AI is also making strides in radiology by assisting in tasks such as drug development and improving the quality of radiological image data. Prototypes already show huge potential to support the work of radiologists by handling complex tasks more efficiently. For instance, at University Hospital Essen in Germany, generative AI is being used to create interfaces that allow for efficient communication with databases, enhancing the ability to extract patient information and aiding in developing improved treatments [97]. Projects in collaboration with companies like Siemens Healthineers are underway to develop software assistants for radiological diagnosis, reflecting the integration of generative AI into clinical workflows [98].

While the exact dollar figures for the cost savings and revenue generation attributed to LLMs and generative AI in radiology are still emerging, the growth of the AI medical imaging market— valued at $1.9 billion in 2022 and projected to reach $29.8 billion by 2032—reflects the significant economic impact these technologies are expected to have on the healthcare industry. Moreover, companies such as RadNet Inc. have reported substantial increases in revenue from their AI initiatives, which, despite initial losses, suggest a promising trend for the financial viability of AI in radiology [99].

Challenges

The fast-paced advancement in AI technology necessitates continuous investment in the latest AI tools and systems. Radiology departments must regularly update their AI software and hardware to keep pace with technological advancements, ensuring they harness the full potential of AI in medical imaging [100]. This requirement for ongoing investment represents a significant financial commitment but is crucial for maintaining a cutting-edge radiology service.

The implementation of AI in radiology must adhere to stringent ethical and regulatory standards, particularly concerning patient data privacy and the transparency of AI algorithms. Ensuring compliance with these standards involves additional costs related to data security, legal consultancy, and the development of ethical AI frameworks [101]. Despite these costs, adherence to ethical and regulatory guidelines is paramount to maintain patient trust and prevent potential legal challenges.

CONCLUSION

In conclusion, the financial impact of AI in radiology is profound and multifaceted. While the initial investment and ongoing costs present challenges, the long-term benefits of AI in terms of cost savings, operational efficiencies, and revenue generation offer a compelling case for its integration into radiological practices [102]. The economic benefits of AI in radiology extend well beyond the immediate healthcare setting, influencing the broader healthcare system by reducing long-term treatment costs and improving overall patient care efficiency.

Radiology, as a key diagnostic tool in medicine, is at the forefront of experiencing the transformative effects of AI. The advancements in AI technology promise not only to enhance the quality and accuracy of radiological services but also to redefine the financial models within which these services operate. As AI continues to evolve and integrate more deeply into radiological practices, it is expected to unlock even greater efficiencies and cost-saving opportunities, while also posing new challenges in terms of ethical considerations and the need for continuous technological updates [103].

The future of AI in radiology is a landscape of both opportunities and challenges. Navigating this landscape requires a balanced approach, leveraging AI's potential to enhance service delivery and patient care, while also addressing the financial, ethical, and regulatory implications that come with such advanced technology [104].

REFERENCES

1. "AI Index Report 2023." Artificial Intelligence Index, Stanford University, 2023. Retrieved from https://aiindex.stanford.edu/report/.

2. "Stanford: Fintech Maintains Position as Third Biggest AI Investment Focus Area." Fintechnews Switzerland, Fintech Schweiz Digital Finance News - FintechNewsCH, 25 Apr. 2023. Retrieved from https://fintechnews.ch/aifintech/stanford-fintech-maintains-position -as-third-biggest-ai-investment-focus-area/59671/.

3. "US Leads World on Gen AI Investment, Innovation, Implementation." PYMNTS, 24 Oct. 2023. Retrieved from https://www.pymnts.com/news/artificial-intelligence/2023/united -states-leads-world-generative-ai-investment-innovation-implementation/.

4. "AI Spending Will Jump to $154 Billion Worldwide in 2023." Insider Intelligence, 19 Apr. 2023. Retrieved from https://www.insiderintelligence.com/content/ai-spending-will-jump-billion -worldwide-2023.

5. "Worldwide Spending on AI-Centric Systems Will Pass $300 Billion by 2026." International Data Corporation (IDC), 12 Sept. 2022. Retrieved from https://www.idc.com/getdoc.jsp?containerId=prUS49670322.

6. "Worldwide Spending on AI-Centric Systems Forecast to Reach $154 Billion in 2023." International Data Corporation (IDC), 7 Mar. 2023. Retrieved from https://www.idc.com/ getdoc.jsp?containerId=prUS50454123.

7. Loucks, Jeff, Susanne Hupfer, David Jarvis, and Timothy Murphy. "AI Investment by Country – Survey." Deloitte Insights, 1 May 2019. Retrieved from https://www2.deloitte.com/us/en/ insights/focus/cognitive-technologies/ai-investment-by-country.html.

8. "Hyper Automation Market Size & Share, Growth Forecasts 2035." Research Nester. Retrieved from https://www.researchnester.com/reports/hyper-automation-market/3960#:~ :text=Hyper%20Automation%20Market%20size%20is,automation%20was%20USD%2020 %20Billion.

9. "Inside Amazon's Robotics Ecosystem." The Robot Report. Retrieved from https://www .therobotreport.com/inside-amazons-robotics-ecosystem/.

10. "A Look At Amazon's (AMZN) Initiatives in the Field of Robotics." Nasdaq. Retrieved from https://www.nasdaq.com/articles/a-look-at-amazons-amzn-initiatives-in-the-field-of-robot-ics-2020-07-29.

11. "Hitting the Books: How Amazon's Aggressive R&D Push Made it an e-Commerce Giant." Engadget. Retrieved from https://www.engadget.com/hitting-the-books-the-exponential -age-azeem-azhar-diversion-books-153033895.html.

12. "UPS: Driving Performance by Optimizing Driver Behavior." Harvard University. Retrieved from https://d3.harvard.edu/platform-digit/submission/ups-driving-performance-by-opti-mizing-driver-behavior/#:~:text=UPS%20cleverly%20framed%20this%20driver,results%20in %20a%20faster%2C%20more.

13. "Fresh Food Solutions." Lineage Logistics. Retrieved from https://www.onelineage.com/ news-stories/lineage-announces-lineage-fresh-expands-fresh-produce-offering-us.

14. "The Self-Awarehouse: Unleashing Lineage's Algorithms in Food Logistics." Lineage. Retrieved from https://www.onelineage.com/news-stories/self-awarehouse-unleashing -lineages-algorithms-warehouse-operations.

15. "Lineage Logistics Announces Lineage Fresh, Expands Fresh Produce Solutions." Business Wire. Retrieved from https://www.businesswire.com/news/home/20230207006000/en/ Lineage-Logistics-Announces-Lineage-Fresh-Expands-Fresh-Produce-Offering-In-Europe.

16. "2021 YEAR IN REVIEW: Feeding a Healing World While Achieving Safety Milestones." Lineage. Retrieved from https://www.onelineage.com/news-stories/2021-year-review-feeding-healing-world-while-achieving-new-heights-sustainability.

17. "Getting There: Lineage Logistics Adjust to a post-COVID World." Supermarket Perimeter. Retrieved from https://www.supermarketperimeter.com/articles/10103-getting-there-lineage-logistics-adjust-to-a-post-covid-world.

18. "Artificial Intelligence [AI] Market Size, Share & Forecast, 2030." Fortune Business Insights. Retrieved from https://www.fortunebusinessinsights.com/industry-reports/artificial-intelligence-market-100114.

19. "AI In Healthcare Market Size, Share & Growth Report, 2030." Grandview Research. Retrieved from https://www.grandviewresearch.com/industry-analysis/artificial-intelligence-ai-healthcare-market.

20. "AI in Healthcare Market Size, Share & Industry Report, 2029." Fortune Business Insights. Retrieved from https://www.fortunebusinessinsights.com/industry-reports/ai-in-healthcare-market-100534.

21. "Artificial Intelligence in Healthcare Market." Vantage Market Research. Retrieved from https://www.vantagemarketresearch.com/industry-report/artificial-intelligence-in-healthcare-market-1141.

22. "Artificial Intelligence (AI) in Healthcare Market Size, Share, Industry ..." Emergen Research. Retrieved from https://www.emergenresearch.com/industry-report/artificial-intelligence-in-healthcare-market.

23. "AI Augmentation Creating $2.9 Trillion of Biz Value | Gartner." Gartner. Accessed 3 Jan. 2024. Retrieved from https://gartner.com/en/newsroom/press-releases/2019-08-05-gartner-says-ai-augmentation-will-create-2point9-trillion-of-business-value-in-2021#:~:text=Decision%20Support%20and%20AI%20Augmentation,%2C%20according%20to%20Gartner%2C%20Inc.

24. "AI Augmentation Will Create Trillions of Business Value in 2021 — Gartner." Information Age. Accessed 3 Jan. 2024. Retrieved from https://www.information-age.com/ai-augmentation-trillions-business-value-2021-gartner-14404/.

25. "Exploring Opportunities in the Gen AI Value Chain | McKinsey." McKinsey & Company. Accessed 3 Jan. 2024. Retrieved from https://www.mckinsey.com/capabilities/quantumblack/our-insights/exploring-opportunities-in-the-generative-ai-value-chain.

26. "AI in Fintech Market - Size, Trends, Report & Growth - Mordor Intelligence." Mordor Intelligence. Accessed 3 Jan. 2024. Retrieved from https://www.mordorintelligence.com/industry-reports/ai-in-fintech-market/market-size.

27. "Artificial Intelligence In Fintech Market Size Report, 2030." Grand View Research. Accessed 3 Jan. 2024. Retrieved from https://www.grandviewresearch.com/industry-analysis/artificial-intelligence-in-fintech-market-report.

28. "AI in Fintech Market Size, Industry Share & Trend – 2032." Future Market Insights. Accessed 3 Jan. 2024. Retrieved from https://www.futuremarketinsights.com/reports/ai-in-fintech-market.

29. "AI in Financial Services: Key Trends & Opportunities for 2022." Wizeline. Accessed 3 Jan. 2024. Retrieved from https://www.wizeline.com/ai-in-financial-services-key-trends-opportunities-for-2022/.

30. "AI Augmentation Will Create $2.9 Trillion of Business Value in 2021 ..." CIOL. Accessed 3 Jan. 2024. Retrieved from https://www.ciol.com/artificial-intelligence-job-creators-not-killers/.

31. "Artificial Intelligence [AI] Market Size, Share & Forecast, 2030." Fortune Business Insights. Retrieved from https://www.fortunebusinessinsights.com/industry-reports/artificial-intelligence-market-100114.

32. "Artificial Intelligence Market Size, Share, Growth Report 2030." Grandview Research. Retrieved from https://www.grandviewresearch.com/industry-analysis/artificial-intelligence-ai-market.

33. "Artificial Intelligence Market Size & Trends, Growth Analysis, Forecast ..." MarketsandMarkets. Retrieved from https://www.marketsandmarkets.com/Market-Reports/artificial-intelligence-market-74851580.html.

34. "Artificial Intelligence Market Size, Share & Growth Report 2030." GMI Research. Retrieved from https://www.gmiresearch.com/report/global-artificial-intelligence-ai-market/.

35. "Artificial Intelligence Industry Overview | North America to Capture 56% Market Share by 2025." PR Newswire. Retrieved from https://www.prnewswire.com/news-releases/artificial-intelligence-industry-overview--north-america-to-capture-56-of-total-ai-market-share--17-000-technavio-research-reports-301377871.html.

36. "Can AI Transform Small Businesses and Grow Prosperity Across North America?" VentureBeat. Retrieved from https://venturebeat.com/2021/06/26/can-ai-transform-small-businesses-and-grow-prosperity-across-north-america/.

37. "The Total Economic Impact™ Of IBM Services And Watson Cognitive ..." IBM. Retrieved from https://www.ibm.com/downloads/cas/MXWB3D6P.

38. "The Total Economic Impact™ Of IBM Watson Natural Language Processing ..." IBM. Retrieved from https://www.ibm.com/downloads/cas/XMRMP7XK.

39. "IBM Security MaaS360 With Watson | ITPro." ITPro. Retrieved from https://www.itpro.com/the-total-economic-impact-of-ibm-maas360-with-watson.

40. "IBM's Watson: Can it Improve Returns on R&D?" ZDNET. Retrieved from https://www.zdnet.com/article/ibms-watson-can-it-improve-returns-on-r-d/.

41. "The Total Economic Impact Of IBM OpenPages - Final - 2020109." IBM. Retrieved from https://www.ibm.com/downloads/cas/8ZDXNKQ4.

42. "The Total Economic Impact of IBM Watson Studio and Watson Knowledge ..." ADTmag. Retrieved from https://adtmag.com/whitepapers/2018/08/ibm-the-total-economic-impact.aspx.

43. "Artificial Intelligence in Business: 9 Functions to Watch in 2023 ..." RediMinds. Retrieved from https://rediminds.com/blog/artificial-intelligence-in-business/.

44. "15 Top Applications of Artificial Intelligence in Business." TechTarget. Retrieved from https://www.techtarget.com/searchenterpriseai/tip/9-top-applications-of-artificial-intelligence-in-business.

45. "How Artificial Intelligence Is Transforming Business ..." Business News Daily. Retrieved from https://www.businessnewsdaily.com/9402-artificial-intelligence-business-trends.html.

46. "Use Cases for Computer Vision in Analytics" Roboflow. Retrieved from https://blog.roboflow.com/computer-vision-in-data-analytics/.

47. "Google Cloud Launches AI-powered Solutions to Safely Accelerate Drug ..." PR Newswire. Retrieved from https://www.prnewswire.com/news-releases/google-cloud-launches-ai-powered-solutions-to-safely-accelerate-drug-discovery-and-precision-medicine-301825674.html

48. "Google Cloud Adds New Features to Vertex AI Search for Healthcare and ..." PR Newswire. Retrieved from https://www.prnewswire.com/news-releases/google-cloud-adds-new-features-to-vertex-ai-search-for-healthcare-and-life-science-companies-301950223.html#:~:text=The%20new%20features%20announced%20today,retain%20control%20over%20their%20data.

49. "Google Cloud Launches AI Solutions for Drug ..." HealthITAnalytics. https://healthitanalytics.com/news/google-cloud-launches-ai-solutions-for-drug-discovery-precision-medicine.

50. "Artificial Intelligence in Healthcare Market worth $102.7 Billion by 2028." Yahoo Finance. Retrieved from https://finance.yahoo.com/news/artificial-intelligence-healthcare-market-worth-140000146.html.

51. "AI in Healthcare Market Size, Share & Growth Report, 2030." Grand View Research. Retrieved from https://www.grandviewresearch.com/industry-analysis/artificial-intelligence-ai-healthcare-market.

52. "AI in Healthcare Market Size, Share, Growth Report to 2028." Market Data Forecast. Retrieved from https://www.marketdataforecast.com/market-reports/artificial-intelligence-in-healthcare-market.

53. "AI in Healthcare Market Size, Share & Industry Report, 2029." Fortune Business Insights. Retrieved from https://www.fortunebusinessinsights.com/industry-reports/artificial-intelligence-in-healthcare-market-100534.

54. "To Pay or Not to Pay for Artificial Intelligence Applications in Radiology." Nature. Accessed 3 Jan. 2024. Retrieved from https://www.nature.com/articles/s41746-023-00861-4#:~:text=radiology%20AI%20applications.-,As%20a%20rule%20of%20thumb%2C%20we%20propose%20that%20radiology%20AI,diagnostic%20performance%20leads%20to%20improved.

55. "How Does Artificial Intelligence in Radiology Improve Efficiency and ..." Springer. Accessed 3 Jan. 2024. Retrieved from https://link.springer.com/article/10.1007/s00247-021-05114-8#:~:text=We%20describe%20six%20clinical%20objectives,accuracy%20and%20more%20personalized%20diagnostics.

56. "The Cost of AI in Radiology: Is It Really Worth it?" European Society of Radiology. Accessed 3 Jan. 2024. Retrieved from https://academic.oup.com/jbi/article/4/3/302/6548154.

57. "The Economic Impact of AI on Breast Imaging." Oxford Academic. Accessed 3 Jan. 2024. Retrieved from https://academic.oup.com/jbi/article/4/3/302/6548154.

58. "Applications of Artificial Intelligence (AI) in Diagnostic Radiology: A ..." PubMed Central. Accessed 3 Jan. 2024. Retrieved from https://pubmed.ncbi.nlm.nih.gov/32945967/.

59. "Artificial Intelligence in Radiology: Opportunities and Challenges." Radiologic Clinics. Accessed 3 Jan. 2024. Retrieved from https://www.radiologic.theclinics.com/article/S0033 -8389(21)00117-2/fulltext.

60. "Impact of the Rise of Artificial Intelligence in Radiology: What do ..." ScienceDirect. Accessed 3 Jan. 2024. Retrieved from https://www.sciencedirect.com/science/article/pii/ S2211568419300907.

61. "Stakeholders' Perspectives on the Future of Artificial ... - Springer." Springer. Accessed 3 Jan. 2024. Retrieved from https://link.springer.com/article/10.1007/s00330-021-08214-z.

62. "AI Can Help Reduce Radiology Errors by Learning the Features of a High-Quality Image, Applying Computational Strategies to Increase the Odds of Producing That Image, and Automatically Compensating for Any Distortions." Patient Safety Network. Agency for Healthcare Research and Quality, 2024. Retrieved from https://psnet.ahrq.gov/perspective/ artificial-intelligence-and-diagnostic-errors.

63. "Implementing Processes to Learn from Diagnostic Error Is Key to Reducing Errors." RadioGraphics. Radiological Society of North America, 2024. Retrieved from https://pubs .rsna.org/doi/abs/10.1148/rg.2018180021#:~:text=Implementing%20processes%20to%20learn %20from,errors%20involving%20imaging%20annually%20worldwide.

64. "The Current and Future State of AI Interpretation of Medical Images." The New England Journal of Medicine. Massachusetts Medical Society, 2024. Retrieved from https://www.nejm .org/doi/10.1056/NEJMra2301725.

65. "Artificial Intelligence in Radiology." Nature Reviews Cancer. Nature Publishing Group, 2024. Retrieved from https://www.nature.com/articles/s41568-018-0016-5.

66. "Diagnostic Errors in Radiology - Healthcare AI." Aidoc. Aidoc Medical Ltd., 2024. Retrieved from https://www.aidoc.com/blog/diagnostic-errors-in-radiology/.

67. "Imaging AI in Practice: A Demonstration of Future Workflow." Radiological Society of North America. Radiological Society of North America, 2024. Retrieved from https://pubmed .ncbi.nlm.nih.gov/34870224/.

68. "Streamlining the Radiology Workflow to Improve Efficiency and Capacity." GE Healthcare. General Electric Company, 2024. Retrieved from https://www.gehealthcare.com/article/ streamlining-the-radiology-workflow-to-improve-efficiency-and-capacity.

69. "Rad AI Showcases New Features in AI-Driven Workflow Tools." Applied Radiology. Anderson Publishing Ltd., 2024. Retrieved from https://appliedradiology.com/articles/rad-ai -showcases-new-features-in-ai-driven-workflow-tools.

70. "AI in Radiology: Essentials of An Integrated Strategy." What's Next blog. Nuance Communications, Inc., 2024. Retrieved from https://whatsnext.nuance.com/healthcare/ai-in -radiology-essentials-of-an-integrated-strategy.

71. "Optimization of Radiology Workflow with Artificial Intelligence." Radiology Key. Radiology Key, 2024. Retrieved from https://radiologykey.com/optimization-of-radiology -workflow-with-artificial-intelligence.

72. "To Pay or Not to Pay for Artificial Intelligence Applications in Radiology." Nature. Nature Publishing Group, 2024. Retrieved from https://www.nature.com/articles/s41746-023-00861 -4.

73. "Rad AI to Unveil Next-Generation Intelligent Radiology Reporting." Rad AI. RadNet, 2024. Retrieved from https://www.prnewswire.com/news-releases/rad-ai-to-unveil-next-genera-tion-intelligent-radiology-reporting-solution-at-launch-event-301905959.html.

74. "RadNet Expects to Log Upward of $18M in Revenue from its AI Division." Radiology Business. Radiology Business News, 2024. Retrieved from https://radiologybusiness.com/topics/artificial-intelligence/radnet-expects-log-upward-18m-revenue-its-ai-division-year.

75. "Artificial Distance: AI in Teleradiology." Aidoc. Accessed 3 Jan. 2024. Retrieved from https://www.aidoc.com/blog/artificial-distance-ai-in-teleradiology.

76. "Artificial Intelligence for the Future Radiology Diagnostic Service." Frontiers in Medicine. Accessed 3 Jan. 2024. Retrieved from https://www.frontiersin.org/articles/10.3389/fmolb .2020.614258/full.

77. "How Does Artificial Intelligence in Radiology Improve Efficiency and ..." SpringerLink. Accessed 3 Jan. 2024. Retrieved from https://link.springer.com/article/10.1007/s00247-021 -05114-8.

78. "AI and Teleradiology Impact Healthcare." HealthManagement.org. Accessed 3 Jan. 2024. Retrieved from https://healthmanagement.org/c/imaging/news/ai-and-teleradiology -impact-healthcare.

79. "Positive Economic Impact of AI in Healthcare." PubMed Central, National Center for Biotechnology Information. Accessed 3 Jan. 2024. Retrieved from https://www.ncbi.nlm.nih .gov/pmc/articles/PMC7059082/.

80. "The Potential Impact of Artificial Intelligence on Healthcare Spending." National Bureau of Economic Research. Accessed 3 Jan. 2024. Retrieved from https://www.nber.org/papers/ w30857.

81. "Economics of Artificial Intelligence in Healthcare." PubMed Central, National Center for Biotechnology Information. Accessed 3 Jan. 2024. Retrieved from https://www.ncbi.nlm.nih .gov/pmc/articles/PMC9777836/.

82. "The Rise of Artificial Intelligence in Healthcare Applications." PubMed Central, National Center for Biotechnology Information. Accessed 3 Jan. 2024. Retrieved from https://www .ncbi.nlm.nih.gov/pmc/articles/PMC7325854/.

83. "To Pay or Not to Pay for Artificial Intelligence Applications in Radiology." Nature. Accessed 3 Jan. 2024. Retrieved from https://www.nature.com/articles/s41746-023-00861-4.

84. "Artificial Intelligence ROI Considerations in Radiology." Radiology Business. Accessed 3 Jan. 2024. Retrieved from https://radiologybusiness.com/topics/artificial-intelligence/artifi-cial-intelligence-roi-considerations-radiology#:~:text=Radiology%20practice%20ROI%20for %20quality%20and%20efficiency&text=AI%20tools%2C%20particularly%20in%20computer ,pathologies%2C%20significantly%20benefiting%20patient%20outcomes.

85. "Streamlining the Radiology Workflow to Improve Efficiency and Capacity." GE Healthcare. Accessed 3 Jan. 2024. Retrieved from https://www.gehealthcare.com/article/streamlining -the-radiology-workflow-to-improve-efficiency-and-capacity.

86. "Trending AI and Digital Solutions in Radiology." GE Healthcare. Accessed 3 Jan. 2024. Retrieved from https://www.gehealthcare.com.au/insights/article/trending-ai-and-digital -solutions-in-radiologycurrent-adoption-applications-and-continuing-developme#:~:text=An %20AI%20tool%20created%20by,a%20single%20technologist%5B3%5D.

87. "Workflow Applications of Artificial Intelligence in Radiology." ScienceDirect. Accessed 3 Jan. 2024. Retrieved from https://www.sciencedirect.com/science/article/abs/pii/S1546144020308760.

88. "Artificial Intelligence in Radiology." PubMed Central, National Center for Biotechnology Information. Accessed 3 Jan. 2024. Retrieved from https://www.ncbi.nlm.nih.gov/pmc/articles/PMC6616181.

89. "Collecting Earned Revenue: How AI Maximizes Radiology Payments." RadiologyBusiness.com. Accessed 3 Jan. 2024. Retrieved from https://radiologybusiness.com/sponsored/35131/infinx-healthcare/topics/healthcare-management/healthcare-economics/medical-billing.

90. "How AI Can Help Radiology Groups Address Reimbursement Changes." RADAI.com. Accessed 3 Jan. 2024. Retrieved from https://www.radai.com/blogs/how-ai-can-help-radiology-groups-address-reimbursement-changes.

91. "The Impact of AI in Radiology: Stats to Know." MammoScreen.com. Accessed 3 Jan. 2024. Retrieved from https://www.mammoscreen.com/ai-radiology-stats-2021.

92. "Johns Hopkins Radiology Explores the Potential of AI in the Reading Room." HopkinsMedicine.org. Accessed 3 Jan. 2024. Retrieved from https://www.hopkinsmedicine.org/news/articles/2023/11/johns-hopkins-radiology-explores-the-potential-of-ai-in-the-reading-room.

93. "Radiology-GPT: A Large Language Model for Radiology." arXiv.org. Retrieved from https://arxiv.org/abs/2306.08666.

94. "Radiology-Llama2: Best-in-Class Large Language Model for Radiology." Retrieved from https://arxiv.org/abs/2309.06419.

95. "RadBERT: Adapting Transformer-based Language Models to Radiology." Radiological Society of North America. Retrieved from https://pubs.rsna.org/doi/full/10.1148/ryai.210258.

96. "[2312.13103] Exploring Multimodal Large Language Models for Radiology ..." arXiv.org. Retrieved from https://arxiv.org/abs/2312.13103.

97. "Large Language Models and Structured Reporting: Never Stop Chasing ..." Radiology Medical, link.springer.com. DOI: 10.1007/s11547-023-01651-4. Retrieved from https://link.springer.com/article/10.1007/s11547-023-01711-9.

98. Mallio, C.A., et al. "Large Language Models for Structured Reporting in Radiology." Radiology Medical, 2023. Retrieved from https://pubmed.ncbi.nlm.nih.gov/37248403/.

99. "RadNet Reports Second Quarter Financial Results, with Record Quarterly Revenue and Adjusted EBITDA(1), and Updates 2023 Financial Guidance Ranges." RadNet. Retrieved from https://www.radnet.com/about-radnet/news/radnet-reports-second-quarter-financial-results-record-quarterly-revenue-and.

100. "An Economic Overview of the Value of AI in Radiology." Aidoc, www.aidoc.com. Accessed 2024. Retrieved from https://www.aidoc.com/blog/value-artificial-intelligence-radiology/.

101. "Artificial Intelligence in Radiology—Ethical Considerations." Radiological Society of North America et al., 2019. Retrieved from https://www.ncbi.nlm.nih.gov/pmc/articles/PMC7235856/.

102. "The Economic Impact of AI on Breast Imaging." Oxford Academic. Retrieved from https://academic.oup.com/jbi/article/4/3/302/6548154.

103. "To Pay or Not to Pay for Artificial Intelligence Applications in Radiology." Nature. Retrieved from https://www.nature.com/articles/s41746-023-00861-4.

104. "Who Will Pay for AI?" Radiology: Artificial Intelligence. Retrieved from https://pubs.rsna .org/radiology/doi/10.1148/ryai.2021210030#:~:text=In%20the%20past%20year%2C%20reim- bursement,Prospective%20Payment%20System%20(IPPS).

PART IV

CURRENT STATE OF AI IN RADIOLOGY

13 Clinical View

Alexander E. Jacobs, Elisa Hofmeister, Jordan Helmer, and Christopher Prokosch

INTRODUCTION

In the past decade, there have been hundreds of firms founded and thousands of papers published exploring AI algorithms that aim to enhance diagnostic accuracy, make radiologist workflows more efficient, and improve patient outcomes [1]. Many of these algorithms show promise in controlled settings, though their performance is typically observed to deteriorate when deployed clinically [1]. Before achieving widespread clinical adoption, radiologists and developers will need to construct tools that produce robust, reproducible outcomes while remaining flexible enough to accommodate inputs that may vary significantly from the data on which the software was trained. And they must do this all while operating within frequently shifting clinical guidelines and regulatory environments [1].

AREAS OF CURRENT DEPLOYMENT

Image processing and quantification algorithms of the chest, breast, and brain constitute a significant majority of current uses of clinical AI in radiology [2]. Currently available intrathoracic algorithms encompass a wide range of conditions, from pneumothorax triage to classification of chronic lung disease to lesion detection. In the brain, applications focus on "detection and triage of intracranial hemorrhage, large vessel obstruction, and for brain segmentation on MR imaging" [3]. The variety of possible applications extends to every region of the body. However, there has yet to be a publicly available AI/ML algorithm that works across systems—nearly all available applications focus on a single body region [3].

Lessons can be drawn from the current areas of industry focus. For instance, AI-assisted detection of neoplastic changes using lung CT and two-dimensional mammography are an areas of concerted investigation not only because of the high disease burden of lung and breast cancers, but also because the binary nature of lesion detection and the large size of available datasets make these pathologies attractive targets for AI. According to one 2020 review, algorithms have focused on diagnosing nodules on chest CT scans and mammography scans because there are large datasets available and are binary classification tasks (present vs. absent) [2].

While artificial intelligence is widely seen as a major component of radiology's future, it is still in its early stages. Venture capital continues to embrace the industry, though no significant breakthrough software firm has yet to emerge. One analysis, based on publicly available records, projects that the market capitalization of all FDA-approved AI/ML applications to approach approximately $2.0 billion by 2022 [2]. As a monetary total encompassing an entire industry, this number stands in contrast to the much larger market capitalizations of firms in mature adjacent industries (such as medical devices) or broader healthcare expenditures in the United States, which totaled $4.1 trillion in 2020 [4].

The potential transformations that AI will bring to radiology are significant, yet there remains, at the time of this writing, fewer than 200 FDA-approved AI applications, according to one widely used and regularly updated list of AI medical products [5]. The reasons for the apparent gap between AI's significant clinical potential and the paucity of market-ready products are primarily due to process-based and regulatory limitations. There are few theoretical limitations on AI's potential role in radiology, but first developers need to navigate challenges regarding data access and management, implementation costs, and privacy/regulatory concerns. The collective challenge is to translate theoretical capability into clinical outcomes because AI in radiology has not yet had a proven direct improvement on patient care or clinical practice [6].

CURRENT LIMITATIONS

Scale of Available Data

A significant bottleneck in developing clinically deployable AI is the availability and size of relevant datasets. For instance, lung CT and mammography processing algorithms are the most common targets for AI/ML tools today, in part because they can rely on large pools of pre-existing radiologic studies to train their algorithms. This underscores the central importance of large and accurately annotated datasets for training new AI/ML applications. However, appropriately labeling and curating medical images for use in ML training requires expertise and is costly [1].

DOI: 10.1201/9781003095279-17

Dataset size and quality is limited by more than economic considerations. Because medical images often reside within different picture archiving and communication systems (PACS) and electronic health records (EHR), coordination between different entities is constrained by privacy concerns and technological incompatibilities. This further limits the size of datasets available to developers. Data use permissions are another concern; as one review of clinical AI applications noted, "FDA-approved products use much smaller datasets compared with open-source AI tools, because the terms of use of public datasets are limited to academic and noncommercial entities, which precludes their use in commercial products" [1].

In a 2021 review of 118 FDA-approved machine learning products, the authors quantified the limited scope of datasets used in many of the algorithms. Out of 66 AI algorithms with stated patient and/or imaging data for validation studies, 68% had a total sample size of less than 500 patients. Only 2 out of the 66 algorithms had over 4,500 patients prior to gaining FDA approval [3].

In part due to the underpowered size of these datasets, as well as the lack of publicly available validation studies, it is difficult to justify clinical application because presence of bias cannot be discerned [3]. Publicly available summaries of FDA-approved software typically lack internal validation data (often withheld for proprietary reasons) that are critical for an objective assessment of software effectiveness [3]. Given the risks of adopting any new technology into clinical practice, this lack of transparency makes clinical adoption less likely. Economic stakeholders—hospital administrators, radiologists, insurers—will likely require much more comparative data before investing in new software and clinical workflows and accepting their attendant risks.

Internal Validation vs. Clinical Performance

A related issue is that AI/ML algorithms are often limited in their generalizability across clinical settings. There is often a decrease in performance when algorithms are applied to the clinic [1, 3]. This deterioration in performance is due to overlapping circumstances, from varying image quality between training and clinical datasets, ever-changing clinical guidelines, and different imaging equipment and protocols used to produce underlying data.

There are technical and regulatory reasons for clinical AI applications to be limited to a single clinical use with little broader generalizability. On the technical side, no firm has successfully demonstrated a generalized AI/ML application that can work across multiple organ systems. On the regulatory side, it is challenging for companies to build software that is suitably flexible to accommodate the kinds of varying inputs that are typical in clinical radiology without risking FDA certification. In order to retain FDA certification, such algorithms cannot stray far from the state in which they were approved [1]. The kinds of on-the-job adjustments that radiologists routinely make when working across different imaging modalities, EHR/PACS systems, and pathologies are, at the moment, difficult for software to accomplish without sacrificing accuracy or regulatory approval.

Data Accessibility

In the United States, medical data are protected by the Health Insurance Portability and Accountability Act of 1996 (HIPAA) and similar protections exist in many other countries. A report from the 2019 meeting of the International Society for Strategic Studies in Radiology emphasized that for "effective and responsible data sharing, it is critical that ethical standards are followed and that privacy concerns are addressed." They note that privacy is of special importance for image-based media, as "image data sharing … has a rather high potential for data re-identification … this is especially true for images that include facial features that could be reconstructed allowing subject identification" [7].

Even when data are successfully depersonalized, the landscape of American healthcare is influenced by a focus on privacy as a predominant concern, making institutions reluctant to share clinically relevant data. Marketplace rivalries exacerbate this trend between private firms, whether they are vendors of competing EHR/PACS systems or developers of AI/ML software. As innovation in the field of AI/ML and radiology is centered in the United States, these trends affect progress worldwide. Given the significant investments required to construct secure, high-quality datasets and the potential financial penalties for a significant privacy breach, incentives currently favor institutions and companies limiting access to shared data [7]. Although there exist promising new methods for securely conducting machine learning via decentralized, cloud-based applications that allow for institutions to protect their data while benefiting from scale, these applications are not yet ready for clinical use [8].

BUILDING TRUST IN PERFORMANCE OF AI ALGORITHMS

Building trust in algorithms is crucial for clinical acceptance. Many radiologists are reluctant to embrace any AI that cannot explain the process by which it reaches conclusions. This is a point of contention with deep learning algorithms that possess the well-known "black box" property, in which it is impossible to recreate the internal state of a neural network that corresponds to a specific result [9]. While some radiologists may be skeptical of algorithms whose inner workings they cannot parse, others may uncritically trust any AI model's output because of the veneer of objectivity ascribed to computerized systems. This problem, known as automation bias, is well-established in medicine and in many other technical fields [10]. Across domains, there is a human tendency to defer to conclusions generated by automated systems, even in the presence of contra-dictory information [10]. Whether radiologists treat algorithms skeptically due to their "black box" nature or accept automated results uncritically because they are perceived as more "objective" than humans, there is a clear need for radiologists to construct frameworks for deciding which clinical AI algorithms to trust and be empowered to provide feedback as a part of continuous quality monitoring and improvement.

A prominent area of inquiry for increasing trust in clinical AI is explainable AI (XAI). The concept of explainability "can be viewed as an active characteristic of a model, denoting any action or procedure taken by a model with the intent of clarifying or detailing its internal functions" [9]. XAI can take many forms, such as heat maps that overlay an image and indicate, via color-coding, the degree to which different regions of the image were weighted by the algorithm in generating its output [11].

While enhancing an algorithm's explainability may appear to be a self-evident benefit, it is not without drawbacks. The underlying premise of XAI is that for an algorithm's output to be trusted it must be explained, yet algorithms that maximize explainability may sacrifice accuracy, which in turn jeopardizes trust [9]. While models that co-maximize explainability and performance are a target of XAI research, they are far from being integrated into contemporary clinical workflows [9, 12].

The adage in ML circles that "a model is only as good as its data" underlines the final issue with explainable AI: if the inputs are flawed, corrupted, miscategorized, or otherwise inaccurate, an algorithm may generate explicable yet inaccurate results. Thus, explainability is not sufficient for generating algorithmic trust. These issues with explainability lead some radiologists and com-puter scientists to question the usefulness of explainability as a proxy for trust altogether [11].

If XAI is not the answer, or at least not the sole answer, then how can radiologists learn to trust AI? In the short term, trustworthiness may be aided by adopting checklists for evaluation of new AI applications, such as the Checklist for AI in Medical Imaging (CLAIM). CLAIM evalu-ates algorithmic performance based on reproducibility and transparency of the underlying data upon which models are trained, providing a formal structure for assessing new AI applications [13]. Ultimately, however, there is likely no single technical or administrative solution to generate trust in AI. Rather, trust will build over time as radiologists grow comfortable with algorithmic aids in clinical settings [11]. Trust is not a function of knowing "why" an algorithm works (i.e., explainability), but a relationship between user and technology that forms if a technology reliably does what it claims to do [14]. Once an algorithm is shown to be effective, is subject to continual oversight and updating, and exists within an environment where data standards and workflow standards are held to the same level of scrutiny and revision as the algorithm itself—only then can radiologists trust clinical AI to the same extent that they trust the other technologies they rely on clinically.

CURRENT WORKFLOWS AND THEIR EVIDENCE

Few practicing radiologists today, outside of research and software development contexts, interact with technologies that could meaningfully be described as artificial intelligence, especially the kinds of transformational algorithms that proponents hope for.

The translational gap between AI's promise and implementation is reflected in the lack of clinical studies assessing AI in radiology. One analysis found that while a PubMed query for "radiology" + "AI" returned around 2,000 results, fewer than 2% of the results discussed clinical implementations [1]. According to the authors, a closer examination of that 2% revealed most of the results were not studies but "opinions and not reports of actual implementation." The authors of this analysis also conducted a search of the ClinicalTrials.gov database, in which they found 71 registered trials within the field AI and radiology. Of these trials, 13 dealt with clinical application

Table 13.1 Eighteen AI Applications Demonstrating Evidence of Clinical Impact According to Van Leeuwen et al.

Company Name	Product Name	On Market Since	Disease Targeted	Modality	Use	FDA Approved?
QView Medical	QVCAD	September 2016	Breast cancer	Ultrasound	Abnormality detection and analysis	Yes, Class III
ScreenPoint Medical	Transpara	September 2015	Breast cancer	2D and 3D mammography	Detection with analysis, exam, and risk scores	Yes, Class II
Volpara Solutions	VolparaDensity	October 2010	Breast cancer	2D and 3D mammography	Breast density quantification	Yes, Class II
HeartFlow	HeartFlow FFRCT Analysis	January 2016	Coronary artery disease	CT	FFR score estimation	Yes, Class II
Lunit	Lunit INSIGHT CXR	November 2019	Lung pathologies	X-ray	Abnormality detection and score with interpretation	Yes, Class II
MeVis Medical Solutions AG	Veolity	March 2015	Lung cancer	CT	Nodule detection and comparison with report generation	Yes, Class II
Riverain Technologies	ClearRead CT - Detect	November 2015	Lung cancer	CT	Nodule detection	Yes, Class II
Riverain Technologies	ClearRead CT - Vessel Suppress	November 2015	Lung cancer	CT	Vessel suppression	Yes, Class II
Riverain Technologies	ClearRead Xray - Bone Suppress	November 2008	Lung cancer	X-ray	Bone suppression	Yes, Class II
Riverain Technologies	ClearRead Xray - Detect	November 2008	Lung cancer	X-ray	Lung nodule detection	Yes, Class III
Thirona	CAD4TB	October 2014	Tuberculosis	X-ray	Tuberculosis risk score with heatmap	No
ImageBiopsy Lab	IB Lab KOALA	August 2019	Knee osteoarthritis	X-ray	Osteoarthritis detection and guideline classification	Yes, Class II
VUNO	VUNO Med®-BoneAge	May 2018	Growth disorders	X-ray	Bone age assessment with report generation	No

(Continued)

Table 13.1 (Continued)   Eighteen AI Applications Demonstrating Evidence of Clinical Impact According to Van Leeuwen et al.

Company Name	Product Name	On Market Since	Disease Targeted	Modality	Use	FDA Approved?
Brainomix	e-ASPECTS	March 2015	Stroke	CT	Occlusion and hyperdensity detection, quantification and scoring with report generation	No
Brainomix	e-CTA	May 2018	Stroke	CT	Collateral assessment, occlusion detection with report generation	Yes, Class II
Cortechs.ai	NeuroQuant	August 2006	Neurodegenerative brain disorders	MR	Brain region quantification with report generation	Yes, Class II
Future Processing	Sens.ai	March 2020	Glioblastoma	MR	Lesion segmentation and volume calculation	No
AmCad BioMed	AmCAD-UT®	2017	Thyroid cancer and nodules	Ultrasound	Nodule recognition with risk analysis	Yes, Class II

Data extracted from publicly available data collated at "AI for Radiology: An Implementation Guide" https://grand-challenge.org/aiforradiology/

of AI but these trials were listed as recruiting, not yet recruiting, or active but not recruiting [1]. This trend is further underscored by the small number of FDA-approved applications on the market, as has been discussed elsewhere.

Today, arguments in favor of clinical AI largely rely on its theoretical potential, but there remains a lack of evidence in support of it in the clinic today. One 2021 review of commercially available radiology algorithms assessed the scientific evidence underlying various AI/ML applications. They found that, out of a pool of 100 programs, 64/100 lacked peer-reviewed evidence of efficacy. Analyzing the remaining 36/100, they found that only half (18/100) were supported by robust enough evidence to convincingly demonstrate whether the algorithms positively influence diagnosis, costs, or other patient outcomes [15].

An overview of the 18 technologies meeting the authors' highest evidentiary standards is a useful illustration of factors that radiologists and developers should consider when evaluating which applications are robust enough for current clinical practice. Publicly available summary data for these 18 applications are reproduced in Table 13.1. Notably, only three of the included applications were released after 2018, indicating that more mature product offerings likely afford more time for peer-reviewed analysis—a promising sign that some recently approved technologies currently lacking independently validated data may eventually have it. Furthermore, all of the applications have discrete clinical goals, aimed at classification and quantification problems limited to a single organ or set of interrelated pathologies. Evidence currently favors applications with a narrow focus.

The lack of clinical data leads some to conclude that "AI for medical imaging has been characterized by hype, with exaggerated claims of superhuman performance compared with clinicians" [1]. At a minimum, there exists debate between those who feel an AI revolution is imminent, and those that feel that the adoption of clinical AI will proceed in a piecemeal fashion over years and decades. In either scenario, there will need to be an increase in clinical studies demonstrating application efficacy before widespread adoption is possible.

CONCLUSION

In recent years, there has been growing promise and speculation regarding the role of artificial intelligence in clinical radiology. Predictions that radiologists would imminently be replaced by computers have yet to come to fruition. To the extent artificial intelligence has infiltrated radiology workflows, it is as an adjunct to radiologist-centered processes, such as triage or second-read applications.

While AI/ML applications in radiology are in their infancy, the theoretical underpinning for these technologies remains strong. Before that promise can be realized, stakeholders need to address barriers to implementation, from data-sharing and privacy concerns, to conducting trials that demonstrate the clinical utility of artificial intelligence in radiology.

REFERENCES

1. Tariq, A., Purkayastha, S., Padmanaban, G. P., Krupinski, E., Trivedi, H., Banerjee, I., & Gichoya, J. W. Current Clinical Applications of Artificial Intelligence in Radiology and Their Best Supporting Evidence. *Journal of the American College of Radiology: JACR* (2020). https://doi.org/10.1016/j.jacr.2020.08.018

2. Tadavarthi, Y., Vey, B., Krupinski, E., Prater, A., Gichoya, J., Safdar, N., & Trivedi, H. The State of Radiology AI: Considerations for Purchase Decisions and Current Market Offerings. *Radiology. Artificial Intelligence* (2020). https://doi.org/10.1148/ryai.2020200004

3. Ebrahimian, S., Kalra, M. K., Agarwal, S., Bizzo, B. C., Elkholy, M., Wald, C., Allen, B., & Dreyer, K. J. FDA-regulated AI Algorithms: Trends, Strengths, and Gaps of Validation Studies. *Academic Radiology* (2022). https://doi.org/10.1016/j.acra.2021.09.002

4. Hartman, M., Martin, A. B., Washington, B., Catlin, A., & The National Health Expenditure Accounts Team. National Health Care Spending In 2020: Growth Driven By Federal Spending In Response To The COVID-19 Pandemic: National Health Expenditures Study Examines US Health Care Spending in 2020. *Health Affairs* (2022). https://doi.org/10.1377/hlthaff.2021.01763

5. AI Central. (n.d.). Retrieved April 29, 2022, from https://aicentral.acrdsi.org/

6. Mongan, J., Vagal, A., & Wu, C. C. Imaging AI in Practice: Introducing the Special Issue. *Radiology. Artificial Intelligence* (2022). https://doi.org/10.1148/ryai.220039

7. Recht, M. P., Dewey, M., Dreyer, K., Langlotz, C., Niessen, W., Prainsack, B., & Smith, J. J. Integrating Artificial Intelligence into the Clinical Practice of Radiology: Challenges and Recommendations. *European Radiology* (2020). https://doi.org/10.1007/s00330-020-06672-5

8. Warnat-Herresthal, S., Schultze, H., Shastry, K. L., Manamohan, S., Mukherjee, S., Garg, V., Sarveswara, R., Händler, K., Pickkers, P., Aziz, N. A., Ktena, S., Tran, F., Bitzer, M., Ossowski, S., Casadei, N., Herr, C., Petersheim, D., Behrends, U., Kern, F., … Schultze, J. L. Swarm Learning for Decentralized and Confidential Clinical Machine Learning. *Nature* (2021). https://doi.org/10.1038/s41586-021-03583-3

9. Barredo Arrieta A, Diaz-Rodriguez N, Del Sera J, Bennetot A.,…Herrera F. Explainable Artificial Intelligence (XAI): Concepts, Taxonomies, Opportunities and Challenges toward Responsible AI (2020). Inf fusion. https://doi.org/10.1016/j.inffus.2019.12.012

10. Goddard K., Roudsari A., Wyatt J. Automation Bias: A Systematic Review of Frequency, Effect Mediators, and Mitigators. *Journal of the American Medical Informatics Association* (2012). https://doi.org/10.1136/amiajnl-2011-000089

11. Kitamura F. & Marques O. Trustworthiness of Artificial Intelligence Models in Radiology and the Role of Explainability. *Journal of the American College of Radiology* (2021). https://doi.org/10.1016/j.jacr.2021.02.008

12. Blazek P. & Lin M. Explainable Neural Networks That Simulate Reasoning. *Nature* (2021). https://doi.org/10.1038/s43588-021-00132-w

13. Mongan J., Moy L., Kahn C.E. Checklist for Artificial Intelligence in Medical Imaging (CLAIM): A Guide for Authors and Reviewers. *Radiology: Artificial Intelligence* (2020). https://doi.org/10.1148/ryai.2020200029

14. Shah N.H. How to Trust, But Verify, in Healthcare. Retrieved May 20, 2022: https://ncats-nih-gov.ezp3.lib.umn.edu/files/4_Shah_Nigam_NCATS_meeting_2019_wide_version_updated.pdf

15. van Leeuwen, K. G., Schalekamp, S., Rutten, M. J. C. M., van Ginneken, B., & de Rooij, M. Artificial Intelligence in Radiology: 100 Commercially Available Products and Their Scientific Evidence. *European Radiology* (2021). https://doi.org/10.1007/s00330-021-07892-z

14 Technological View

Mireia Crispin-Ortuzar

TECHNICAL APPROACHES

Most instances of AI in radiology are based on the application of a machine learning framework to automate or augment a given task involving some degree of image analysis. In this section we will discuss the differences between radiology AI frameworks based on engineered features, and those that are completely data-driven.

Radiomics

Radiomics refers to the extraction of a large number of handcrafted, quantitative features from a region of interest within an image, which are then used to train a machine learning model. Features typically include intensity descriptors, shape characteristics, and textural or higher-order features, which quantify the spatial co-occurrence of certain intensity patterns [1]. Features can also be designed to capture specific hypotheses, for example by integrating the signal of multiple imaging modalities [2]; or they can be partially data-driven, by using deep neural networks as feature extractors [3]. Although radiomics features can be used to train any class of machine learning model, a significant fraction of the literature is still dominated by simpler algorithms such as logistic regression or traditional machine learning methods [4].

Deep Learning

Deep learning algorithms are a particular class of machine learning models characterized by the use of data processing networks with a large number of layers. They do not rely on predefined features; they learn them by themselves. Each layer is composed of a series of building blocks that represent certain mathematical operations: the nature of such operations, and the way they are all interconnected, is what defines a given network architecture. Depending on the training mechanism, models can be classified into supervised, if the model learns from ground truth data; unsupervised, if it does not have access to any ground truth; or weakly supervised, when the ground truth is limited or imprecise.

The most popular type of deep learning algorithm is the convolutional neural network (CNN), composed of convolution layers that act as feature detectors. CNNs have been hugely successful in the analysis of all types of images thanks to their ability to automatically learn the spatial hierarchies of features [5]. Fully convolutional networks (FCNs) [6] are a special type of CNN in which all layers are convolutional, which enables them to take images of different size as input.

CNNs can also be used as elements of a bigger structure. For example, spatial transformer networks (STNs) are designed to transform feature maps spatially and are therefore widely used for registration tasks [7]. STNs contain a CNN that learns the parameters of the transformation, followed by a grid generator and a sampler. Another common example are generative adversarial networks (GANs), where two parallel structures (usually two CNNs) learn, respectively, to synthesize new data and to distinguish real from synthesized samples [8].

Although there are many different versions of network architectures, one category is particularly relevant for medical imaging applications: the encoder–decoder structure. The encoder is itself a network that reduces the input data to a certain internal representation; this encoded representation is read by another network, the decoder, and mapped onto the desired output. U-net, a very popular choice for image segmentation tasks, is based on this type of structure [9].

Some applications, such as natural language processing, require a different class of network. Recurrent neural networks (RNNs) act by performing the same task on the elements of a sequence, and the output is dependent on the previous computations—in other words, they are said to have "memory." This makes them particularly attractive to work with text inputs.

APPLICATIONS

AI can be applied to almost every step in the radiological journey, from the reconstruction of an image all the way to clinical decision-making (Figure 14.1).

Reconstruction

Generally, the data acquired by medical imaging sensors does not live in the spatial domain, and therefore transforming it into an image is a highly complex process. Traditional methods involve

DOI: 10.1201/9781003095279-18

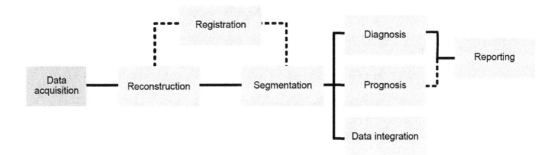

Figure 14.1 Simplified schematic of steps in the radiology pathway where AI tools have been developed. Relevant steps after data acquisition are shown in grey and correspond to the different sections of this chapter. Continuous (dashed) lines between the steps represent typical (possible) radiology pathways. Other connections and pathways not illustrated here are also possible.

either analytic reconstruction, based on the mathematical inversion of an idealized model of the data generation process; or numerical iterative methods, where inversion is unknown or too complex [10].

Deep learning models can be used as replacements for individual steps in the reconstruction pipeline, or as single end-to-end solutions. For example, encoder-decoder networks have been used to remove artefacts from reconstructed images [11, 12], with improvements such as interactive training with a radiologist in the loop [13]. So-called unrolling or unfolding methods exploit the power of iterative methods, for example, by mapping each step to a layer of a deep neural network [14]. Another popular family of methods, called "plug-and-play," use deep neural networks as a prior term in the iterative procedure—for example, a CNN [15]. In parallel, GANs have shown promise in situations where ground truth images are either unknown or impossible to obtain [16], or for image synthesis [17].

Registration

Image registration is the process of finding a transformation that maps two or more images onto the same spatial coordinate frame. The process of registration generally involves three key components: a transformation model; a similarity metric to evaluate how close the resulting images are; and an optimization algorithm to improve the result. Traditional algorithms proceed iteratively and are computationally intensive and time consuming.

Deep learning approaches have been proposed to improve the accuracy and accelerate the different elements of the registration process. The similarity metric can be replaced with CNNs or GANs, providing flexibility that is particularly useful for multi-modal registration without compromising the performance; however, these approaches still rely on an iterative process and therefore do not improve running time [18]. Deep learning can also be used for the transformation models. Supervised methods significantly accelerate the process but are hampered by the difficulty to obtain ground-truth samples to train on. Unsupervised methods do not have this problem and have received significant attention, particularly those based on STNs [19]. An intermediate solution is the use of weakly supervised models, in which some information such as anatomical segmentations are used to improve the performance with respect to fully unsupervised approaches [20].

Segmentation

Segmentation refers to the task of delineating the boundaries of anatomical structures of interest on a medical image. Traditionally done manually by radiologists, the field has moved towards automatic or semi-automatic approaches based on computer vision methods, which have been dominated by deep learning in recent times.

Supervised deep learning segmentation methods are the most popular and achieve a high degree of accuracy. U-net has been successful across segmentation tasks thanks to its ability to fuse low-level and high-level features and is considered the benchmark for many medical image segmentation tasks [9]. A common method to improve segmentation accuracy is to employ a succession ("cascade") of networks that provide increasingly accurate results [21]. nnU-net, a framework that exploits network cascades, has recently received significant attention due to its ability to automatically configure itself while achieving state-of-the-art performance [22].

The biggest challenge for supervised approaches is the scarcity of segmentation ground truth. Several methods have been proposed to deal with this limitation, including data augmentation, the generation of synthetic datasets using methods such as GANs, and the development of interactive segmentation approaches [23, 24].

Disease Detection and Diagnosis

The diagnosis of some diseases, including many types of cancer, relies on the initial radiological detection of abnormal features on a radiological image. Traditional computational methods based on manual feature engineering, typically referred to by the umbrella term "computer-aided detection" (CADe), have not enjoyed widespread success in part due to their suboptimal performance; for example, some studies showed that the use of CADe had no impact on the diagnostic accuracy of mammographic screening [25, 26]. However, recent studies based on CNNs and FCNs have achieved promising results for the detection of breast masses [27], lung nodules [28], strokes [29], or even Alzheimer's [30].

Beyond disease detection, AI is also used to classify a disease into a particular diagnostic category. This type of task is based on the quantitative evaluation of the visual features that characterize the diseased region, using radiomics or deep learning approaches. These types of studies have boomed in recent years, with a yearly increase of almost 200% in the number of papers, mostly focusing on CT imaging [4]. Thoracic radiology applications are at the forefront, including the classification of cancer vs. benign pulmonary nodules, and molecular classification of lung cancer [31]. Although a recent review found that most existing radiomics studies were based on logistic regression or LASSO, deep learning approaches are now becoming more popular [4].

Prognosis and Prediction

AI models can be flexible and powerful enough to be optimized directly on endpoints that can directly support clinical decision-making, such as prognosis of treatment response. Lung cancer provides a comprehensive example, as it has been extensively studied using radiomics approaches [32]. Examples of prognostic and predictive AI applications developed for lung cancer include the prediction of distant metastasis using radiomics features and consensus clustering [33]; mortality risk stratification using CNNs [34]; or response to chemotherapy [35].

Imaging Genomics and Data Integration

One of the main challenges of radiomics studies, particularly in oncology, is the difficulty of interpreting the meaning of the features involved in the predictions. The field known as imaging genomics, or radiogenomics, is focused on addressing this challenge by finding links between radiomics and molecular features. Crucially, datasets that include imaging as well as molecular data enable the development of joint prediction models in which the power of the various data streams is integrated [36].

The radiogenomics literature is evenly split between correlation and prediction studies, with 41% of the studies being based on simple associations, and only 2% using deep learning [37]. Most of the studies have focused on brain and lung cancer, followed closely by breast cancer; similarly, the most common molecular features studied are mutations in EGFR, RAS, and IDH1 [37]. The observation that some cancers exhibit a high degree of spatial heterogeneity has prompted the development of spatial studies that attempt to correlate regions with a well-defined imaging profile (so-called habitats) with underlying molecular data, for example in ovarian cancer [38, 39].

Predictive models based on data integration are still scarce, with a few examples for glioblastoma [40], melanoma [41], and lung [42], all of them based on traditional analytical methods such as Cox proportional hazards or logistic regression.

Reporting

Radiology reports are the main product of clinical radiological analyses. Deep learning has provided significant advances in both the automatic generation of reports and in the conversion of clinical free text into structured data.

The most common approach to report generation is to link a CNN for image analysis and an RNN for natural language generation, including word-, sentence-, and paragraph-based outputs

[43]. Chest X-ray datasets are the most used to train this type of algorithms [44, 45]. Multi-task algorithms that jointly predict tags and generate paragraphs have been particularly successful [43, 46].

In parallel, manually generated radiology reports have been mined using deep NLP for diagnostic surveillance, cohort building, or quality assessment [47, 48]. RNNs are ideally suited to analyze sentences and have been used for a diverse collection of tasks, from standardizing breast radiology reports [49] to classifying fractures [50]. Attention models are also promising, as they are capable of giving context to words within sentences [51].

CHALLENGES

Dataset Limitations

There are a limited number of public datasets to train radiology AI models. Clinician-led studies have access to local cohorts, but most studies are single-center with a median number of participants of 85—although that number has been growing lately [4]. This results in a significant risk of biases, particularly as datasets are often imbalanced, potentially confounding both the training and the evaluation of the results. In addition, many of the existing public datasets are shared without the corresponding annotations by radiologists.

COVID-19 has incentivized the creation of dynamic national and international imaging database collaborations [52, 53]. The networks and operating procedures set up for this purpose could expedite the development of similar platforms for other radiological applications in the future. An example is the National AI Medical Imaging Platform (NMIP) being set up by the British National Health System after the success of National COVID-19 Chest Imaging Database (NCCID) [54].

Methodological Limitations

The explosion of radiology AI applications has triggered the development of a series of guidelines for adequate analysis and reporting, such as CHARMS [55], TRIPOD [56], or TRIPOD-AI [57]. Their impact is however still limited [58], and there is also a need for a clinically oriented set of guidelines such as STARD-2015 [59] adapted for AI models.

The rapid pace of development triggered by COVID-19 makes these methodological reviews particularly relevant. A recent study highlighted that none of the papers published between January and October 2020 proposing machine learning models for the diagnosis or prognosis of COVID-19 from radiographs or CT images could be used clinically due to methodological flaws and/or underlying biases [60].

In parallel to all this, several studies have highlighted the need to quantify and control the stability of deep learning algorithms, for example in reconstruction algorithms, which may lead to artefacts or missing features [61]. Traditional radiomics approaches can also suffer from lack of robustness and reproducibility [62].

Experimental Limitations

Radiological AI applications are expected to help alleviate one of the most pressing problems in the field—the global shortage of radiologists [63, 64]. However, this very problem is also responsible for the growing workload for radiologists, which in some cases means that they have to interpret one image every 3–4 seconds to meet the productivity requirements [65]. Errors are therefore almost inevitable in such a setting, which without careful curation could negatively affect model training. Models that can deal with noisy labels such as weakly supervised learning can help with these challenges [66].

Even when data are annotated in detail, some tasks are ambiguous by nature. Tasks such as lesion detection or segmentation, or qualitative feature assessments, can present poor interobserver agreement [67, 68], which can even lead to differences in response assessment [69]. This can be further emphasized by differences in experience level [70].

OUTLOOK

The field of AI for radiology is vast and includes advances in almost all areas. Deep learning is starting to dominate, although traditional machine learning methods are still the norm mostly due to dataset and annotation limitations. As the pace of development grows, addressing these as well as the methodological challenges will become essential to guarantee a successful future for the field.

REFERENCES

1. A. P. Apte *et al.*, Medical Physics (2018).

2. M. Crispin-Ortuzar *et al.*, Radiotherapy and Oncology (2017).

3. P. Afshar, A. Mohammadi, K. N. Plataniotis, A. Oikonomou, and H. Benali, IEEE Signal Processing Magazine **36**, 132 (2019), 1808.07954.

4. J. Song *et al.*, European Journal of Radiology **127** (2020).

5. R. Yamashita, M. Nishio, R. K. G. Do, and K. Togashi, Insights into Imaging **9**, 611 (2018).

6. J. Long, E. Shelhamer, and T. Darrell, IEEE Transactions on Pattern Analysis and Machine Intelligence **39**, 640 (2014), 1411.4038.

7. M. Jaderberg, K. Simonyan, A. Zisserman, and K. Kavukcuoglu, Spatial transformer networks, in *Advances in Neural Information Processing Systems*, Vol. 2015–January, pp. 2017–2025, Neural information processing systems foundation, 2015, 1506.02025.

8. I. J. Goodfellow *et al.*, Generative Adversarial Nets, in *Advances in Neural Information Processing Systems*, 2014.

9. O. Ronneberger, P. Fischer, and T. Brox, U-net: Convolutional networks for biomedical image segmentation, in *Lecture Notes in Computer Science (Including Subseries Lecture Notes in Artificial Intelligence and Lecture Notes in Bioinformatics)*, Vol. 9351, pp. 234–241, Springer Verlag, 2015.

10. G. Wang, J. C. Ye, and B. De Man, Nature Machine Intelligence **2**, 737. (2020).

11. K. H. Jin, M. T. McCann, E. Froustey, and M. Unser, IEEE Transactions on Image Processing **26**, 4509 (2017), 1611.03679.

12. D. Lee, J. Yoo, S. Tak, and J. C. Ye, IEEE Transactions on Biomedical Engineering **65**, 1985 (2018), 1804.00432.

13. H. Shan *et al.*, Nature Machine Intelligence **1**, 269 (2019).

14. Y. Yang, J. Sun, H. Li, and Z. Xu, Deep ADMM-Net for Compressive Sensing MRI, in *Advances in Neural Information Processing Systems*, Vol. 29, 2016.

15. J. Liu *et al.*, IEEE Journal on Selected Topics in Signal Processing **14**, 1088 (2019), 1912.05854.

16. J. M. Wolterink, T. Leiner, M. A. Viergever, and I. Iˇsgum, IEEE Transactions on Medical Imaging **36**, 2536 (2017).

17. F. Liu, H. Jang, R. Kijowski, T. Bradshaw, and A. B. McMillan, Radiology **286**, 676 (2018).

18. X. Chen, A. Diaz-Pinto, N. Ravikumar, and A. Frangi, Progress in Biomedical Engineering **3**, 012003 (2020).

19. B. D. de Vos, F. F. Berendsen, M. A. Viergever, M. Staring, and I. Iˇsgum, End-to-end unsupervised deformable image registration with a convolutional neural network, in *Lecture Notes in Computer Science (including subseries Lecture Notes in Artificial Intelligence and Lecture Notes in Bioinformatics)*, Vol. 10553 LNCS, pp. 204–212, Springer Verlag, 2017, 1704.06065.

20. Z. Xu and M. Niethammer, DeepAtlas: Joint Semi-supervised Learning of Image Registration and Segmentation, in *Lecture Notes in Computer Science (including subseries Lecture Notes in Artificial Intelligence and Lecture Notes in Bioinformatics)*, Vol. 11765 LNCS, pp. 420–429, Springer, 2019, 1904.08465.

21. T. Lei *et al.*, Medical Image Segmentation Using Deep Learning: A Survey, 2009.13120v2.

22. F. Isensee, P. F. Jaeger, S. A. Kohl, J. Petersen, and K. H. Maier-Hein, Nature Methods **18**, 203 (2021).

23. G. Wang *et al.*, IEEE Transactions on Medical Imaging **37**, 1562 (2018), 1710.04043.

24. K. B. Girum, G. Cŕehange, R. Hussain, and A. Lalande, International Journal of Computer Assisted Radiology and Surgery **15**, 1437 (2020).

25. E. B. Cole *et al.*, American Journal of Roentgenology **203**, 909 (2014).

26. C. D. Lehman *et al.*, JAMA Internal Medicine **175**, 1828 (2015).

27. Z. Hu *et al.*, Pattern Recognition **83**, 134 (2018).

28. A. Halder, D. Dey, and A. K. Sadhu, Journal of Digital Imaging **33**, 655 (2020).

29. R. Karthik, R. Menaka, A. Johnson, and S. Anand, Computer Methods and Programs in Biomedicine **197**, 105728 (2020).

30. M. A. Ebrahimighahnavieh, S. Luo, and R. Chiong, Computer Methods and Programs in Biomedicine **187**, 105242 (2020).

31. B. Chen, R. Zhang, Y. Gan, L. Yang, and W. Li, Radiation Oncology **12**, 154 (2017).

32. C. Bortolotto *et al.*, Expert Review of Anticancer Therapy **21**, 257 (2021).

33. L. Fan *et al.*, Academic Radiology **26**, 1253 (2019).

34. A. Hosny *et al.*, PLOS Medicine **15**, e1002711 (2018).

35. M. R. Chetan and F. V. Gleeson, European Radiology **31**, 1049 (2021).

36. E. Sala *et al.*, Clinical Radiology **72**, 3 (2017).

37. Z. Bodalal, S. Trebeschi, T. D. L. Nguyen-Kim, W. Schats, and R. BeetsTan, Abdominal Radiology **44**, 1960 (2019).

38. B. Weigelt *et al.*, JCO Precision Oncology **1** (2019).

39. P. Martin-Gonzalez *et al.*, Insights into Imaging **11**, 94 (2020).

40. M. Nicolasjilwan *et al.*, Journal of Neuroradiology **42**, 212 (2015).

41. L. Basler *et al.*, Clinical Cancer Research **26**, 4414 (2020).

42. P. Grossmann *et al.*, eLife **6** (2017).

43. M. M. A. Monshi, J. Poon, and V. Chung, Artificial Intelligence in Medicine **106**, 101878 (2020).

44. X. Wang *et al.*, ChestX-ray8: Hospital-scale chest X-ray database and benchmarks on weakly-supervised classification and localization of common thorax diseases, in *Proceedings - 30th IEEE Conference on Computer Vision and Pattern Recognition, CVPR 2017*, Vol. 2017–January, pp. 3462–3471, Institute of Electrical and Electronics Engineers Inc., 2017, 1705.02315.

45. D. Demner-Fushman *et al.*, Journal of the American Medical Informatics Association **23**, 304 (2016).

46. B. Jing, P. Xie, and E. Xing, ACL 2018 - 56th Annual Meeting of the Association for Computational Linguistics, Proceedings of the Conference (Long Papers) **1**, 2577 (2017), 1711.08195.

47. E. Pons, L. M. Braun, M. G. Hunink, and J. A. Kors, Radiology **279**, 329 (2016).

48. V. Sorin, Y. Barash, E. Konen, and E. Klang, Journal of the American College of Radiology **17**, 639 (2020).

49. S. Miao *et al.*, International Journal of Medical Informatics **119**, 17 (2018).

50. C. Lee, Y. Kim, Y. S. Kim, and J. Jang, American Journal of Roentgenology **212**, 734 (2019).

51. B. Shin, F. H. Chokshi, T. Lee, and J. D. Choi, Classification of radiology reports using neural attention models, in *Proceedings of the International Joint Conference on Neural Networks*, Vol. 2017–May, pp. 4363–4370, Institute of Electrical and Electronics Engineers Inc., 2017, 1708.06828.

52. J. Jacob *et al.*, European Respiratory Journal **56** (2020).

53. COVID-19 imaging datasets EIBIR (https://www.eibir.org/covid-19imaging-datasets/).

54. AI in imaging - NHSX (https://www.nhsx.nhs.uk/ai-lab/ai-labprogrammes/ai-in-imaging/), 2020.

55. K. G. M. Moons *et al.*, PLoS Medicine **11**, e1001744 (2014).

56. K. G. Moons *et al.*, Annals of Internal Medicine **162**, W1 (2015).

57. G. S. Collins and K. G. Moons, The Lancet **393**, 1577 (2019).

58. M. Yusuf *et al.*, BMJ Open **10**, e034568 (2020).

59. J. F. Cohen *et al.*, BMJ Open **6**, e012799 (2016).

60. M. Roberts *et al.*, Nature Machine Intelligence **3**, 199 (2021).

61. V. Antun, F. Renna, C. Poon, B. Adcock, and A. C. Hansen, Proceedings of the National Academy of Sciences of the United States of America **117**, 30088 (2020).

62. J. E. Park, S. Y. Park, H. J. Kim, and H. S. Kim, Korean Journal of Radiology **20**, 1124 (2019).

63. A. Nishie *et al.*, Japanese Journal of Radiology **33**, 266 (2015).

64. The Royal College of Radiologists, Clinical radiology UK workforce census 2019, 2020.

65. R. J. McDonald *et al.*, Academic Radiology **22**, 1191 (2015).

66. N. Natarajan, I. S. Dhillon, P. Ravikumar, and A. Tewari, Learning with Noisy Labels, in *Advances in Neural Information Processing Systems*, Vol. 26, 2013.

67. E. Sala *et al.*, Radiology **257**, 125 (2010).

68. H. A. Vargas *et al.*, Radiology **285**, 482 (2017).

69. T. Lestra *et al.*, Diagnostic and Interventional Imaging **99**, 73 (2018).

70. E. M. Crowe, W. Alderson, J. Rossiter, and C. Kent, Frontiers in Psychology **8**, 1628 (2017).

15 Societal View

Suely Fazio Ferraciolli, André Lupp Mota, and Felipe Campos Kitamura

INTRODUCTION TO SOCIAL VIEW

The concept of artificial intelligence (AI) is frequently associated with expensive equipment and technological complexity. Considering that most of the world's population currently live in regions with limited resources and that two-thirds of the global population face a shortage of radiologists [1], if a large part of humanity is deprived of the benefits of AI, then the social gap between low-/middle-income countries (LMIC) and high-income countries (HIC) would be further increased.

Nevertheless, in several areas of healthcare, AI is already established as an important tool with the potential of bringing improvements in the quality of care to less-resourced communities, with positive impact even in the setting of limited investment and availability of equipment. In this aspect, AI becomes an essential tool for social transformation and equity among different populations worldwide.

In locations with limited resources and high demand for assistance, AI promises to optimize existing structures, reaching the largest possible number of people, with relatively low investment-to-benefit ratio. To achieve this optimization in LMICs, one proposed model is to divide AI in healthcare into three levels of complexity: use of portable devices for diagnosis (as an aid for remote health professionals and to perform triage); clinical decision support systems (diagnostic tools, support for specialists and sub-specialists); and, finally, algorithms for use in public healthcare (epidemiological data analysis, resource distribution) [2].

The use of portable devices with AI algorithms has allowed the diagnosis of diseases in locations without radiologists. One example is in the experience of using smartphones to diagnose tuberculosis in Peru using convolutional neural network algorithms, thus optimizing care both in urban regions, where there are large numbers of people and few health services, and in rural areas where patients are far from specialized care centers. As a result, correct diagnosis and rapid treatment facilitated by AI were beneficial in helping to control a prevalent disease in poorer localities [3].

In this and other scenarios, such as the COVID-19 pandemic, AI can be used as a tool to reduce the social gap between LMICs and HICs. Another key question is how can LMICs be included in the AI universe, enabling the creation of dedicated radiological services for these locations, using only currently available resources? The RAD-AID group proposes solutions for the implementation of these services, taking into account personnel and location differences (clinical experience, disease pattern, demography, digital infrastructure, and equipment) through a staged implementation with the following steps: clinical education, implementation of the infrastructure, and phased AI-introduction [4]. In addition to benefiting global welfare, the implementation of AI-enabled tools can facilitate the training of professionals, creation of a specialized labor force, and attract investors, together contributing to local socio-economic development.

However, special attention must be paid to potential technical and ethical concerns, such as possible bias related to the AI systems having been trained on populations different from target populations, and to the ownership of the data obtained. One potential source of inequality would come from information obtained in LMICs that contribute to the creation of high-value AI products that only benefit the HICs involved in the development of these technologies [4].

ETHICAL ASPECTS

Like any new technology, AI in medical practice raises ethical issues about its use. Although there are laws and ethical guidelines on how clinical information is shared, the field of AI deserves special attention, as that information used in AI development may transit through new elements that have been traditionally outside the care chain, such as through for-profit companies developing AI algorithms [5].

The development of guidelines based on bioethics principles (Figure 15.1) is a way to regulate legal and ethical limits in AI. Nevertheless, in AI there are some peculiarities that can conflict with these principles, due to factors such as: the individual may be both patient and participant of the clinical research; even if anonymized, patient imaging data may reveal information about the patient; the potential for bias; and, potential conflicts of interest among the developers [6]. One

DOI: 10.1201/9781003095279-19

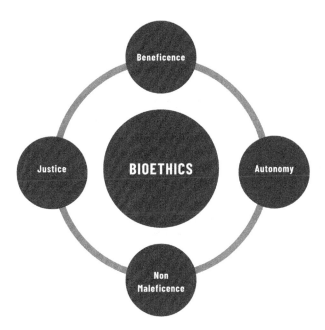

Figure 15.1 Schematic of bioethics principles

solution to ensure ethical boundaries that protect patients is the proposal of ethical obligations addressed to each of the parties involved in the process, as have been proposed by Larson et al. [5].

Another issue that deserves attention is that AI algorithms are built to discriminate patterns and give weight to some factors over others. This way of functioning can predispose to biases, which can be accentuated and perpetuated in certain categories such as gender, age, race, sexual orientation, etc. Minority groups or groups traditionally excluded from population studies may lack representation of their characteristics in the algorithms built based on the predominant groups, which may hinder access and further inequalities. One suggested solution to avoid bias is to require representation of minority groups in the training and test sets of the AI algorithms during development and internal validation prior to algorithms' use in public healthcare systems [7].

Another relevant ethico-legal issue is in who assumes responsibility for errors involving AI algorithms (the doctor, the manufacturer or the seller?). It is up to governments and associations to draft laws and ethical standards to regulate the uses of data, algorithms, and best practices.

REGULATORY FRAMEWORK

It is difficult to establish an ethical environment for AI in healthcare without a well-designed regulatory framework. Given that the use of AI algorithms in medicine and radiology may influence the suggested diagnosis, prognosis, and treatment, we need to carefully evaluate these AI models, so the final result would avoid harm and produce a positive impact in the patient's health.

Larson et al. [5] showed real-world problems that involved the question of who may control and profit from secondary uses of the data. In many discussions, there were two paths drawn: one where the secondary use of data belongs to the patient and another where it belongs to provider organizations. The enlightening insight of this article is to propose a third path, where the secondary use of data could be treated as a public good. From the point-of-view of a public good, data is not "owned" by anyone, should not be sold for profit, or confined under exclusive arrangements. On the contrary, people or organizations that control or interact with data need to ensure that it is used for the benefit of future patients, of the healthcare system, and of broader society. This premise is the cornerstone for answering questions about access, regulation, permission, profit, and exclusivity of data.

There are some conditions that need to be fulfilled when sharing clinical data with outside entities, including: individual privacy is protected; the receiving organization is forbidden to share the data again without the consent of the original provider organization; and both organizations must willingly accept the same fiduciary responsibilities of data stewardship (without intent of re-identifying individuals from the data) and agree to specify the exact purposes for the use of the data.

By establishing this robust ethical framework for clinical imaging data in this article [5], there could be an expansion of the ethical framework's use to other types of data, such as electronic medical records and pathologic data.

Another article by Larson et al. [8] reviewed regulatory frameworks for Software as a Medical Device (SaMD) applications, including the FDA in the United States, the Global Harmonization Task Force, and its predecessor, the International Medical Device Regulators Forum (IMDRF). These groups proposed key definitions [9], risk categories [10], a quality management system [11], and standards for clinical evaluation and investigation [12, 13]. In these frameworks, various fundamental aspects were defined to ensure the safety, effectiveness, and performance of SaMD applications, primarily focusing on manufacturers' responsibilities. However, many gaps were identified and described in this article [8], such as the integration of the algorithm with its diagnostic task, a superficial approach to defining the diagnostic task, the lack of an effective comparison mechanism for similar algorithms, inadequate characterization of performance and safety elements, the absence of assessment of the model's performance at the installed sites (often due to limited resources), and inherent conflicts of interest.

In order to overcome these gaps, some suggested measures are:

1. To separate the diagnostic task from the algorithm, with each task having its specifications (background, purpose, definitions, limitations, measurements, ground-truth, and description of their desired output) [8].

2. The performance of the model should be evaluated not only in the laboratory but also in the real-world, with its response to different sites and datasets being analyzed before clinical deployment. After initial deployment, a constant and continuous monitoring of algorithms' performance should be in place [8].

3. The evaluation process could be divided into five steps: 1. diagnostic task definition; 2. capability (related to the algorithm performance in the AI laboratory compared with other algorithms); 3. performance in the real-world; 4. validation at other sites; and 5. durability (performance over time) [8].

4. Third-party evaluation should be encouraged [8].

5. All of these steps should also be incorporated into the manufacturers' development process, divided into these four phases: feasibility, capability, effectiveness, and durability. These steps are similar to the ones applied to pharmaceuticals and proposed for software applications. These specifications could ensure prime performance of the algorithms installed at all sites [8].

All these excellent points brought by this article [8] make us recognize that although there has been impressive progress with regulation of AI algorithms in healthcare, there are still many aspects that remain uncovered, and much research remains to be done in the near future to ensure accuracy, reliability, and safety of these AI models.

Following this lead, the FDA in 2021 has released an AI/ML Action Plan [14] in response to the stakeholder feedback received from the 2019 discussion paper "Proposed Regulatory Framework for Modifications to AI/Machine Learning-Based Software as a Medical Device" [15]. In this action plan, the FDA delineates five future actions:

1. To continuously develop the regulatory framework to achieve a Tailored Regulatory Framework for AI/ML-based SaMD, especially for the learning algorithms that are not "frozen" [14].

2. To participate in collaborative communities and efforts to develop consensus standards, to support the fulfillment of good machine learning practices (GMLP) to evaluate and improve machine learning (ML) algorithms (Figure 15.2) [14].

3. To support a patient-centered approach, including a future public workshop to increase the transparency and trust of the users related to ML algorithms in healthcare [14].

4. To develop methods of evaluation and refinement of ML algorithms to precisely define the regulatory science and methods to overcome algorithm bias and ensure robustness [14].

5. To advance pilots of real-world performance monitoring [14].

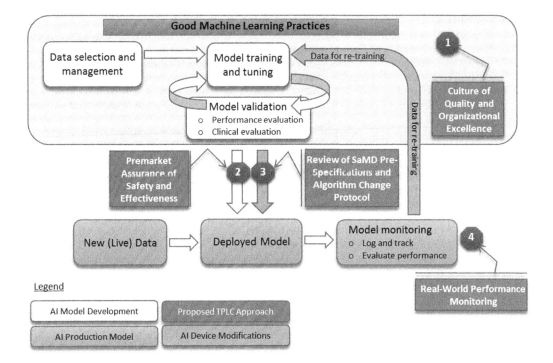

Figure 15.2 Overlay of FDA's total product lifecycle (TPLC) approach on AI/ML workflow. Source: FDA (https://www.fda.gov/files/medical%20devices/published/US-FDA-Artificial -Intelligence-and-Machine-Learning-Discussion-Paper.pdf)

Also in 2020, the Digital Health Center of Excellence was created as part of the evolution of the Digital Health Program in the Center for Devices and Radiological Health, which will work within the framework of the FDA's regulation. It will strategically advance science and evidence for digital health technologies by supporting responsible and high-quality digital health innovation [16].

All of these initiatives, suggestions, and actions are based on the complex view of the areas involved in the development, deployment, and monitoring of AI algorithms that directly or indirectly affect patient care. Regulatory frameworks worldwide should contemplate all these broad aspects to guarantee algorithms that safely and accurately deliver healthcare. The fulfillment of all these points described above would have substantial societal impact in the years to come, positively influencing the construction of more ethical and trustworthy AI-based healthcare tools to helping overcome existing challenges.

DATA SECURITY

Data security and privacy protection regulations exist in the form of HIPAA (Health Insurance Portability and Accountability Act) in the US, GDPR (general data protection regulation) in the EU, and similar regulations in many countries, such as LGPD (general law of data protection) in Brazil. These are essential steps to make the future of AI in radiology safer by guaranteeing data de-identification. Another aspect that should be considered is possible cybersecurity breaches in the healthcare system. The hacking of medical data may have huge social impact, with numerous ways of potential harm for the patients under our care. In this light, it is crucial that cybersecurity measures are in place in medical systems like a picture archiving and communication system (PACS), a radiological information system (RIS), and electronic health records (EHR), embedded in electronic devices.

As stated by Chu et al. [17], we need to be aware of potential bad actors that may abuse AI to their own benefit, creating chaos or harming healthcare systems and our patients. In the following years, we need to search, recognize, and develop new ways of addressing critical vulnerabilities in data security to protect data, patients, and healthcare systems from attack.

RESPONSIBILITY

Another important issue that needs to be explored and clarified is the liability of radiology reports and decision-making based on AI produced results. Neri et al. [18], addressed this question at the core of their article, discussing the actions/consequences and responsibilities inherent to AI use in patient care. They state that typically, when human beings make decisions, the responsibility stems from the action of the individual who generated the decision. When an AI algorithm is used, it becomes difficult but essential to be able to know where to impute the responsibility when something goes wrong.

There are some important statements in this manuscript [18], such as, by using AI, radiologists are responsible for the diagnosis and need to be trained on the use of AI. Also, radiologists who support the development of AI are responsible for reducing bias and improving machine learning explainability, aiming to create trustworthy AI. There is a risk that radiologists who work with non-explainable ML algorithms could validate the unknown (black box), which may lead to attesting wrong outputs as true. Also, automation bias, which is the tendency for humans to favor machine-generated decisions, thus ignoring contrary data or conflicting human decisions, may lead to errors in AI use. All these points must be considered when drafting legislations that define the professional's responsibility and accountability.

Who will pay for use of AI/ML tools [19] is also a fundamental question with significant social impact as the lack of economic power of a patient, institution, or country may limit access to beneficial AI algorithms in healthcare. An unbalanced way of payment for AI algorithms would only increase the disparity between LMICs and HICs. This interesting topic will be further explored in the next chapters.

CONCLUSION

Use of AI in radiology presents multifaceted social impacts. Since AI promises to provide a broad range of benefits, ranging from promoting health equity by reducing the difference between LMICs and HICs, through optimizing the use of economic resources and improving social equity, to having a direct impact in the diagnosis, treatment, and outcome of our patients, AI algorithms should be critically analyzed before being put into practice in order to maximize benefit. Considerations driving the ethical use of AI should guide the regulatory frameworks, legislation, and cybersecurity protections that are being developed and that will need to be adapted in the years to come.

REFERENCES

1. PAHO/WHO. (2012). <https://www.paho.org/hq/index.php?option=com_content&view=article&id=7410:2012-dia-radiografia-dos-tercios-poblacion-mundial-no-tiene-acceso-dia gnostico-imagen&Itemid=1926(=en>

2. Hosny A, Aerts HJWL. Science (2019). <https://doi.org/10.1126/science.aay5189>

3. Curioso WH, Brunette MJ. Revista Peruana de Medicina Experimental y Salud Pública (2020). <https://doi.org/10.17843/rpmesp.2020.373.5585>

4. Mollura DJ, Culp MP, Pollack E, Battino G, Scheel JR, Mango VL, et al. Radiology (2020). <https://doiorg/101148/radiol2020201434>

5. Larson DB, Magnus DC, Lungren MP, Shah NH, Langlotz CP. Radiology (2020). <https://doiorg/101148/radiol2020192536>

6. D'Antonoli TA. Diagn Interv Radiol (2020). <https://doi.org/10.5152/dir.2020.19279>

7. Schönberger D. International Journal of Law and Information Technology (2019). <https://doi.org/10.1093/ijlit/eaz004>

8. Larson DB, Harvey H, Rubin DL, Irani N, Tse JR, Langlotz CP. Journal of the American College of Radiology (2020). <https://doi.org/10.1016/j.jacr.2020.09.060>

9. IMDRF (2013). <http://www.imdrf.org/docs/imdrf/final/technical/imdrf-tech-131209-samd-key-definitions-140901.pdf >

10. IMDRF (2014). <http://www.imdrf.org/docs/imdrf/final/technical/imdrf-tech-140918-samd-framework-risk-categorization-141013.pdf>

11. IMDRF (2015). <http://www.imdrf.org/docs/imdrf/final/technical/imdrf-tech-151002-samd-qms.pdf>

12. IMDRF (2017). <http://www.imdrf.org/docs/imdrf/final/technical/imdrf-tech-170921-samd-n41-clinical-evaluation_1.pdf>

13. IMDRF (2019). <http://www.imdrf.org/docs/imdrf/final/technical/imdrf-tech-191010-mdce-n57.pdf>

14. FDA (2021). <https://www.fda.gov/medical-devices/software-medical-device-samd/artificial-intelligence-and-machine-learning-software-medical-device>

15. FDA (2019) <https://www.fda.gov/media/122535/download>

16. FDA (2021). <https://www.fda.gov/news-events/press-announcements/fda-releases-artificial-intelligencemachine-learning-action-plan>

17. Chu LC, Anandkumar A, Shin HC, Fishman EK. Journal of the American College of Radiology (2020). <https://doi.org/10.1016/j.jacr.2020.04.010>

18. Neri E, Coppola F , Miele V, Bibbolino C, Grassi R. La Radiologia Medica (2020). <https://doi.org/10.1007/s11547-020-01135-9>

19. Chen MM, Golding LP, Nicola GN. Radiological Society of North America (2021). https://doi.org/10.1148/ryai2021210030

16 Financial View

Florian Dubost

MARKET SIZE AND PROJECTIONS

The market in artificial intelligence (AI) is exponentially growing, with a 30 to 40% forecasted compound annual growth rate (CAGR) from 2021 to 2026. Over the last two years, those estimations have even been corrected upward. In 2019, Brand Essence, a British market research company, estimated the market size of AI in radiology to be $180M in 2025 with a CAGR of 35.9% [1]. This year, in 2021, Signify Research, a market search company specialized in healthcare technologies, estimated the AI radiology market to grow to $1.2B in 2025, with a CAGR of 39.5% from 2021 to 2025 [2]. Yet, Signify Research estimated the 2019 market to be $300M, a sharp increase from Brand Essence's estimation of the 2018 market at $25.1M. Throughout our meta-analysis, we identified substantial differences in the dollar evaluation of the market of AI in radiology. Besides potential differences in methodology in market-size evaluation, we think that estimation differences mostly arise from the loose definition of AI. Does it refer to the recent developments in deep learning? To machine learning in general? Or any statistical algorithm? Signify Research estimates the proportion of deep learning–based AI medical imaging products to substantially increase in the coming years. In 2020, Signify Research estimated other computer vision methods to be in higher proportion in AI radiology products, while deep learning takes the lead in 2021.

Despite their differences in the evaluation of the dollar value of the market size, market search companies seem to agree about the CAGR being from 30 to 40% from 2021 to 2025, which is much higher than the return of top-quartile VCs (15–27%), making it very attractive for investors [3].

Brand Essence predicts the overall medical imaging market to be as high as $35.8Bn in 2027, with a CAGR of 3.99% from 2021 to 2027, a CAGR ten times lower than that of radiology AI, confirming that AI is currently the key innovator in the medical imaging market.

MARKET SEGMENTATION

AI is not expected to influence every radiology subfield to the same extent. Four disciplines are clearly ahead of the pack: neurological, cardiovascular, breast, and lung imaging, each projected to capture 15 to 20% of the market according to Signify Research studies [4]. In this chapter, we do not include AI for pathology images, which also captures a substantial share of the AI-powered medical imaging market. This is demonstrated by the amount of private investment in startups such as Freenome—which is performing early cancer detection from blood cells and raised $513M since its creation in 2014—as well as recent scientific interest and funding with an increasing number of published articles or conference workshops organized around the topic [5].

Neurological and cardiovascular imaging lead the way in the forecasted market shares of AI radiology. For example, Heartflow, a startup founded in 2007, leverages standard CT images to build an AI-powered model of the coronary arteries and their blood flood to help doctors in the visualization and prevention of potential heart diseases. Heartflow has received $577M of funding and is planned to go public via a SPAC deal in the fourth quarter of 2021 at a valuation of $2.4Bn. With this amount funding and target valuation, Heartflow is among the leading startups in AI radiology.

The market can also be segmented w.r.t. imaging modalities. Above, we gave the example of a successful startup in CT (Heartflow). Butterfly Network (BFLY) is another big player, with its disruptive hand-held ultrasound device. Butterfly Network was founded in 2011 and went public on February 16, 2020, with a current market capitalization of $2.2Bn. Exo, another startup founded in 2015, developed a similar product and raised this summer 2021 another $220M, bringing it a total of $320M in funding.

The sum of the valuations and funding raised for the startup mentioned above (Heartflow, Butterfly Network, and Exo) is already at $5Bn. This is far above the market size forecasted by Brand Essence and Signify Research for 2025 to 2026, suggesting that investors expect the AI radiology market to be only at its beginning.

DOI: 10.1201/9781003095279-20

PRODUCTS AND BUSINESS MODELS

We identified three major types of business models for current AI radiology startups: software on-premise, cloud-based software-as-a-service, and model 3.

The first model is software on-premise. Multinational corporations, or startups partnering with multinationals, usually embed their AI software to existing imaging systems. For example, that is the case of Quantib BV., a Dutch startup founded in 2012, which partners with GE, Philips, and IBM Watson. The advantage of such a business model for start-ups is that they have a customer base before developing the product, consequently leading to a lower economical and sale risk.

The second business model is cloud-based software-as-a-service. Patients or their healthcare providers upload radiographs to the company's infrastructure for analysis. Examples of start-ups with such business models are Heartflow, see above, or Thirona, a Dutch company founded in 2014, which proposes AI analysis of CTs, chest X-rays, and retinal images for COPD, asthma, ILD-19, and COVID. This business model has the advantage of having more stable, constant, and uncapped revenues. Moreover, the company has access to imaging data to further improve its AI software, creating a positive network effect: more customers/patients leads to more imaging data, which improves the AI software, which attracts new customers. As with all networks effect types of business models, the most difficult is the initial growth of the customer base.

The last type of business model is the sale of a physical product. The company creates a new imaging device and sells it to either healthcare providers or directly to patients. Innovation of the device usually involves a lower price than existing devices; a smaller, portable version of existing devices; a more informative and precise image acquisition; imaging a body part or biological function less invasively than previously possible. These innovations are made possible through the combinations of either new or more subtle physical measurements and AI, which can easily learn to cope with missing information or process new radiological modalities. For example, Butterfly Network (BFLY) and Exo propose a hand-held ultrasound device, at a lower price than machines. This business model can be combined with the second business model, cloud-based software-as-a-service. Often, startups with this model also propose additional advanced AI features by allowing the users to send the acquired images to the company. By combining both models, the company can ensure early steady revenue through the sales of its physical imaging device while growing its customer base and prepare future constant revenue through the positive network effects of the cloud-based software-as-a-service model.

CURRENT INVESTMENTS AND FUNDING IN THE INDUSTRY

Philips, GE Healthcare, Siemens Healthineers, IBM, Genentech are among the largest multinational companies in medical imaging. These companies are spending part of their research and development budget on AI radiology or partnering with startups to develop new products. For example, Genentech is developing AI for lung CT, GE a software suite for chest X-rays with application to pneumonia, COVID-19, tuberculosis, and lung nodules. Budgets are often not disclosed and estimating them is challenging.

Many AI radiology startups have been founded in the last decades. We already covered some of the largest ones above, namely Butterfly Network (BFLY), Heartflow, Exo, but there are many more, smaller, startups addressing the AI radiology market. Across our meta-search, we did not find a strong correlation between founding date and funding raise or valuation. Figure 16.1 shows a chart comparing funding/valuation of AI radiology startups vs. their founding date.

SUMMARY

The AI radiology market is in exponential expansion, with an estimated compound annual growth rate of 30 to 40% from 2021 to 2026. The major impacted radiology fields are neurological, cardiovascular, lung and breast imaging. Multinational giants such as Siemens Healthineers, Philips, GE Healthcare are actively investing in AI radiology R&D. The sum of valuation and funding for radiology AI startups exceeds $5.5Bn. AI extends the offer of radiology imaging from new imaging modalities to more advanced information retrieval. This is an attractive opportunity for investors, and, most importantly, is helping to improve patient care by improving diagnosis and prevention.

Disclosure: Dr. Florian Dubost works for a company of the 4Catalyzer group in September 2021.

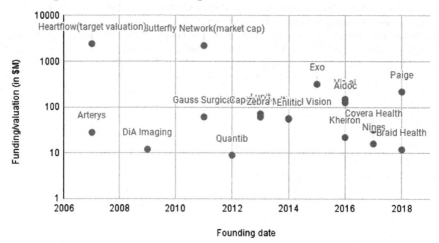

Figure 16.1 Funding/valuation vs. founding date. This is a non-exhaustive chart of current AI radiology startups. Data was gathered from crunchbase.com

REFERENCES

1. https://www.medgadget.com/2019/12/artificial-intelligence-in-radiology-market-2019-size-shar e-upcoming-trends-segmentation-and-forecast-to-2025-cagr-of-35-9.html

2. https://www.signifyresearch.net/medical-imaging/medical-imaging-ai-market-projected-reach- 1-2-billion-2025/

3. https://seraf-investor.com/compass/article/dividing-pie-how-venture-fund-economics-work-part-ii

4. https://www.signifyresearch.net/medical-imaging/ai-medical-imaging-top-2-billion-2023/

5. Solorzano, L., Almeida, G.M., Mesquita, B., Martins, D., Oliveira, C., Wählby, C., Stoyanov, D., Taylor, Z., Ciompi, F., Xu, Y. and Martel, A., 2018. *Computational Pathology and Ophthalmic Medical Image Analysis.*

PART V

AI APPLICATIONS IN DEVELOPMENT

17 Clinical View

Leonid L. Chepelev

INTRODUCTION

The demonstration of AlexNet by Krizhevsky, Sutskever, and Hinton at the University of Toronto in 2012 forever changed the landscape of artificial intelligence research, exemplifying not only the tremendous potential of deep convolutional neural networks (CNNs), but also demonstrating the feasibility of CNN implementation utilizing affordable hardware [1]. Encouraged by this success, Dr. Hinton famously opined that new radiologists should no longer be trained, as their jobs would be made obsolete by the numerous artificial intelligence–based software solutions. However, nearly a decade later, these words still could not be further from the truth as deep learning–based tools seem to be barely able to help facilitate improved radiologist productivity in the face of growing workloads and strained workforces. Radiologists are beginning to participate in the development and adoption of AI-based solutions in daily practice, harnessing the potential of artificial intelligence in radiology, from study requisition, workload/schedule balancing, image quality improvement, cost optimization, study prioritization, to assistance in study interpretation (Table 17.1). Novel approaches to image segmentation and AI-enriched radiomics applications that were previously too time-consuming or altogether not envisioned prior to the widespread availability of CNN-based solutions stand poised to further extend the scope and value of radiology.

While a wide array of applications is available to address numerous clinical needs across multiple disciplines, attempts at practical and clinically meaningful implementation of otherwise impressive algorithms have encountered an equally wide array of limitations. Clinical experience has demonstrated that software appraisal for the purpose of frontline clinical deployment is a much more complex process than initially anticipated. To begin appreciating the scope of this multifaceted problem, we must address every stage of the software lifecycle. First, not every algorithm that is generally available is assessed using metrics that are clinically meaningful. Secondly, algorithms that attain an acceptable level of clinically meaningful performance may not have been trained on data that is representative of the entirety of the patient population or representative of various subgroups, including racial minority subpopulations, which may adversely impact treatment decisions for these underrepresented subgroups. This was demonstrated by the capability of neural networks to correctly infer, for example, patient race from otherwise objectively anonymized data, such as chest radiographs [2]. Third, algorithms that are trained on fair and representative data likely suffer from generalizability deficiencies, given temporal, regional, and institutional disparities in patient population compositions, radiology hardware, imaging protocol differences, and the use of additional software within the imaging pipeline that may tangibly impact the input data, including: AI-based super-resolution, contrast signal boosting, acquisition acceleration, or image denoising algorithms. Fourth, algorithms that are practically deployable and validated at a specific time point shall not necessarily remain so as time progresses and local practices and population compositions evolve, as has been demonstrated with networks that exhibit performance degradation over time [3]. Fifth, algorithms that satisfy all of the above requirements may still not necessarily be safe and effective in frontline clinical applications and may require continuous careful evaluation as to cost–benefit tradeoffs and severity of impact on clinical workflows and decision-making.

Radiologists occupy a unique role as medical imaging specialists and clinical experts, being able to centralize all aspects of medical imaging within a single dedicated department in order to support further positive transformation of the scope, accessibility, and throughput of radiology, in a fashion similar to that facilitated by digital PACS systems. This transformation, however, will not occur without maintaining a sober clinical outlook on the impact and nature of the developing software solutions in the nascent algorithmic age of radiology. This chapter first provides a non-comprehensive overview of several major directions of ongoing AI related research and the results achieved so far, so as to ground a discussion of the future of clinically relevant AI-based applications. This chapter concludes with a discussion of an idealized future seamless clinical AI deployment system. Key to this future is the transformation of the current model of distribution of AI solutions through numerous independent vendors as standalone products to a more uniform distribution of AI solutions within a unified vendor neutral computational fabric that natively supports novel algorithms, along with built-in capabilities for ongoing quality monitoring.

DOI: 10.1201/9781003095279-22

Table 17.1 An Overview of Selected Major Directions in Radiology AI Application Development

Major Axis of AI Development	Specific Directions
Image Quality/Cost Improvement	Noise Reduction
	Radiation Dose Reduction
	Contrast Dose Reduction
	MRI Acquisition Acceleration
	On-Scanner Image Acquisition Support
Image Postprocessing	Automated Anatomic Segmentation
	Automated Pathology Quantification
	Malignancy Follow-Up Automation
	Radiomics-Based Imaging Biopsy
Diagnostic Assistance	Prediction of Disease Progression/Stratification
	Workflow Prioritization
	Automated Diagnosis
	Imaging Dilemma Resolution
Non-Voxel Radiology AI	Appropriate Study Requisition Support
	Automated Follow-Up
	Automated Electronic Chart Synthesis
	Report Pre-Dictation
	Workload Balancing
	CME/Learning Optimization
AI Quality Improvement/Quality Assurance	Universal AI Evaluation Frameworks
	Localization of Continuous AI QA/QI Efforts
	Dataset Bias Identification

IMAGE QUALITY AND COST IMPROVEMENT

A large subset of deep learning applications has been focused on the optimization of image quality in adverse environments. The two principal approaches to this problem include the direct mapping of the detector data to high-quality image outputs, or alternatively, reconstruction of high-quality image outputs from low quality inputs. The most notable and one of the earliest representatives of direct mapping of image data to sensor data was AUTOMAP [4], which replaced physics-based image reconstruction from k-space with a neural network–based reconstruction. The benefit of this approach is the ability to fully leverage sensor data in the setting of various levels of k-space undersampling and accommodating a wide range of imaging artifacts without having to alter the theoretical approach; new combinations of sensor data and optimal corresponding images could be used to learn to correct for artifact, noise, or undersampling. Similar approaches have been deployed by others, including to accelerate knee magnetic resonance imaging (MRI) [5], with the aim of bringing MR image acquisition time to the point where other traditional methods such as radiographs would no longer be the recommended entry point of imaging workups due to decreased MRI costs. Additional recent developments in improving the quality of low-field MRI could improve instrument affordability with only limited degradation in image quality [6], which represent just one example of AI-enabled novel imaging instrument development. An alternative approach focuses on the use of encoder–decoder architectures to train networks capable of image denoising, contrast signal boosting, and image super-resolution based on pairs of suboptimal and optimized quality images. For example, such approaches have been used to improve image resolution in cardiac imaging [7], boosting reduced-dose contrast signal [8], to create synthetic contrast-enhanced images thereby bypassing contrast administration altogether [9], as well as predicting the appearance of lesions on entirely different modalities. Additionally, assistive algorithms have been deployed to facilitate image acquisition and planning in cardiac MRI [10, 11] where such algorithms may eventually be involved in automated quality evaluation as part of a semiautomated workflow [12].

IMAGE POSTPROCESSING

Image postprocessing tasks, including the precise segmentation of pathology and salient anatomy, are key to supporting precision medicine applications, for example by enabling 3D printing applications in a range of clinical scenarios [13] where segmentation architectures such as nnU-Net are employed with excellent Dice similarity coefficients [14]. In lung cancer screening and follow-up, where nodule detection, measurement, and accurate comparison are some of the most common and time-consuming tasks, clinically integrated solutions have emerged not only to facilitate the daily activities of nodule detection and characterization, but to transform the decision-making process in lung cancer screening based on a new AI-enabled strategy [15], producing improved specificity and significant cost savings. Besides significantly facilitating the menial tasks of radiology, deep learning has been applied to dramatically transform traditionally highly time-consuming tasks such as cardiac strain analysis [16] as well as to add tremendous value to medical imaging by effortlessly providing prognostic value through opportunistic computed tomography (CT) analysis of sarcopenia [17] or using a wide range of markers [18]. By facilitating advanced image postprocessing, AI-enabled software could add tremendous value to radiology by enabling analyses that are known to add value but are too time-consuming to be practical in the frontline clinical setting. This could not only enable entirely novel analyses but could democratize access to evaluations currently only performed in tertiary centers due to resource constraints in smaller centers. Beyond the improved access to advanced image postprocessing, AI-based solutions have been called on to enable the concept of "digital biopsy"—the AI-enabled radiomic evaluation of lesions of interest to more accurately establish malignant potential based on imaging alone [19]. In addition to general categorical prognostication, such as effective prognosis of COVID-19 using deep learning based on chest radiographs [20], one may envision the application of encoder–decoder networks trained on pairs of studies that are temporally separated with the aim of predicting the appearance of the future study, for example with respect to stroke evolution, with the aim of modifying current management [21].

DIAGNOSTIC ASSISTANCE AND SOFTWARE EVALUATION

Recent literature provides a constant array of studies that not only demonstrate the feasibility of diagnosis based on deep learning, but also demonstrate superiority of automated approaches when compared with radiologist performance and have been addressed elsewhere in this book. To urge caution, some large-scale practical evaluations of such reportedly well-performing algorithms sometimes fall short of expectations, as was the case in a large-scale analysis of breast cancer screening on digital mammograms where 94% of the 36 tested AI systems were found to be less accurate than even a single radiologist [22]. In another study, none of the published models to detect and prognosticate for COVID-19 on chest radiographs and CT scans were of clinical use due to methodological flaws and biases [23]. There are ample reasons for such limitations, including differences between instruments employed in the published studies and used in subsequent large-scale evaluations, different imaging techniques, differences in patient population compositions in terms of clinical circumstances, age, and race, fraction of imaging artifact, presence of imaging data in the testing set that was outside the training set thus resulting in a confident silent failure, and many other technical factors. Such validation studies highlight the importance of ongoing model training and tuning to the local implementation site, and a culture of quality and organizational excellence that fosters ongoing real-world performance monitoring for AI-based solutions. Such goals were highlighted by the United States Food and Drug Administration [24]. Conscious effort must be made to overcome implicit known biases such as those introduced by selection of only the highest-quality data rather than truly representative data for training; as well as possible unanticipated data biases, such as from racial bias [2] which may be overcome using explicit subgroup analyses. While the topic of algorithm appraisal is beyond the scope of this chapter, numerous papers have been published on the evaluation of AI-based software solutions and practical implementation of such solutions in the clinical setting [25, 26]. Additionally, the American College of Radiology has initiated the Assess-AI initiative for algorithm monitoring [27] that may be helpful in objective evaluation of algorithm performance in the future.

NON-VOXEL–BASED CLINICAL AI

Finally, another major direction of ongoing development is in improving radiologist productivity through automated integration of clinical data and automated assistive functions. Such functions include identification of cases for prompt communication [28], estimation of pretest probability, study protocolling, predicting missed imaging appointments, radiology workflow management, automated radiology–pathology correlation, identification of similar studies based on image similarity search to inform differential diagnoses, and workload distribution, among numerous others [29]. On a practical level, it is these technologically simpler applications which may ultimately have the most profound impact on radiologist workflow and quality of life.

THE FUTURE OF CLINICALLY READY AI

AI-enabled solutions in medical imaging encompass a broad spectrum of software tools with goals of improving image acquisition quality and affordability, extending the scope of radiology, supporting clinical diagnosis, and improving everyday workflows in medical imaging clinical practice and research. As this chapter has amply described, AI-based software solutions hold tremendous potential for meaningful transformation of their respective domains. Unfortunately, algorithm generalizability, vendor-neutral integration, result explanation/transparency, and built-in support for ongoing quality assurance and improvement to account for changes in local practice and patient populations remain significant challenges that can limit broad adoption of artificial intelligence in clinical practice [30].

Currently, radiological AI is facing a multilayered computational Tower-of-Babel problem propagated by the ever-growing number of the substantially disparate software vendors and local implementations in various stages of true clinical readiness. The sheer challenge of seamless computational integration further interacts with a complex layer of the management of the clinical and practical perceptions and expectations regarding the use of radiological AI. Currently, an AI solution that is published is not necessarily immediately available for clinical use [23]; and if available, such a solution is not necessarily immediately deployable in a specific computational environment; and if deployable, such a solution is not necessarily validated on a specific scanner in a specific clinical setting or is ready for ongoing validation. If objectively validated from the standpoint of accuracy, such a solution may not be capable of failing safely, is not necessarily overall safe and effective, or meaningfully improves daily workflows. Finally, if all of the above requirements are satisfied, a solution may still be limited in its clinical use by radiologists' and referring providers' implicit or explicit perception of the value of its contribution to patient care [31].

To address some of these issues, a vendor-neutral general computational fabric similar in nature to the Internet and supported by common ontologies such as RadLex and computational service integration standards laid out by the Digital Imaging and Communications in Medicine workgroups would likely be needed to support a more facile integration of smaller independently developed solutions. Alternatively, larger software/instrument vendors could develop vendor-specific and specific instrument-optimized "App Marketplaces" similar to major smartphone platforms to enable independent developers to create and deploy their solutions in a standardized environment, with well-defined inputs, outputs, workflows, and lists of compatible/cleared imaging devices.

In either pathway, the transformation of an AI-based solution from a rigid and centrally updated software product into a self-optimizing framework would likely be necessary to support clinically meaningful implementations. Additionally, the nature of clinical records in AI-supported work would likely need to be dramatically revised to ensure that all algorithms employed in a specific pipeline are completely traceable for medicolegal and quality improvement purposes. Crucial information in such a setting would include algorithm version, provenance, modifications, and order within a workflow that could contain numerous other algorithms (e.g., prognostication based on segmentation of images that are first denoised). Collaborative, widely accessible, anonymized, and expertly curated data commons would likely need to be established to ensure fair representation of patient populations and to support accurate algorithm validation and ongoing research efforts. Importantly, such data commons should contain not only carefully selected high-quality images, but a wide selection of suboptimal image studies to reflect the true clinical image datasets. A routine participation in ongoing quality monitoring by key clinical members of radiology departments will be necessary to gain and maintain trust in all of these computational solutions, ensure tangible added value, and avoid scenarios where computer assistance is of limited practical utility [32].

CONCLUSIONS

This chapter describes the many current directions of AI development in radiology, but regardless of the precise trajectory of the evolution of deep learning applications in the clinical setting, radiologists should expect novel AI applications to transform the scope and nature of their work in unprecedented ways. It is the responsibility of radiologists to meticulously leverage their clinical expertise to drive AI implementations that are safe, effective, amenable to ongoing quality monitoring, transparent, and fair. Supporting a culture of quality and organizational excellence while transforming the nature of AI-based software solution distribution will be key to positively transforming the clinical scope of radiology in the nascent algorithmic age.

REFERENCES

1. Krizhevsky A et al., Advances in Neural Information Processing Systems (2012).

2. Banerjee I et al. (2021) https://arxiv.org/abs/2107.10356

3. Pianykh OS et al., Radiology (2020) PMID: 32840473

4. Zhu B et al., Nature (2018) PMID: 29565357

5. Recht MP et al., AJR Am J Roentgenol (2020) PMCID: PMC8209682

6. Koonjoo N et al., Sci Rep (2021) PMCID: PMC8050246

7. Masutani EM et al., Radiology (2020) PMCID: PMC7263289

8. Gong E et al., J Magn Reson Imaging (2018) PMID: 29437269

9. Kim SW et al., Sci Rep. (2021) PMCID: PMC8516935

10. Blansit K et al., Radiol Artif Intell (2019) PMCID: PMC6884027

11. Bahrami N et al., Magn Reson Med (2019) PMCID: PMC7962153

12. Piccini D et al., Radiol Artif Intell (2020) PMCID: PMC8082371

13. Chepelev L et al., 3D Print Med (2018) PMCID: PMC6251945

14. Isensee F et al., Nat Methods (2021) PMID: 33288961

15. Adams SJ et al., J Am Coll Radiol (2021) PMID: 33482120

16. Ghadimi S et al., J Cardiovasc Magn Reson (2021) PMCID: PMC7949250

17. Boutin RD, Lenchik L, AJR Am J Roentgenol (2020) PMID: 32755187

18. Pickhardt PJ et al., Lancet Digit Health (2020) PMCID: PMC7454161

19. Sahiner B et al., Med Phys (2019) PMID: 30367497

20. Jiao Z et al., Lancet Digit Health (2021) PMCID: PMC7990487

21. Yu Y et al., AJNR Am J Neuroradiol (2021) PMCID: PMC8191664

22. Freeman K et al., BMJ (2021) PMCID: PMC8409323

23. Roberts M et al., Nat Mach Intell (2021) DOI: 10.1038/s42256-021-00307-0

24. The United States Food and Drug Administration (2021) https://www.fda.gov/files/medical%20devices/published/US-FDA-Artificial-Intelligence-and-Machine-Learning-Discussion-Paper.pdf

25. Omoumi P et al., Eur Radiol (2021) PMCID: PMC8128726

26. Tadavarthi Y et al., Radiol Artif Intell (2020) PMCID: PMC8082344

27. American College of Radiology (2021) https://www.acrdsi.org/DSI-Services/Assess-AI

28. Meng X et al., J Biomed Inform (2019) PMCID: PMC6506378

29. Lakhani P et al., J Am Coll Radiol (2018) PMID: 29158061

30. Yan W et al., Radiol Artif Intell (2020) PMCID: PMC8082399

31. Gaube S et al., NPJ Digit Med (2021) PMCID: PMC7896064

32. Lehman CD et al., JAMA Intern Med (2015) PMCID: PMC4836172

18 Technological View

Tyler Gathman, Adrianna M. Rivera-León, Thomas Kane,
Ranveer Vasdev, Ibrahim Abdalla, and Amy Song

CONCEPTUAL FOUNDATIONS

With the rapid development of robust computational architectures, AI has shown the potential to transform radiology. The development of models that can analyze patient information, imaging, and other medically relevant features which can then make predictions and interpretations based on the data has been well-demonstrated. Traditionally, machine learning (ML) algorithms, including decision trees, artificial neural networks, and support vector machines, have been utilized in prediction models due to their computational efficiency and superior performance with limited data [1]. However, with the advent of cost-effective computing resources and large datasets, notably ImageNet which consists of hand-annotated images, deep learning, also referred to as deep neural networks, has provided major advancements while transfer (or continuous) learning has enabled AI applications to be efficiently developed and updated with new information or objectives [1, 2].

Deep learning involves an architecture with multiple layers consisting of interconnected nodes, or "neurons." For example, convolutional neural networks (CNNs) utilize convolution and pooling layers, which allow algorithms to make extensive associations and connections among complex data as opposed to conventional ML methods. AlexNet, one of the first technically successful CNN architectures, demonstrated a significant leap in performance over prior years during the 2012 ImageNet challenge, where models compete to classify natural color images into 1,000 classes [3]. Additional advances have led to residual networks (ResNets) which have more layers, can be more efficiently trained, and lead to better accuracy as measured by area under the curve (AUC) [2, 4]. Deep learning, especially in the context of medical image analysis, has been utilized for classification, detection, and segmentation applications [2]. Studies have demonstrated the utility of deep neural networks for identifying malignancies, fractures, and neurological pathology from imaging among other radiological objectives [2, 5–7].

In tandem with deep learning, transfer learning has increased the efficiency of the training process and enabled smaller datasets to be utilized [5, 8]. Previous applications in radiology employed static algorithms which ingest information and undergo a comprehensive learning process [9]. However, static algorithms tend to perform worse over time as the data and environment change and cannot be easily adapted to changing objectives [9]. As radiology and the broader medical field continuously evolve with changing treatment regimens and diagnostic protocols, AI applications must undergo the same process. Transfer learning enables the model to be efficiently updated as additional data becomes available or better associations between features and output are uncovered. Novel objectives, including new diagnostic targets, can also be trained effectively by maintaining a portion of the deep learning architecture such as only training the terminal layer of a CNN after pre-training the algorithm.

Various studies have reinforced the benefits of deep neural networks and transfer learning. Antropova et al. utilized a pre-trained CNN on ImageNet to characterize breast lesions [2, 5]. The deep learning CNN algorithm was shown to outperform existing computer-aided diagnosis (CADx) methods despite utilizing a relatively small dataset due to transfer learning with ImageNet pretraining [2, 5].

CLINICALLY RELEVANT APPLICATIONS

AI methodologies can be exploited for various radiological applications that range from image acquisition and optimization to ease of workflow and decision-making. The implementation of AI for computerized imaging procedures has shown considerable success in improving image quality. This has been achieved through the development of algorithms that allow accurate positioning of patients for image taking, also resulting in marginal benefits such as reducing the amount and number of times that patients need to be scanned for imaging procedures including CTs and MRIs [1]. Moreover, AI systems can be used for recognition and extraction of quantitative image attributes that can aid in disease characterization, potentially improving patient outcomes [10]. This AI-based approach to medical imaging, referred to as radiomics, demonstrates particular promise

DOI: 10.1201/9781003095279-23

for individualized and precision-based medicine, as it could be integrated with non-pixel information such as electronic medical records and patient-specific or signature data for achieving optimal care and treatment plans [11]. Provided with reliable training data sets, these applications could decrease error rates and even help ease radiologist's workflow by taking care of ordinary and routine tasks (e.g., organization, sequencing, localization) and allowing radiologists to focus on the clinical interpretation of data.

Recently, a deep transfer learning approach purely based on CT images was proposed for the detection of COVID-19. This method, which was pretrained with the ImageNet dataset, showed high detection accuracy when predicting COVID-19 cases [12]. Although this method did not provide information as to the severity or progression state of COVID-19 cases, its high accuracy in disease detection could certainly prove useful in the development of additional approaches for similar or future time-sensitive instances where clinician time, effort, and performance are a top priority.

In addition to image acquisition, optimization, and processing, AI based systems could be implemented to provide clinical assistance in decision-making. Clinical decision support systems (CDSS) are technologies that provide clinicians with support in the process of making decisions. To do so, traditional CDSS make use of a so-called knowledge base that contains compiled and associated data as input, an inference engine that can integrate the knowledge base with patient information, and a communication mechanism to generate and deliver an output [13]. AI-based systems could be exploited for developing CDSS that do not rely on the system structure described above; that is to say, systems that do not rely on a knowledge base or an associated set of rules as input. In contrast, AI-CDSS can learn from past experiences and find patterns in clinical data without strict supervision. AI-CDSS algorithms allow the computer to carry out a stream of tasks and learn from its own outputs through reinforcement learning and optimization.

De Fauw et al. developed a deep neural network for interpreting optical coherence tomography images for retinal disease and referring patients for additional care visits as needed [14]. Their final method's performance was comparable to that of experts in the field, surpassing one of the major challenges in AI-based systems. This and similar technologies have led to the generation of so-called "intelligent-agents," which can execute functions in real time, adapt to changing sets of rules, analyze themselves for error and success rates, learn from previous performance, develop memory, and build upon their own output for optimal results [13].

PRACTICAL CONSIDERATIONS AND LIMITATIONS

Major challenges in implementing clinical AI technologies, however, lie in the availability of patient data and images, generalizability of training data, and the creation of efficient algorithms that are accurate and reliable [2]. Although transfer learning by pretraining can reduce the need for large medical datasets, the complexity of deep neural networks may result in overfitting [2]. Overfitting occurs when deep learning models are tuned to the training data and are not generalizable to unseen, but related, test data. Moreover, the notion of interpretability has raised concerns due to the increasing complexity of deep learning systems which are not always explainable or illustratable to humans. This phenomenon, particularly within unsupervised deep learning systems, is referred to as the "black box," in which an algorithm with a training data set is developed but with no clear understanding of how it generates a given output.

To aid and circumvent the "black box" phenomenon, several interpretability models have been developed. Collectively, these comprise varying approaches and methods of interpretability to accurately understand and discern data predictions in AI-based systems. One such method relies on the development of saliency maps. Essentially, this approach highlights features within an image that contribute significantly to the system's prediction and output [15]. Similarly, a different approach that makes use of influential functions attempts to leave out and retain individual images within a training dataset to determine which images within said dataset are most influential in driving a prediction model [16]. A different non-feature attribution reliant method is the use of semantics through concept activation vectors [6]. This approach generates a textual explanation that provides contextual information when an output or probabilistic or statistical value is generated. All these methods allow for data to become more interpretable to humans while providing a comprehensive framework to evaluate and improve AI systems. Importantly, these approaches could even be applied in unison to achieve maximal interpretability.

CASE STUDY: Deep Learning for Lung Cancer Screening

Investigative efforts into radiomics, deep learning, and transfer learning have demonstrated promising findings when applied to clinical problems. Literature on lung cancer screening and management illustrates the potential of the modalities. The discussions below highlight current applications of AI in that identification, management, and longitudinal evaluation of lung malignancies.

Identification and assessment of pulmonary nodules on CT imaging remain a difficult but important task for practicing radiologists. Even without AI, clinical trials such as the National Lung Screening Trial (NLST) for cancer, have demonstrated a 20% reduction in mortality by screening for lung cancer in high-risk populations [17]. Nevertheless, human error and inherent variability in image evaluation can confound the assessment of pulmonary nodules. Armato et al. demonstrated that variance exists even in individual radiologists' evaluations of CT images of the same nodules, resulting in different reported numbers of nodules for identical images [18]. To this end, multiple studies have investigated the ability of radiomics to classify lung nodules as benign or malignant. Chen et al. trained a support vector machine model on a sample of CT images from 72 patients with known benign and malignant lung nodules, achieving 84% accuracy [19]. Deep learning algorithms have also shown encouraging results. For instance, Ardila et al. developed a deep learning CNN algorithm with pretraining on the ImageNet dataset that compared current and prior imaging to screen for lung cancer [20]. While the model performed similar to humans when prior imaging was available, in the absence of prior imaging, the model outperformed radiologists, leading to an absolute reduction of 11% false positive and 5% false negative reads [20].

Applications of radiomics to lung cancer go beyond screening; several studies have investigated histological classification and response to treatment of lung cancer. In a study of patients with non-small cell lung cancer (NSCLC) on definitive chemotherapy, deep learning models based on pretreatment and posttreatment CT scans were able to predict survival and outcomes including metastasis or progression with statistical significance [21]. Another study by Canuzwa et al. was able to train a CNN to discriminate between adenocarcinoma and squamous cell carcinoma, the two most common NSCLC subtypes, based on CT data [16]. The evaluation and prognostic prediction of lung cancer remains an important objective and illustrates an example of the mounting applications for AI at the leading edge of technology in radiology.

REFERENCES

1. C. Malamateniou, K. M. Knapp, M. Pergola, N. Woznitza, and M. Hardy, "Artificial Intelligence in Radiography: Where Are We Now and What Does the Future Hold?," *Radiography*, vol. 27, pp. S58–S62, Oct. 2021, doi: 10.1016/j.radi.2021.07.015.

2. M. A. Mazurowski, M. Buda, A. Saha, and M. R. Bashir, "Deep Learning in Radiology: An Overview of the Concepts and a Survey of the State of the Art with Focus on MRI," *J. Magn. Reson. Imaging*, vol. 49, no. 4, pp. 939–954, Apr. 2019, doi: 10.1002/jmri.26534.

3. A. Krizhevsky, I. Sutskever, and G. E. Hinton, "ImageNet Classification with Deep Convolutional Neural Networks," *Commun. ACM*, vol. 60, no. 6, pp. 84–90, May 2017, doi: 10.1145/3065386.

4. K. He, X. Zhang, S. Ren, and J. Sun, "Deep Residual Learning for Image Recognition," 2015, doi: 10.48550/ARXIV.1512.03385.

5. N. Antropova, B. Q. Huynh, and M. L. Giger, "A Deep Feature Fusion Methodology for Breast Cancer Diagnosis Demonstrated on Three Imaging Modality Datasets," *Med. Phys.*, vol. 44, no. 10, pp. 5162–5171, Oct. 2017, doi: 10.1002/mp.12453.

6. B. Kim *et al.*, "Interpretability Beyond Feature Attribution: Quantitative Testing with Concept Activation Vectors (TCAV)." *arXiv*, Jun. 07, 2018. Accessed: May 16, 2022. [Online]. Available: http://arxiv.org/abs/1711.11279

7. for the Alzheimer's Disease Neuroimaging Initiative, A. Ortiz, J. Munilla, F. J. Martínez-Murcia, J. M. Górriz, and J. Ramírez, "Learning Longitudinal MRI Patterns by SICE and Deep Learning: Assessing the Alzheimer's Disease Progression," in *Medical Image Understanding and Analysis*, vol. 723, M. Valdés Hernández and V. González-Castro, Eds. Cham: Springer International Publishing, 2017, pp. 413–424. doi: 10.1007/978-3-319-60964-5_36.

8. J. Yosinski, J. Clune, Y. Bengio, and H. Lipson, "How Transferable Are Features in Deep Neural Networks?," 2014, doi: 10.48550/ARXIV.1411.1792.

9. O. S. Pianykh *et al.*, "Continuous Learning AI in Radiology: Implementation Principles and Early Applications," *Radiology*, vol. 297, no. 1, pp. 6–14, Oct. 2020, doi: 10.1148/radiol.2020200038.

10. G. Choy *et al.*, "Current Applications and Future Impact of Machine Learning in Radiology," *Radiology*, vol. 288, no. 2, pp. 318–328, Aug. 2018, doi: 10.1148/radiol.2018171820.

11. B. Saboury, M. Morris, and E. Siegel, "Future Directions in Artificial Intelligence," *Radiol Clin N Am*, vol. 59, no. 6, pp. 1085–1095, Nov. 2021, doi: 10.1016/j.rcl.2021.07.008.

12. W. Zhao, W. Jiang, and X. Qiu, "Deep Learning for COVID-19 Detection Based on CT Images," *Sci Rep*, vol. 11, no. 1, p. 14353, Dec. 2021, doi: 10.1038/s41598-021-93832-2.

13. F. Shaikh *et al.*, "Artificial Intelligence-Based Clinical Decision Support Systems Using Advanced Medical Imaging and Radiomics," *Curr Probl Diagn Rad*, vol. 50, no. 2, pp. 262–267, Mar. 2021, doi: 10.1067/j.cpradiol.2020.05.006.

14. J. De Fauw *et al.*, "Clinically Applicable Deep Learning for Diagnosis and Referral in Retinal Disease," *Nat Med*, vol. 24, no. 9, pp. 1342–1350, Sep. 2018, doi: 10.1038/s41591-018-0107-6.

15. M. Reyes *et al.*, "On the Interpretability of Artificial Intelligence in Radiology: Challenges and Opportunities," *Radiol Artif Intell.*, vol. 2, no. 3, p. e190043, May 2020, doi: 10.1148/ryai.2020190043.

16. P. W. Koh and P. Liang, "Understanding Black-box Predictions via Influence Functions," arXiv, Dec. 29, 2020. Accessed: May 16, 2022. [Online]. Available: http://arxiv.org/abs/1703.04730

17. B. S. Kramer, C. D. Berg, D. R. Aberle, and P. C. Prorok, "Lung Cancer Screening with Low-Dose Helical CT: Results from the National Lung Screening Trial (NLST)," *J Med Screen*, vol. 18, no. 3, pp. 109–111, 2011, doi: 10.1258/jms.2011.011055.

18. S. G. Armato *et al.*, "Assessment of Radiologist Performance in the Detection of Lung Nodules: Dependence on the Definition of 'Truth,'" *Acad Radiol*, vol. 16, no. 1, pp. 28–38, Jan. 2009, doi: 10.1016/j.acra.2008.05.022.

19. C.-H. Chen *et al.*, "Radiomic Features Analysis in Computed Tomography Images of Lung Nodule Classification," *PLoS One*, vol. 13, no. 2, p. e0192002, Feb. 2018, doi: 10.1371/journal.pone.0192002.

20. D. Ardila *et al.*, "End-to-end Lung Cancer Screening with Three-Dimensional Deep Learning on Low-Dose Chest Computed Tomography," *Nat Med*, vol. 25, no. 6, pp. 954–961, Jun. 2019, doi: 10.1038/s41591-019-0447-x.

21. Y. Xu *et al.*, "Deep Learning Predicts Lung Cancer Treatment Response from Serial Medical Imaging," *Clin Cancer Res*, vol. 25, no. 11, pp. 3266–3275, Jun. 2019, doi: 10.1158/1078-0432.CCR-18-2495.

19 Societal View

Charles Lu, Ken Chang, Jay B. Patel, Praveer Singh, and Jayashree Kalpathy-Cramer

INTRODUCTION

Recent breakthroughs in artificial intelligence (AI) have renewed expectations for AI clinical decision support tools in a broad range of application areas in healthcare. Deep learning (DL), a specific subset of techniques in the field of machine learning (ML), has demonstrated impressive predictive capability across a wide range of tasks using large amounts of data such as images, text documents, and audio recordings (Esteva 2019), and have been applied to a broad range of biomedical applications such as drug discovery, hospital operations, and bioinformatics (Senior 2020, Zou 2019, Singh 2020, Nelson 2019).

The field of radiology has changed substantially over the years as newly developed technologies are incorporated into the existing clinical workflow, and several technical and occupational factors make radiology primed for integration of AI tools. Most medical imaging data is standardized into the Digital Imaging and Communication in Medicine (DICOM) format and stored in a picture archiving and communication system (PACS) infrastructure. Additionally, radiologists are usually consulted by a referring physician and do not need to be physically situated near the patient to interpret a scan, which allows for the possibility of teleradiology. As such, tools that are able to interface with the PACS and the radiologist's workstation have the potential to improve current patient workflows. The steady increase in the numbers of medical imaging exams ordered every year has led to radiologist burnout and increased interest in investigating AI-tools for high-volume, low disease-incidence applications such as mammography screening (McKinney 2020). Thus, radiologists have the opportunity to lead efforts in adoption and standardization of best practices that demonstrate the increased efficiency gained from new AI workflows to other areas of medicine.

One could envision numerous ways AI technology could transform healthcare over the coming decades, leading to lower costs and improved patient outcomes to alleviate rising healthcare costs and diseases from aging populations. However, uncertainties remain on how to best incorporate AI into existing clinical workflows and infrastructure in the face of complex social dynamics and clinician–AI interaction. In this chapter, we discuss specific opportunities and challenges for AI in radiology to better inform the development and deployment of AI for healthcare institutions.

AI APPLICATIONS IN DEVELOPMENT

The number of FDA clearances granted to "artificial intelligence" medical devices for applications in medical imaging have increased significantly over the previous decade, according to data from the American College of Radiology (ACR).

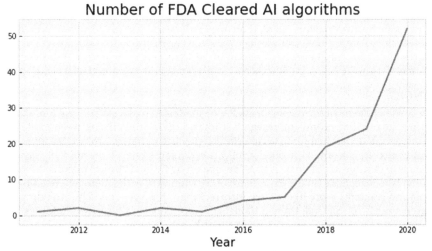

This recent growth in the number of available AI tools for radiology could greatly accelerate existing clinical workflows by automating tasks that were previously performed manually to

DOI: 10.1201/9781003095279-24

increase efficiency and improve interpretation turnaround times. Similar to Topol (Topol 2019), we do not foresee AI tools replacing radiologists wholesale. Rather, we believe these tools will empower radiologists and radiological technicians by automating tedious and routine tasks, thus alleviating burnout and helping to reduce overall workload. Indeed, much of medicine consists of tasks which may be time-consuming but are not difficult relative to the gamut of higher-level cognitive tasks that modern radiologists perform. Recent studies have shown AI to have high predictive performance in common tasks such as determining the presence of findings on routine chest radiographs, delineating brain metastases, and detecting lung nodules (Rajpurkar 2018, Majkowska 2020, Grøvik 2020, Ardila 2019). Automating these types of tasks will allow the clinician to focus on more challenging and clinically important tasks such as determining optimal and personalized treatment regimens. Opportunities for AI also exist in improving specific components of the overall clinical workflow—such as improving image quality, better triage of urgent cases, or acting as a second read to reduce error and misses (Titano 2018, Annarumma 2019, Harvey 2019).

While currently most AI tools are designed to utilize only radiologic imaging, newer methods will be able to incorporate contextual information from a host of clinically relevant data such as that from electronic health records, genomics, and pathology (Rajkomar 2018, Zou 2019, Campanella 2019). This ability to leverage multi-modal sources of clinical data will provide more holistic assessments and predictions in the diagnosis of certain diseases (Huang 2020). Another promising application is in assessment of disease trajectories and prediction of future outcomes. For example, deep learning can be utilized to assess and quantify changes in disease severity between patient visits, as well as to assess disease stability and progression (Li 2020a, Li 2020b). Other studies demonstrate predictive ability for prognostic measures, such as the need for intubation, treatment response, and patient survival (Li 2020c, Jin 2021, Lao 2017). These are just some of the ways AI may fundamentally change the role of radiology to be more customizable and tailored to each individual patient, a recurring trend in the era of precision medicine.

A final opportunity for AI in healthcare is the development of techniques and formation of partnerships to leverage disparate silos of medical data to train and validate ML algorithms in a distributed and continuous manner (Pianykh 2020, Lee 2020, Gupta 2021, Zhang 2021). Just as dictionaries are continuously updated and refined as language evolves, AI algorithms must also be adaptive and periodically recalibrated to account for changing disease prevalence, patient populations, data acquisition methodologies, and clinical guidelines. It is widely known that training DL models on data from diverse sources leads to increased model performance and robustness. Unfortunately, due to patient privacy issues, it is difficult to acquire a centralized repository of multi-institutional data. Presently, much research is being done toward developing distributed and federated learning paradigms so that AI models may be trained on data from multiple healthcare institutions while preserving patient privacy (Mahajan 2018; Jared A. Dunnmon 2019; Roth 2020). Distributed approaches would potentially enable AI algorithms to be trained on hundreds of thousands of cases, and such algorithms may prove invaluable for rare diseases or in regions where clinical experience and expertise is limited. For instance, even if a developing country had access to adequate healthcare equipment, there may still be too few fully trained radiologists to provide satisfactory care for their population. We see AI algorithms developed in collaboration with high-resource countries being deployed in these settings, thus helping overburdened healthcare personnel to provide better patient care to underrepresented and underserved populations.

CHALLENGES

Radiology is one of the most promising areas for AI models to make a positive social impact. However, in spite of many promising research studies, a divide remains between model development and successful deployment into actual clinical practice in the real world. Machine learning in healthcare differs from many other application domains in a few prominent ways. First, labeled data is expensive to obtain as it must be labeled by trained clinical experts. Moreover, the quality of the labels depends on the experience of the clinician, with labels generated by medical students and residents more likely to be incorrect than those generated by attending radiologists (Rauschecker 2020). Second, trained models must be deployed directly into legacy hospital systems, requiring tight vendor integration in a highly regulated environment in most countries. Finally, promised clinical utility may not be achieved due to several known challenges such as generalization, model biases, and gaps in engaging relevant stakeholders during the design and development of AI applications. Additionally, given that most AI medical devices are commercially developed, companies tend to focus on high-prevalence and critical diseases over rare diseases or non-critical conditions, in which it may not be commercially feasible to recoup development and regulatory costs.

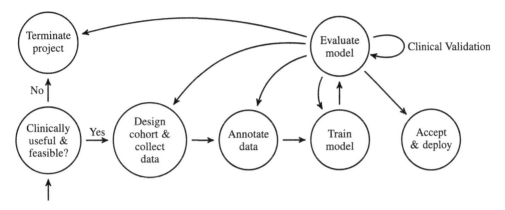

Many AI models are developed and tested on data from only a single institution, but are intended to make predictions on data from external institutions (Wu 2021). In principle, it is desirable for the model's performance to generalize to new data, but in practice, a generalization gap usually exists as the training data is never truly representative of the real-world data distribution. Performance often declines when applied to new data from different scanners, acquisition parameters, and patient populations (AlBadawy 2018; Chang 2020; Zech 2018; Mårtensson 2020; Gupta 2020). This is an issue that is not often addressed in the literature due to the lack of external testing sets (Liu 2019; Kim 2019). However, as journals move toward expecting studies to perform external validation (Mongan 2020), the brittleness of models across a wide range of applications is becoming more apparent. Compounding the issue, most algorithms only use training datasets from limited geographic regions (Kaushal 2020). More processes and controls need to be developed to improve the generalizability of models while identifying conditions to mitigate model failures (Chen 2021; Subbaswamy 2021).

Another open question is what impact clinical AI tools will have on clinical decision-making and user behavior. Several studies demonstrate that a clinician's diagnostic performance may be improved with the help of AI (Sung 2021), resulting in increased efficiency (as measured by time advantage) and consistency (as measured by reader variability). However, the over-utilization of AI can result in automation bias from overreliance by deferring decision-making in some tasks (Lyell 2017). Another study found that incorrect AI predictions negatively impacted clinicians' performance across skill levels and case difficulties for evaluation of pathology slides (Kiani 2020). Similar negative impact was reported in another study regarding performance of radiologists, internists, and emergency medicine physicians receiving inaccurate external advice on the interpretation of chest X-rays (Gaube 2021). These studies demonstrate the need for further study of clinician–AI interactions and highlights the importance of training clinicians about both the limitations of AI in general, and the particular limitations of each specific model that clinicians may be utilizing.

Interpretability and explainability are also major challenges in the applicability of current AI algorithms (Xie 2020). Commonly used deep learning algorithms are often described as "black-box," meaning that they do not admit any explanation for how or why they produced any specific prediction. Therefore, it would be difficult for a clinician to trust an algorithm prediction without adequate justification on how it came to its decision. Non-obvious false-positive AI predictions may even slow down or degrade radiological workflow. Explainability is especially important in light of recent work showing that neural networks often utilize "shortcut" learning, i.e., associating spurious imaging features (laterality markers, arrows, and other forms of human annotation) with the label of interest (DeGrave 2021). Saliency maps are a commonly used approach to make neural networks interpretable (Reyes 2020). However, these methods are limited in that they provide no explanation beyond localizing where the network is looking (Rudin 2019). Furthermore, there is emerging evidence that the localization capacity of these saliency maps is extremely limited and far underperforms detection/segmentation models (Arun 2021). There are several other promising methods for explainability that have been developed, but rigorous validation under a wide range of clinical scenarios still remains to be seen (Chen 2020; Kim 2018).

Inequality with regard to access to emerging technology such as AI has also attracted more attention recently. Disability-adjusted life years (DALY) is a simple metric which accounts for both mortality and morbidity and measures how many years of health are lost due to premature death from disease or disability. DALY is often used to compare the burden of disease amongst different countries. Much of the healthcare burden as measured by DALY is concentrated in developing countries (Roser and Ritchie 2016).

Burden of disease, 2017

Disability-Adjusted Life Years (DALYs) per 100,000 individuals from all causes.
DALYs measure the total burden of disease – both from years of life lost due to premature death and years lived with a disability.
One DALY equals one lost year of healthy life.

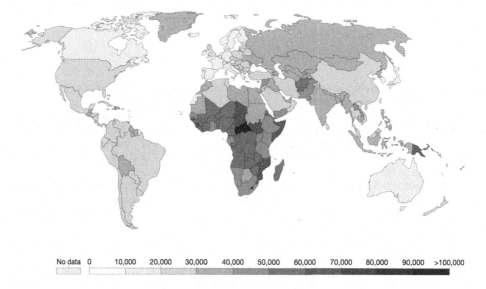

No data | 0 | 10,000 | 20,000 | 30,000 | 40,000 | 50,000 | 60,000 | 70,000 | 80,000 | 90,000 | >100,000

However, most private funding for AI applications (medical or otherwise) are concentrated in just two countries: the United States and China (Zhang 2021). Along with different population demographics and disease prevalence, developing countries have lower access to healthcare than developed countries and lack much of the infrastructure necessary to develop or deploy new healthcare technologies. As current AI techniques such as the transformer language models require large amounts of computing resources and access to data, which is biased towards English-speaking developed countries (e.g., Wikipedia, social media, Western tech companies), there is a concern that developing countries will be further left behind and unlikely to capture much of the potential benefit from technological advances (Bender 2021).

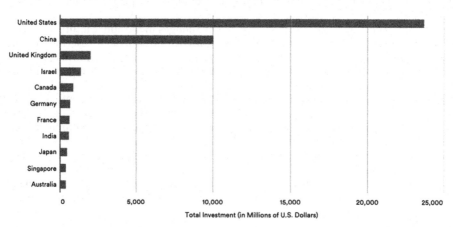

Other issues relate to regulation of clinical AI software as medical devices (FDA 2019). AI software differs from traditional medical software in its capability to continuously learn from data, which presents challenges from a verification standpoint. Additional training for new technologies should be practiced so clinicians will use AI tools as intended. Post-market monitoring of AI applications should enforce transparency of failure modes and potential biases, and new frameworks or platforms for auditing should capture how AI tools are being used within current guidelines (FDA 2021). Similar to evidence-based medicine, the frequency and severity of errors should be recorded and reported to the manufacturer of AI models to ensure proper management of risk and liability within future regulatory frameworks (Pasquale 2020). Changing populations and disease epidemiology will also affect the efficacy of AI models trained on retrospective data and may require periodic calibration or fine tuning to be integrated while ensuring adherence to privacy protections and integrity with regard to safety-critical clinical infrastructure.

Many other questions are yet to be satisfactorily answered in detail and remaining uncertainties present exciting opportunities for interdisciplinary collaborations between clinicians, researchers, developers, patients, regulators, and policymakers to shape the future direction of healthcare, which we believe is to ensure a future where technology positively benefits society and advances the standard of care for everyone.

CONCLUSION

In this chapter, we explored several opportunities and challenges for medical AI that highlight the current state of research and development. Specifically, we discussed the future of AI in the field of radiology and the challenges faced in bringing these tools to the clinic. While there are many reasons to be optimistic for successful and beneficial deployment of AI to healthcare applications, we emphasize the importance of tackling both technical and social challenges that have been core principles to the medical community such as safety, fairness, equity, privacy, and well-being.

REFERENCES

Addressing catastrophic forgetting for medical domain expansion. https://arxiv.org/pdf/2103.13511.pdf

AlBadawy, E.A., Saha, A., & Mazurowski, M.A. Deep learning for segmentation of brain tumors: Impact of cross-institutional training and testing. *Med Phys* **45,** 1150–1158 (2018). https://doi.org/10.1002/mp.12752

Ardila, D., Kiraly, A.P., Bharadwaj, S. *et al.* End-to-end lung cancer screening with three-dimensional deep learning on low-dose chest computed tomography. *Nat Med* **25**, 954–961 (2019). https://doi.org/10.1038/s41591-019-0447-x

Arun, N., Gaw, N., Singh, P., Chang, K., Aggarwal, M., Chen, B., Hoebel, K., Gupta, S., Patel, J., Gidwani, M., Adebayo, J., Li, M.D., & Kalpathy-Cramer, J. Assessing the trustworthiness of saliency maps for localizing abnormalities in medical imaging. *Radiol Artif Intell* **3**, 6 (2021).

Bender, Emily M., Gebru, Timnit, McMillan-Major, Angelina, & Shmitchell, Shmargaret. On the dangers of stochastic parrots: Can language models be too big? Proceedings of the 2021 ACM Conference on Fairness, Accountability, and Transparency (FAccT '21). Association for Computing Machinery, New York, NY, USA, 610–623. https://doi.org/10.1145/3442188.3445922

Campanella, G., Hanna, M.G., Geneslaw, L. *et al.* Clinical-grade computational pathology using weakly supervised deep learning on whole slide images. *Nat Med* **25**, 1301–1309 (2019). https://doi.org/10.1038/s41591-019-0508-1

Cecilia, S. Lee, & Lee, Aaron Y. Clinical applications of continual learning machine learning. *The Lancet Digital Health* **2**(6), e279–e281 (2020).

Chang, Ken, Beers, Andrew L., Brink, Laura, Patel, Jay B., Singh, Praveer, Arun, Nishanth T., Hoebel, Katharina V., Gaw, Nathan, Shah, Meesam, Pisano, Etta D., Tilkin, Mike, Coombs, Laura P., Dreyer, Keith J., Allen, Bibb, Agarwal, Sheela, & Kalpathy-Cramer, Jayashree. Multi-Institutional assessment and crowdsourcing evaluation of deep learning for automated classification of breast density. *J Am Coll Radiol* **17**(12), 1653–1662 (2020), ISSN 1546–1440. https://doi.org/10.1016/j.jacr.2020 .05.015

Chen, Emma, Kim, Andy, Krishnan, Rayan, Long, Jin, Ng, Andrew Y., & Rajpurkar, Pranav. CheXbreak: Misclassification identification for deep learning models interpreting chest X-rays. *arXiv preprint arXiv:2103.09957* (2021).

Chen, Z., Bei, Y., & Rudin, C. Concept whitening for interpretable image recognition. *Nat Mach Intell* **2**, 772–782 (2020). https://doi.org/10.1038/s42256-020-00265-z

DeGrave, A.J., Janizek, J.D., & Lee, S.I. AI for radiographic COVID-19 detection selects shortcuts over signal. *Nat Mach Intell* **3**, 610–619 (2021). https://doi.org/10.1038/s42256-021-00338-7

Dunnmon, Jared A., Yi, Darvin, Langlotz, Curtis P., Ré, Christopher, Rubin, Daniel L., & Lungren, Matthew P. Assessment of convolutional neural networks for automated classification of chest radiographs. *Radiology* **290**(2), 537–544 (2019).

Esteva, A., Robicquet, A., Ramsundar, B. *et al.* A guide to deep learning in healthcare. *Nat Med* **25**, 24–29 (2019). https://doi-org.ezp-prod1.hul.harvard.edu/10.1038/s41591-018-0316-z

FDA. Artificial intelligence and machine learning in software. U.S. Food and Drug Administration (2021 Sept 22).. Retrieved June 9, 2022, from https://www.fda.gov/medical-devices/software-medical-device-samd/artificial-intelligence-and-machine-learning-software-medical-device

Gaube, S., Suresh, H., Raue, M. *et al.* Do as AI say: Susceptibility in deployment of clinical decision-aids. *NPJ Digit Med* **4**, 31 (2021). https://doi.org/10.1038/s41746-021-00385-9

Grøvik, E., Yi, D., Iv, M., Tong, E., Rubin, D., & Zaharchuk, G. Deep learning enables automatic detection and segmentation of brain metastases on multisequence MRI. *J Magn Reson Imaging* **51**, 175–182 (2020). https://doi.org/10.1002/jmri.26766

Harvey, H., Karpati, E., Khara, G. *et al.* The role of deep learning in breast screening. *Curr Breast Cancer Rep* **11**, 17–22 (2019). https://doi.org/10.1007/s12609-019-0301-7

Huang, S.C., Pareek, A., Zamanian, R. *et al.* Multimodal fusion with deep neural networks for leveraging CT imaging and electronic health record: a case-study in pulmonary embolism detection. *Sci Rep* **10**, 22147 (2020). https://doi.org/10.1038/s41598-020-78888-w

Jin, C., Yu, H., Ke, J. *et al.* Predicting treatment response from longitudinal images using multi-task deep learning. *Nat Commun* **12**, 1851 (2021). https://doi.org/10.1038/s41467-021-22188-y

Kaushal, A., Altman, R., & Langlotz, C. Geographic Distribution of US Cohorts Used to Train Deep Learning Algorithms. *JAMA* **324**(12), 1212–1213 (2020). https://doi.org/10.1001/jama.2020.12067

Kiani, A., Uyumazturk, B., Rajpurkar, P. *et al.* Impact of a deep learning assistant on the histopathologic classification of liver cancer. *NPJ Digit Med* **3**, 23 (2020). https://doi.org/10.1038/s41746-020 -0232-8

Kim, D.W., Jang, H.Y., Kim, K.W., Shin, Y., & Park, S.H. Design characteristics of studies reporting the performance of artificial intelligence algorithms for diagnostic analysis of medical images: Results from recently published papers. *Korean J Radiol* **20**(3), 405–410 (2019 Mar). https://doi.org/10 .3348/kjr.2019.0025

Kim, B., Wattenberg, M., Gilmer, J., Cai, C., Wexler, J., Viegas, F., & Sayres, R. Interpretability beyond feature attribution: Quantitative Testing with Concept Activation Vectors (TCAV). Proceedings of the 35th International Conference on Machine Learning, in Proceedings of Machine Learning Research **80**, 2668–2677 (2018). https://proceedings.mlr.press/v80/kim18d.html

Lao, J., Chen, Y., Li, Z. C. *et al.* A Deep Learning-based radiomics model for prediction of survival in glioblastoma multiforme. *Sci Rep* **7**, 10353 (2017). https://doi.org/10.1038/s41598-017-10649-8

Larrazabal, A.J., Nieto, Nicolás, Peterson, Victoria, Milone, Diego H., & Ferrante, Enzo. Gender imbalance in medical imaging datasets produces biased classifiers for computer-aided diagnosis. *Proceedings of the National Academy of Sciences* **117**(23), 12592–12594 (2020 Jun). https://doi.org/10.1073/pnas.1919012117

Li, M. D., Chang, K., Bearce, B. *et al.* Siamese neural networks for continuous disease severity evaluation and change detection in medical imaging. *NPJ Digit Med* **3**, 48 (2020). https://doi.org/10.1038/s41746-020-0255-1

Li, Matthew D., Thumbavanam Arun, Nishanth, Gidwani, Mishka, Chang, Ken, Deng, Francis, Little, Brent P., Mendoza, Dexter P., Lang, Min, Lee, Susanna I., O'Shea, Aileen, Parakh, Anushri, Singh, Praveer, & Kalpathy-Cramer, Jayashree. *Radiol Artif Intell* **2**, 4 (2020).

Liu, Xiaoxuan *et al.* A comparison of deep learning performance against health-care professionals in detecting diseases from medical imaging: A systematic review and meta-analysis. *The Lancet Digital Health* **1**(6), e271–e297 (2019).

Lu, Charles, Lemay, Andreanne, Chang, Ken, Hoebel, Katharina, & Kalpathy-Cramer, Jayashree. Fair conformal predictors for applications in medical imaging. 36th AAAI Conference on Artificial Intelligence. AAAI, 2022.

Lyell, D., & Coiera, E. Automation bias and verification complexity: A systematic review. *J Am Med Inform Assoc* **24**(2), 423–431 (2017). https://doi.org/10.1093/jamia/ocw105

Mahajan, D. *et al.* Exploring the limits of weakly supervised pretraining. In: Ferrari, V., Hebert, M., Sminchisescu, C., & Weiss, Y. (eds.) *Computer Vision – ECCV 2018. ECCV 2018. Lecture Notes in Computer Science*, vol. 11206. Springer, Cham, 2018. https://doi.org/10.1007/978-3-030-01216-8_12

Majkowska, Anna, Mittal, Sid, Steiner, David F., Reicher, Joshua J., McKinney, Scott Mayer, Duggan, Gavin E., Eswaran, Krish, Chen, Po-Hsuan Cameron, Liu, Yun, Raju Kalidindi, Sreenivasa, Ding, Alexander, Corrado, Greg S., Tse, Daniel, & Shetty, Shravya. Chest radiograph interpretation with deep learning models: Assessment with radiologist-adjudicated reference standards and populationadjusted evaluation. *Radiology* **294**(2), 421–431 (2020).

Mauro, Annarumma, Withey, Samuel J., Bakewell, Robert J., Pesce, Emanuele, Goh, Vicky, & Montana, Giovanni. Automated triaging of adult chest radiographs with deep artificial neural networks. *Radiology* **291**(1), 196–202 (2019).

Mårtensson, Gustav, Ferreira, Daniel, Granberg, Tobias, Cavallin, Lena, Oppedal, Ketil, Padovani, Alessandro, Rektorova, Irena, Bonanni, Laura, Pardini, Matteo, Kramberger, Milica G., Taylor, John-Paul, Hort, Jakub, Snædal, Jón, Kulisevsky, Jaime, Blanc, Frederic, Antonini, Angelo, Mecocci, Patrizia, Vellas, Bruno, Tsolaki, Magda, Kłoszewska, Iwona, Soininen, Hilkka, Lovestone, Simon, Simmons, Andrew, Aarsland, Dag, & Westman, Eric. The reliability of a deep learning model in clinical out-of-distribution MRI data: A multicohort study. *Medical Image Analysis* **66**, 101714 (2020), ISSN 1361–8415. https://doi.org/10.1016/j.media.2020.101714.

McKinney, S.M., Sieniek, M., Godbole, V. *et al.* International evaluation of an AI system for breast cancer screening. *Nature* **577**, 89–94 (2020). https://doi.org/10.1038/s41586-019-1799-6

Mongan, John, Moy, Linda, & Kahn, Jr. Charles E. Checklist for Artificial Intelligence in Medical Imaging (CLAIM): A guide for authors and reviewers. *Radiol Artif Intell* **2**, 2 (2020).

Morrison, Cecily, Huckvale, Kit, Corish, Bob, Dorn, Jonas, Kontschieder, Peter, O'Hara, Kenton. ASSESS MS Team, Criminisi, Antonio, & Sellen, Abigail. Assessing multiple sclerosis with kinect: Designing computer vision systems for real-world use. *Human–Computer Interaction* **31**(3–4), 191–226 (2016). https://doi.org/10.1080/07370024.2015.1093421

Nelson, A., Herron, D., Rees, G. *et al.* Predicting scheduled hospital attendance with artificial intelligence. *NPJ Digit Med* **2**, 26 (2019). https://doi.org/10.1038/s41746-019-0103-3

Pasquale, F. When medical robots fail: Malpractice principles for an ERA of automation (2020 Nov 09). Retrieved June 9, 2022, from https://www.brookings.edu/techstream/when-medical-robots-fail-malpractice-principles-for-an-era-of-automation/

Pianykh, O.S., Langs, Georg, Dewey, Marc, Enzmann, Dieter R., Herold, Christian J., Schoenberg, Stefan O., & Brink, James A. Continuous learning AI in radiology: Implementation principles and early applications. *Radiology* **297**, 1, 6–14 (2020).

Pierson, E., Cutler, D.M., Leskovec, J. *et al.* An algorithmic approach to reducing unexplained pain disparities in underserved populations. *Nat Med* **27**, 136–140 (2021). https://doi.org/10.1038/s41591-020-01192-7

Rajkomar, A., Oren, E., Chen, K. *et al.* Scalable and accurate deep learning with electronic health records. *npj Digital Med* **1**, 18 (2018). https://doi.org/10.1038/s41746-018-0029-1

Rajpurkar, P., Irvin, J., Ball, R.L., Zhu, K., Yang, B. *et al.* Deep learning for chest radiograph diagnosis: A retrospective comparison of the CheXNeXt algorithm to practicing radiologists. *PLoS Med* **15**(11), e1002686 (2018). https://doi.org/10.1371/journal.pmed.1002686

Rauschecker, A.M., Rudie, J.D., Xie, L., Wang, J., Duong, M.T., Botzolakis, E.J., Kovalovich, A. M., Egan, J., Cook, T.C., Bryan, R.N., Nasrallah, I.M., Mohan, S., & Gee, J.C. Artificial intelligence system approaching neuroradiologist-level differential diagnosis accuracy at brain MRI. *Radiology* **295**(3), 626–637 (2020, Jun). https://doi.org/10.1148/radiol.2020190283. Epub 2020 Apr 7. PMID: 32255417; PMCID: PMC7263320.

Reyes, Mauricio, Meier, Raphael, Pereira, Sérgio, Silva, Carlos A., Dahlweid, Fried-Michael, Tengg-Kobligk, Hendrik von, Summers, Ronald M., & Wiest, Roland. On the interpretability of artificial intelligence in radiology: Challenges and opportunities. *Radiol Artif Intell* **2**, 3 (2020).

Roser, M., & Ritchie, H. Burden of disease. *Published online at OurWorldInData.org* (2016). https://ourworldindata.org/burden-of-disease' [Online Resource]

Roth, H., Chang, K., Singh, P., Neumark, N., Li, W., Gupta, V., Gupta, S., Qu, L., Ihsani, A., Bizzo, B., Wen, Y., Buch, V., Shah, M., Kitamura, F., Mendonça, M., Lavor, V., Harouni, A., Compas, C., Tetreault, J., & Kalpathy-Cramer, J. Federated learning for breast density classification: A real-world implementation. (2020). https://doi.org/10.1007/978-3-030-60548-3_18.

Rudin, C. Stop explaining black box machine learning models for high stakes decisions and use interpretable models instead. *Nat Mach Intell* **1**, 206–215 (2019). https://doi.org/10.1038/s42256-019-0048-x

Senior, A.W., Evans, R., Jumper, J. *et al.* Improved protein structure prediction using potentials from deep learning. *Nature* **577**, 706–710 (2020). https://doi.org/10.1038/s41586-019-1923-7

Seyyed-Kalantari, L., Liu, G., McDermott, M.B., & Ghassemi, M. CheXclusion: Fairness gaps in deep chest X-ray classifiers. *Pacific Symposium on Biocomputing* **26**, 232–243 (2021).

Seyyed-Kalantari, L., Zhang, H., McDermott, M.B.A. *et al.* Underdiagnosis bias of artificial intelligence algorithms applied to chest radiographs in under-served patient populations. *Nat Med* **27**, 2176–2182 (2021). https://doi.org/10.1038/s41591-021-01595-0

Singh, A., Haque, A., Alahi, A., Yeung, S., Guo, M., Glassman, J.R., Beninati, W., Platchek, T., Fei-Fei, L. & Milstein, A. Automatic detection of hand hygiene using computer vision technology. *J Am Med Inform Assoc* **27**(8), 1316–1320 (2020 Aug 1). doi: 10.1093/jamia/ocaa115. PMID: 32712656; PMCID: PMC7481030.

Sung, Jinkyeong Sohee Park, Sang Min Lee, Woong Bae, Beomhee Park, Eunkyung Jung, Joon Beom Seo, & Kyu-Hwan Jung. Added value of deep learning–based detection system for multiple major findings on chest radiographs: A randomized crossover study. *Radiology* **299**(2), 450–459 (2021).

Subbaswamy, A., Adams, R., & Saria, S. Evaluating model robustness and stability to dataset shift. Proceedings of The 24th International Conference on Artificial Intelligence and Statistics, in Proceedings of Machine Learning Research **130**, 2611–2619 (2021). https://proceedings.mlr.press/v130/subbaswamy21a.html.

Titano, J.J., Badgeley, M., Schefflein, J. *et al.* Automated deep-neural-network surveillance of cranial images for acute neurologic events. *Nat Med* **24**, 1337–1341 (2018). https://doi.org/10.1038/s41591-018-0147-y

Tomašev, N., Glorot, X., Rae, J.W. *et al.* A clinically applicable approach to continuous prediction of future acute kidney injury. *Nature* **572**, 116–119 (2019). https://doi.org/10.1038/s41586-019-1390-1

Topol, E.J. High-performance medicine: The convergence of human and artificial intelligence. *Nat Med* **25**, 44–56 (2019). https://doi.org/10.1038/s41591-018-0300-7

Wu, E., Wu, K., Daneshjou, R. *et al.* How medical AI devices are evaluated: limitations and recommendations from an analysis of FDA approvals. *Nat Med* **27**, 582–584 (2021). https://doi.org/10.1038/s41591-021-01312-x

Yao, X., Chen, M., Kao, D., Gao, G., & Xiang, 'Anthony' C. CheXplain: Enabling physicians to explore and understand data-driven, AI-enabled medical imaging analysis. Proceedings of the 2020 CHI Conference on Human Factors in Computing Systems (CHI '20). Association for Computing Machinery, New York, NY, 2020, 1–13. https://doi.org/10.1145/3313831.3376807

Zech, J.R., Badgeley, M.A., Liu, M., Costa, A.B., Titano, J.J. *et al.* Variable generalization performance of a deep learning model to detect pneumonia in chest radiographs: A cross-sectional study. *PLOS Med* **15**(11), e1002683 (2018). https://doi.org/10.1371/journal.pmed.1002683

Zhang, Daniel, Mishra, Saurabh, Brynjolfsson, Erik, Etchemendy, John, Ganguli, Deep, Grosz, Barbara, Lyons, Terah, Manyika, James, Carlos Niebles, Juan, Sellitto, Michael, Shoham, Yoav, Clark, Jack, & Perrault, Raymond. The AI index 2021 annual report. AI Index Steering Committee, Human-Centered AI Institute, Stanford University, Stanford, CA, 2021 Mar.

Zou, J., Huss, M., Abid, A. *et al.* A primer on deep learning in genomics. *Nat Genet* **51**, 12–18 (2019). https://doi-org.ezp-prod1.hul.harvard.edu/10.1038/s41588-018-0295-5

20 Financial View

Ribhav Gupta, Katelyn Rypka, and Saurin Kantesaria

MARKET LANDSCAPE AND FUNDING

The use of AI in healthcare is vast and quickly evolving, from prioritizing the reading of patient radiologic images and improving imaging clarity, to optimizing workflow and providing physicians with support of their clinical decisions [1]. According to Deloitte NWE LLP, a global financial advisory firm, improvements in AI medical imaging applications could lead to improved healthcare outcomes and a potential cost savings of 2020 €16.1–18.2 billion (2021 $~19.1–22.0 USD billion) by healthcare systems nationally when accounting for opportunity costs [2, 3].

Increased opportunities for AI applications in healthcare, compounded by improved technology and a clear potential for large healthcare cost savings, have incentivized many start-ups to enter the market [1, 3, 4], more than tripling the number of competing companies, and driving others to invest in the ever-evolving technology. This has resulted in more than three times the number of companies competing in the radiology/medical imaging space in 2019 (N=113) as compared to 2017 (N=32) [1]. Estimates now indicate nearly 200 start-ups may be working in the space [4]. Many established medical technology companies have also begun investing heavily in the field, such as the recent development and approval of Critical Care Suite, an AI algorithm that can flag chest X-rays indicative of a pneumothorax, by the industry titan GE Healthcare [1, 5].

Following a similar trend, investment in the development of AI for medical imaging has generally increased in the past decade, with a notable resilience to the financial crises of the era [1, 4]. Investment in AI-based medical imaging companies specifically have doubled since 2014, reaching 2021 $1.22 USD billion by 2019 with increased acceleration in investment seen through 2020 [1, 2, 4]. Despite the proliferation in AI-based medical imaging companies—a likely result of increased financial investment in the space—the median investment per company has decreased by 34% from $7.6 million to $5.0 million after accounting for inflation with the average number of investments per AI-based company decreasing by a similar 34% [1, 2]. Interestingly, these trends have not homogeneously affected all AI-based medical imaging companies. Investments in late-stage (Series B or later) companies have seen the greatest absolute increase in venture capital funding offered compared to early stage (Seed or Series A) companies. These trends suggest investors may currently be looking to build out existing companies and earn back investments as opposed to incubating early-stage companies which can carry higher risks [3]. Ultimately these fragmentations are very likely contributing to the slowed progression and adoption of AI medical imaging in medicine [4].

REFORMS TO REGULATORY APPROVAL POLICIES

Despite current support for the development of AI-based medical imaging companies as signaled by private-sector investors, much of the future investment in development is dependent upon reforms to US Food and Drug Administration (FDA) market approval processes and adoption of new insurance coverage strategies [1–3]. Restricted by regulatory practices and government policies previously unequipped to evaluate AI-based softwares as medical devices (SaMDs), the FDA approved 16 AI-based applications over four years between 2016 and 2019 [1]. Estimates as of 2019 suggest the regulatory agency has increased their approval rate to approximately one AI-based imaging algorithm per month [6]. However, these numbers are difficult to validate without a public list tracking approved AI-based medical technologies [6].

Diagnostic devices, such as AI-based medical imaging technologies, are generally subject to one-time technical validation tests to assess accuracy with the optionality for advanced clinical trials [8]. Until recently, once a 510(k) pre-market approval was filed, the algorithm was considered "locked," thus restricting developers from improving models without additional filing [9, 10]. Although this strategy offers many advantages for traditional SaMDs and interventional devices, it impedes upon the primary advantage of AI-based technology: the ability to continuously learn from new information and self-improve [9]. Starting in 2020, the FDA created a clear pathway for AI-based companies to file "Predetermined Change Control Plans," thus allowing for routine monitoring and refinement of existing AI models to complete the intended lifecycle [10]. The American College of Radiology in a letter to the FDA indicates more steps that allow for continuous, real-time monitoring of deployed products will ultimately prompt clinical confidence in the

DOI: 10.1201/9781003095279-25

efficacy and safety of these technologies [9]. Overall, as AI-based applications continue to evolve, legislative frameworks must be developed that are flexible yet provide clarity and guidance to ensure sustained future innovation [2].

REIMBURSEMENT AND LIABILITY POLICIES

While AI-based medical technologies, particularly in radiology, continue to proliferate rapidly, widespread adoption is largely stalled by reimbursement and liability concerns. In 2020, the Centers for Medicare & Medicaid Services (CMS) approved the first set of autonomous AI-based medical imaging tools for the diagnosis of diabetic retinopathy (IDx-DR by Digital Diagnostics) and large-vessel occlusion strokes (ContactCT by Viz.ai), a major leap from current physician-augmented systems [7–10]. While these breakthrough technologies have ushered in a new era of AI in medicine, integrating feasible reimbursement strategies for these new technologies into existing systems remains difficult.

To understand how the Center for Medicare & Medicaid Services (CMS) merges AI-based tools into their infrastructure, it would be helpful to know the current reimbursement model. There are three main payment systems under CMS: Medicare Physician Fee Schedule (MPFS—office services), Hospital Outpatient Payment System (HOPPS—outpatient services), and Inpatient Prospective Payment System (IPPS—inpatient services) [8]. We can best illustrate the differences with two distinct case studies.

ContactCT by Viz.ai is currently under New Technology Add-on Payment (NTAP; under IPPS) which covers a portion of reimbursement based on cost of care (Medicare Severity-Diagnosis Related Group; MS-DRG) and the new technology. As a NTAP, the technology is considered an add-on to the existing treatment as opposed to a stand-alone technology and is therefore only eligible for a partial reimbursement with a set maximum of 65% of the technology cost [11]. For other algorithms to qualify for NTAP they must:

1) "Be considered new and not 'substantially similar' to existing technologies"

2) "[Be] Inadequately paid for under the existing … system"

3) "Substantially improve clinical outcomes more than existing services or technology" [8].

While this sets a standard for coverage of new algorithms, the current structure is only available for three years, with CMS historically reducing reimbursement amounts later on [8, 11]. Furthermore, the first requirement would encourage development of "novel" algorithms rather than improvements to existing ones [8]. This would presumably slow progress in the field by preventing researchers and companies from building off of other's work.

As a second case study, IDx-DR by Digital Diagnostics is currently under Current Procedure Terminology (CPT; under MPFS), which reimburses depending on how much of the given service can be attributed to labor, equipment costs, specific patient encounters, and other details [8]. These rigid criteria for reimbursement do not fully capture the ways AI algorithms and other SaMDs work within healthcare ecosystems. For example, IDx-DR is charged to clients (i.e., clinics) as a system-wide subscription service which is difficult to link with a case-based reimbursement system [8].

There is hope, however, with value-based reimbursement models like the Merit-based Incentive Payment System (MIPS). Here AI-based medical imaging, treatment stratification, and other emerging technologies can prove their worth by decreasing the burden of extracting data and potentially by allowing for early intervention should imaging indicate a patient's condition has a high likelihood of exacerbation and therefore provide improved clinical outcomes and save on substantial downstream costs [8].

Furthermore, gradual introduction of AI-based technologies into radiologists' workflows can help in their universal integration into standard practices if straightforward applications are initially sought out [9]. Having AI-based algorithms perform an initial image-analysis/evaluation of CT images and triage them based on suspected emergencies (i.e., stroke, hemorrhage) is one relatively accessible way for AI to noninvasively assist radiologists. This is a particularly feasible opportunity as reimbursement models take into account door-to-diagnosis turnaround time for emergency cases [9]. In the current system, the majority of wait time is spent queued into a radiologist's reading list to be reviewed as opposed to the actual reading time [10]. One study found that this door-to-diagnosis time could be reduced from 15.75 min to 12.01 minutes if their algorithm was implemented to automatically order images in a given list by whether or not a hemorrhage was identified [10]. In these emergency scenarios every minute can make a tremendous difference in retaining neural function.

Financial liability for errors is another critical caveat for consideration. To date, many AI-based diagnostic companies, such as Digital Diagnostics, have taken on legal liability for diagnostic errors [6]. However, if radiology is to see a proliferation of new technologies, we as a society must reevaluate who to indemnify when mistakes happen. Current alternatives include shared liability across medical systems and AI developers, the development of specialized adjudication systems, and even exempting select AI systems in the future from tort lawsuits if they are central to public health [12].

FRAMEWORK FOR DEVELOPMENT

One significant factor hindering the development of novel artificial intelligence tools for medical imaging has been the difficulty demonstrating a return on investment. As recently as 2019, there was a general perception that companies developing AI for medical imaging applications may not be viable in the long term due to lack of revenue. However, with the first reimbursement approval of AI augmented care in 2020, there has been a resurgence of interest in the field of AI-based medical imaging [8]. As the development and utilization of AI software in radiology is still in its infancy, the precise functionality of AI software can vary widely, thus requiring developers to apply need-finding to isolate what users and key stakeholders will most benefit from. While the technical capability of AI software to perform accurate interpretations of images is central, it is in itself not sufficient to motivate medical systems to adopt the use of artificial intelligence into their workflow [13]. As an example, the reimbursement for the use of computer-aided detection (CAD; a technology from the 1990s) was recently embedded into the price of a mammogram in the United States, meaning it is no longer possible to directly charge for the use of CAD [13]. The implication of this for developers is that for the AI software to be marketable, it must either provide significant improvements in the outcomes achieved by radiologists while using the software, and/or improve the efficiency of the interpretations [14].

Recently, a framework of considerations has been proposed for industries that are interested in purchasing AI software to use when determining which system to invest in. These considerations, coined the ECLAIR guidelines, provide insight for developers into what is relevant for their buyers [15]. Some of these considerations have direct effects on the cost of development for an AI software. For instance, the method and extent of training the algorithms, both in the specificity of diagnosis and the amount of data included in the training, may significantly alter the budget of development. One major factor contributing to the increased cost of developing artificial intelligence programs, particularly in radiology, is the development of quality databases used for training the system [15]. Datasets must be accurately labeled for supervised training. This work is most often completed by radiologists and can be a very time-consuming and labor-intensive task. Moreover, for AI-based technologies to reach their potential, the training datasets and models must be continuously refined, which can pose extensive computational costs. Better-trained models, which are necessary for increased clinical validity, will naturally face higher recurring costs, potentially driving prices up.

While each new generation of artificial intelligence technology provides a higher diagnostic accuracy, there is currently a significant focus and desire for AI software to make more immediate improvements in time and cost savings. This focus on efficiency thus drives companies to consider ways such as reducing the time for acquisition or interpretation [13]. To this end, developers of AI software may focus on features in their software that can reduce the downstream time of interpretation by radiologists. Simple features such as indicating the potential for malignancy in an area of interest on a scale of 1 to 10 can reduce the reading time by radiologists [16].

Additional financial considerations must be considered in addition to development of the algorithms themselves. For example, developers must determine the type of licensing agreement they will utilize with their software, whether that is through purchasing the software or through a subscription. Additionally, how the users of such software will be trained to effectively use the software could impact the desirability of a specific software to be purchased. Finally, whether the software is continually updated, and how maintenance to such programs is implemented, could alter planning in the design [15]. Taken together, these aspects of software development can drastically change the scope of the financial budget and anticipated return on investment.

REFERENCES

1. A. Alexander, A. Jiang, C. Ferreira, and D. Zurkiya, "An Intelligent Future for Medical Imaging: A Market Outlook on Artificial Intelligence for Medical Imaging," *Journal of the American College of Radiology*, vol. 17, no. 1, pp. 165–170, Jan. 2020, doi: 10.1016/j.jacr.2019.07.019.

2. Bureau of Labor Statistics, "CPI Inflation Calculator," 2020. [Online]. Available: https://data.bls.gov/cgi-bin/cpicalc.pl?cost1=113&year1=201001&year2=202002

3. Eliana Biundo, Andrew Pease, Koen Segers, Michael de Groote, Thibault d'Argent, and Edouard de Schaetzen, "The Socio-Economic Impact of AI in Healthcare," *Deloitte*, Oct. 2020. [Online]. Available: https://www2.deloitte.com/content/dam/Deloitte/be/Documents/life-sciences-health-care/Deloitte%20Belgium%20_%20MedTech_Socio-economic%20impact%20of%20AI%20in%20healthcare.pdf

4. Sanjay Parekh, "VC-Funding for Medical Imaging AI Companies Tops $2.6 Billion," *Signify Research*, Mar. 2021. [Online]. Available: https://www.signifyresearch.net/medical-imaging/vc-funding-for-medical-imaging-ai-companies-tops-2-6-billion/

5. U.S. Food & Drug Administration, "K183182," Aug. 12, 2019. [Online]. Available: https://www.accessdata.fda.gov/cdrh_docs/pdf18/K183182.pdf

6. S. Reardon, "Rise of Robot Radiologists," *Nature*, vol. 576, no. 7787, pp. S54–S58, Dec. 2019, doi: 10.1038/d41586-019-03847-z.

7. S. H. Park, J. Choi, and J.-S. Byeon, "Key Principles of Clinical Validation, Device Approval, and Insurance Coverage Decisions of Artificial Intelligence," *Korean Journal of Radiology*, vol. 22, no. 3, p. 442, 2021, doi: 10.3348/kjr.2021.0048.

8. M. M. Chen, L. P. Golding, and G. N. Nicola, "Who Will Pay for AI?," *Radiology: Artificial Intelligence*, vol. 3, no. 3, p. e210030, May 2021, doi: 10.1148/ryai.2021210030.

9. S. D. O'Connor and M. Bhalla, "Should Artificial Intelligence Tell Radiologists Which Study to Read Next?," *Radiology: Artificial Intelligence*, vol. 3, no. 2, p. e210009, Mar. 2021, doi: 10.1148/ryai.2021210009.

10. T. J. O'Neill *et al.*, "Active Reprioritization of the Reading Worklist Using Artificial Intelligence Has a Beneficial Effect on the Turnaround Time for Interpretation of Head CT with Intracranial Hemorrhage," *Radiology: Artificial Intelligence*, vol. 3, no. 2, p. e200024, Mar. 2021, doi: 10.1148/ryai.2020200024.

11. A. T. Clyde, L. Bockstedt, J. A. Farkas, and C. Jackson, "Experience With Medicare's New Technology Add-On Payment Program," *Health Affairs*, vol. 27, no. 6, pp. 1632–1641, Nov. 2008, doi: 10.1377/hlthaff.27.6.1632.

12. George Maliha, Sara Gerke, Ravi B. Parikh, and I. Glenn Cohen, "To Spur Growth in AI, We Need a New Approach to Legal Liability," *Harvard Business Review*. Accessed: Aug. 04, 2021. [Online]. Available: https://hbr.org/2021/07/to-spur-growth-in-ai-we-need-a-new-approach-to-legal-liability

13. Y. Tadavarthi *et al.*, "The State of Radiology AI: Considerations for Purchase Decisions and Current Market Offerings," *Radiology: Artificial Intelligence*, vol. 2, no. 6, p. e200004, Nov. 2020, doi: 10.1148/ryai.2020200004.

14. K. J. Geras, R. M. Mann, and L. Moy, "Artificial Intelligence for Mammography and Digital Breast Tomosynthesis: Current Concepts and Future Perspectives," *Radiology*, vol. 293, no. 2, pp. 246–259, Nov. 2019, doi: 10.1148/radiol.2019182627.

15. P. Omoumi *et al.*, "To Buy or not to Buy—Evaluating Commercial AI Solutions in Radiology (the ECLAIR guidelines)," *European Radiology*, vol. 31, no. 6, pp. 3786–3796, Jun. 2021, doi: 10.1007/s00330-020-07684-x.

16. A. Rodríguez-Ruiz *et al.*, "Detection of Breast Cancer with Mammography: Effect of an Artificial Intelligence Support System," *Radiology*, vol. 290, no. 2, pp. 305–314, Feb. 2019, doi: 10.1148/radiol.2018181371.

PART VI
POTENTIAL OF AI

21 Clinical View

Ali S. Tejani, Bhavya R. Shah, Fang F. Yu, Chandan G. B. Yogananda, and Joseph A. Maldjian

INTRODUCTION

Artificial intelligence (AI) is a disruptive technology with novel imaging applications being introduced to address an expanding array of clinical questions. Exponential growth in AI technology is expanding the scope of applications for medical imaging, with increasing ubiquity of AI in healthcare. Accordingly, we must define AI's role in the clinical context to ensure responsible and effective use of this resource. The future clinical potential of AI in medical imaging will rely on its ability to augment the clinical decision-making process as well as improve the practice of radiology.

Ultimately, successful deployment of AI solutions may benefit from an "AI–physician" interface that promotes AI efficiencies while benefiting from human experience. Insights from this hybrid model, perhaps better termed "augmented intelligence," will accelerate growth in fields that rely on expert assessment of trends in large datasets, such as precision medicine, thereby accelerating advancements in diagnostic, prognostic, and therapeutic capabilities.

In this chapter, we will discuss the potential of AI to address both image-interpretive and non-interpretive scenarios. Examples of image-interpretive AI solutions include identifying pathology, automating provision of descriptive characteristics for risk stratification, and determining molecular phenotype based on underlying imaging feature [1–3]. These examples improve clinical decision-making by detecting pertinent findings, identifying imaging features inaccessible to the human eye, or improving efficiency by automating tedious tasks. Non-interpretive use cases address the *practice* of radiology. Examples of non-interpretive use cases include worklist prioritization of urgent studies, automated study protocoling and scheduling, chatbots for patient education, and enhancing image quality.

IMAGE-INTERPRETIVE USE CASES

Precision Medicine and Radiogenomics

Precision medicine, a term adapted by the National Research Council, refers to the "tailoring of medical treatment to the individual characteristics of each patient … to classify individuals into subpopulations that differ in their susceptibility to a particular disease or their response to a specific treatment" [4]. In other words, precision medicine relies heavily on genomics, molecular profiling, data science, and analytics to maximize the success and minimize side effects of therapeutic intervention by targeting individuals most likely to respond to a given treatment. Artificial intelligence (AI) has to potential to augment precision medicine by improving diagnostic/prognostic capabilities and predict therapeutic outcomes by generating insights from large quantities of imaging and non-imaging data [5].

Radiogenomics represents a novel application of AI to improve diagnostic, prognostic, and therapeutic capabilities. The term "radiogenomics" was initially used in the context of radiation oncology to describe genomic changes resulting from radiotherapy-induced toxicity [6, 7]. However, the definition of radiogenomics has shifted in recent years, now describing the relationship between a quantitative imaging phenotype and a lesion's underlying genotype [8]. Radiogenomics provides opportunities for non-invasive profiling of molecular and genetic data determination based on imaging.

For example, deep-learning networks can accurately and non-invasively determine genetic and molecular subtypes of brain gliomas, a task that currently requires an invasive surgical procedure. Glioma molecular profiling studies have focused on three major genetic markers for prognostic and therapeutic outcomes: *O*-6-methylguanine-DNA methyltransferase (MGMT), isocitrate dehydrogenase (IDH), and 1p/19q co-deletion status [3, 9, 10]. Mutations of the MGMT promoter are associated with improved median survival in patients treated with radiation and temozolomide [11]. Similarly, gliomas with mutant IDH are associated with better prognosis than those with wild-type IDH [12]. Further subtyping allows for identification of IDH mutated tumors with 1p/19q co-deletion, *oligodendroglioma*, which carry better prognosis and predict increased responsiveness to chemotherapy than IDH-mutated non-1p/19q co-deleted tumors [9]. With regard to MGMT, recent studies have shown limited success in non-invasive imaging-based determination

DOI: 10.1201/9781003095279-27

of marker status [10]. Accordingly, determining glioma genotype from characteristic imaging features provides an opportunity to leverage prognostic and therapeutic associations while minimizing procedural risk to the patient.

Similar radiogenomic associations have been reported for other tumors, including breast and other solid organ malignancies [13–16]. Links between imaging phenotypes and specific molecular profiles are documented for hepatic, gastrointestinal, pancreatic, renal, and adrenal/autonomic tumors [17]. Information derived from radiogenomic models, in conjunction with other genomics data, can help train AI applications that predict response to chemotherapy, guiding clinicians to select the most effective treatment option [18, 19].

However, continued advancement in AI-informed radiogenomics requires large, well-curated, and accessible datasets to train and validate deep learning algorithms [20]. Limited data, as is the case for certain gastrointestinal as well as head and neck malignancies, impair stratification between training, validation, and testing datasets, resulting in biased model output [20]. Small datasets also increase susceptibility to overfitting and poor generalization due to inadequate representation of tumor heterogeneity [20]. Accordingly, there is a need for diverse, large, open-source datasets (i.e., The Cancer Genome Atlas) composed of imaging data with expert annotations for reliable and accurate radiogenomic model development.

Automated Lesion Characterization

Manual segmentation of normal and pathologic findings is a time-consuming, though necessary, task for lesion characterization to guide clinical judgment. For instance, while lesion segmentation helps identify high-risk endometrial cancer patients more likely to benefit from invasive procedures, manual segmentation is not practical in high-volume settings [21]. Similarly, objective quantification of disease burden in patients with multiple sclerosis (MS), currently relying on subjective assessment with high intra- and interobserver variability, would require labor-intensive manual segmentation of numerous individual lesions [22]. AI-based automated segmentation methods offer an alternative means of minimizing the time required to obtain accurate, consistent lesion characteristics [21–24]. Information extracted from automated lesion characterization facilitates automated categorization of standardized guidelines, such as the Liver Imaging Reporting and Data System (LI-RADS) and the Thyroid Imaging Reporting and Data System (TI-RADS), promoting increased efficiency in performing tedious but clinically important tasks.

Point-of-Care Imaging

Synergy between wireless telecommunications and AI can augment the use of point-of-care (PoC) imaging devices, especially in resource-constrained settings. PoC modalities, such as portable low-field MRI and handheld ultrasound devices, enable timely screening and diagnosis in emergency and primary-care settings [25–27]. However, hardware limitations, such as low magnet strength and probe size, results in noisy, low-resolution images that may limit diagnostic capabilities. In this context, AI can help to optimize image quality, and thus, diagnostic reliability. For example, portable, low-field MRI systems employing denoising algorithms in the image reconstruction process attain significant gains in image quality that allow for diagnostic capabilities similar to conventional higher field (1.5-3 Tesla) MR scanner [28, 29]. Furthermore, incorporating models for computer vision tasks, such as intracranial hemorrhage and stroke detection, directly on the portable systems allows for an added automated tool for triage of urgent findings [30].

Opportunistic Screening

Opportunistic screening describes extraction of reproducible, objective data from imaging studies performed for alternative clinical indications to assess patient risk for future adverse events [31]. Important imaging features have been described for opportunistic screening of osteoporosis, cardiovascular disease, metabolic syndrome, and sarcopenia on routine CT examinations [31–35]. AI is well-poised to maximize the derivation of clinically pertinent information from imaging data that may otherwise be overlooked without explicit prior mention of "screening" in the clinical indication. Automated methods of quantifying imaging metrics such as aortic calcification, visceral or subcutaneous fat, and bone mineral density show promise in preliminary cohorts [36–39]. However, future work must develop algorithms on more diverse training sets, as current studies feature relatively homogenous cohorts that limit model generalizability and development of widely applicable reference values [31]. Furthermore, cohorts must feature a balance between asymptomatic and symptomatic patients to avoid other sources of bias, including negative or positive set bias. While there remains a need for further work, AI applications leveraging opportunistic data can add further value to imaging examinations with potential to establish robust paradigms for preventative healthcare.

Improving Image-Guided Intervention, Targeting, and Interpretation

The clinical potential of AI applications extends beyond diagnostic use cases to enable more accurate, efficient image-guided interventions. Real-time organ/target segmentation can account for challenging image-guided biopsies, such as transrectal ultrasound-guided prostate biopsies susceptible to artifacts obscuring the target boundary or adjacent structures [40]. Similarly, image-based navigation for more invasive procedures, such as robot-assisted procedures, stand to benefit from machine learning driven improvements to current processes for 2D/3D registration [41]. Furthermore, deep learning models show potential for applications beyond tissue sampling, specifically for histopathological staining and interpretation [42, 43].

Less commonly explored avenues of AI-enhanced image-guided intervention include identifying therapeutic targets. For example, successful magnetic resonance–guided high-intensity focused ultrasound (MRgHIFU) and deep-brain stimulation (DBS) require identification of deep brain nuclei often not clearly identified on standard MRI sequences [44, 45]. Novel, advanced sequences such as diffusion tensor imaging tractography (DTI), quantitative susceptibility mapping, and susceptibility weighted imaging can offer increased clinical efficacy. However, these methods either require preexisting knowledge of target white-matter tracts, extensive post-processing with longer scan times, or higher magnet strengths that limit generalizability and practical use [44]. Current literature has shown promise in automated segmentation of deep-brain nuclei, including thalamic and subthalamic nuclei, with high levels of performance across nuclei, MRI sequences, and magnet field strengths [46–48]. Future work will build on these initial findings, allowing for validation with advanced sequences, such as DTI, and improved model generalizability outside of designated training/testing sets. The future may also expand AI to directly improving image-guided therapy through an intricate closed-loop monitoring system that is able to predict patient response based on treatment thresholds. Even more exciting is the possibility that AI could aid in the targeted delivery of specific therapies to discrete regions of a glioma with magnetic resonance–guided focused ultrasound (MRgFUS) induced transient blood–brain barrier disruption.[49, 50]

NON-INTERPRETIVE USE CASES

Radiation-Dose Reduction

Deep-learning enhanced image reconstruction allows for unique opportunities for radiation-dose reduction in diagnostic and nuclear medicine studies while minimizing noise. Current methods of deep learning reconstruction have shown CT-dose reduction of >50% for delineating both normal anatomy and pathology, including malignancies, in both adult and pediatric populations while maintaining image quality comparable to higher-dose scans [51, 52]. Studies have also validated AI tools featuring convolutional neural networks (CNNs) to reduce radiotracer dose in [18] F-FDG exams to achieve radiation dose reduction and acting synergistically with reduction in radiation exposure from PET/MRI technology as compared to PET/CT [53]. Future work in this domain should aim for further reduction in dose while employing denoising capabilities, creating "vendor-neutral" algorithms, and assessing the integrity of dose-reduction protocols for a wider array of pathology.

"Fast" MRI Protocols

AI-accelerated MRI acquisition represents a vigorous area of research, with the potential to decrease patient discomfort, motion-related artifacts, exam cost, and limitations on exam volume. Currently described methods of AI-accelerated MRI acquisition draw on machine learning techniques for estimation of full resolution images that can operate either in image space or in partially sampled k-space [54]. Additional methods rely on resolution-enhancement algorithms or generative adversarial networks to decrease scan time and provide realistic-appearing images [54, 55]. Ongoing endeavors in this domain will involve creation of publicly available datasets featuring exams from multiple vendors, magnet strengths, and pathologies [56]. Additionally, development of 3D-CNN techniques will further delineate and leverage the correlation between under-sampled and fully-sampled k-space data [54]. Translation of subspecialty targeted approaches will be necessary to realize the full potential of these methods [54, 56, 57].

Worklist Prioritization and Communication of Urgent Findings

The communication of urgent findings to treatment teams is crucial for timely clinical intervention and preventing adverse events. However, increasing exam volumes may delay identification, and thus notification, of these findings. AI worklist prioritization reduces imaging wait and turnaround times, providing a means of preventing delayed critical clinical intervention [58, 59]. However, clinical utility of these algorithms depends on seamless integration with existing workflows, which can vary significantly across practices. Furthermore, algorithms for worklist prioritization feature varying performance-mandating continuous model training on local institution-specific data to minimize false-positive and -negative reads. Notably, these algorithms demonstrate the significance of the "AI–physician" team, where AI provides a preliminary read flagging urgent results to be confirmed by a physician overread.

Beyond flagging worklists, AI applications can assist with communication of confirmed urgent findings to the clinical team. Natural language processing (NLP) models have been proposed to extract critical findings from finalized radiology reports [60–62]. Utilizing NLP models to extract critical findings can inform and initiate contact of the appropriate treatment teams, assisting in coordination of urgent care with both inpatient and outpatient imaging exams. Additionally, NLP models can be used for quality assurance or auditing purposes to track incidence of urgent findings and frequency of successful reporting (i.e., timing, method) to treatment teams [60].

Resource Utilization

With regard to the practice of radiology, resource utilization broadly includes the allocation of scanner, technologists, and radiologists for expected exam volume. AI can help optimize operational and strategic planning to maximize efficiency of scan acquisition and image interpretation. One method of achieving this goal involves future exam forecasting based on prior trends, taking into account factors such as exam acuity, ordering time, type/modality, and subspecialty [63]. Algorithms trained on data from past exams, including scanner usage, can inform future scheduling of subspecialty radiologists and allocation of scanners to optimize available resources. Additional potential of AI in this domain involves processes to account for relatively more unpredictable factors such as "stat," or emergency scans or unavailability of scanners due to maintenance requiring fast reallocation of resources in real-time. Automated protocol assignment for ordered studies utilizing NLP classification models can achieve further gains in efficiency, allowing radiologists to focus on image interpretation rather than tedious upstream tasks [64]. Of note, complete autonomy of these processes is unlikely, as scheduler and technologist overview of model output is necessary to assess for discrepancies in exam scheduling and protocol assignment

---

Select Artificial Intelligence (AI) Solutions in Clinical Radiology Workflow

Step	Potential Clinical AI Intervention
Ordering Imaging Exam	Guide appropriate study selection
	Flag inappropriately ordered studies
Scheduling studies	Optimize scanner and resource utilization
	Optimize staffing with personalized worklists by volume, specialty, speed, and other variables pertinent to interpreting studies
Protocoling studies	Automate protocoling with option for technologist review
Acquiring images	"Fast" scans
	Resolution enhancement, artifact reduction
Interpreting images	Automate reporting of quantitative data (i.e., measurements, lesion characterization, radiogenomic data)
	Recognize reporting errors with option for corrections
Communicating urgent results	Extract actionable findings from reports
	Report urgent findings to clinical teams
Coordinating follow-up imaging	Extract incidental findings from reports and schedule follow-up exams
	Track follow-up imaging and notify ordering physicians if imaging pending

stemming from potential algorithm imperfections. Additional considerations include the creation of vendor-neutral models that can accommodate different institutions' range of various scanners.

Monitoring Follow-Up Imaging for Incidental Findings

Challenges associated with obtaining follow-up imaging for incidental findings are well documented, involving both patient- and broader health system-related factors [65]. Unfortunately, regardless of causative factor, missed follow-up imaging recommendations and exams lead to missed opportunities for timely intervention, such as detecting and treating early malignancy. Automated methods of extracting follow-up recommendations allow for the development of tracking systems that monitor whether patients receive recommended follow-up imaging [65, 66]. Creation of an AI-enabled follow-up imaging tracking system requires formal governance for quality assurance and appropriate identification of a patient's clinical provider. Furthermore, NLP models extracting follow-up recommendation from reports must account for both structured and unstructured reports given heterogeneity of reporting across practices [65].

CONCLUSION

The use cases discussed in this chapter should provide a glimpse of the clinical potential of AI in both image-interpretive and non-interpretive contexts. Each of the highlighted scenarios demonstrates the ability of AI to maximize efficiency and enable shared clinical decisions informed by AI-driven insights. By doing so, AI allows for enhanced prognostic, diagnostic, and therapeutic capabilities that were previously unavailable due to limited time and resources.

Of note, the clinical potential described in this chapter reflects projections derived from the current state of AI. Given the rapidly growing nature of this field, the clinical potential of AI will fluctuate as a function of both the latest technology and changing demands from patient care. Similarly, paradigms of AI integration into clinical workflow will continue to evolve, necessitating conscientious decisions regarding tool governance, especially as AI results become directly available to all care providers moments after image acquisition. Ultimately, achieving the full clinical potential of AI will require targeted AI applications that reflect the evolving state of clinical practice.

REFERENCES

1. Thian YL, Ng D, Hallinan J, et al. Deep Learning Systems for Pneumothorax Detection on Chest Radiographs: A Multicenter External Validation Study. *Radiol Artif Intell* 2021;3:e200190.

2. Wildman-Tobriner B, Buda M, Hoang JK, et al. Using Artificial Intelligence to Revise ACR TI-RADS Risk Stratification of Thyroid Nodules: Diagnostic Accuracy and Utility. *Radiology* 2019;292:112–9.

3. Bangalore Yogananda CG, Shah BR, Vejdani-Jahromi M, et al. MRI-Based Deep Learning Method for Classification of IDH Mutation Status. *Bioengineering* 2023;10:1045.

4. National Research Council. *Toward Precision Medicine: Building a Knowledge Network for Biomedical Research and a New Taxonomy of Disease.* Washington, DC: National Academies Press; 2011.

5. Johnson KB, Wei WQ, Weeraratne D, et al. Precision Medicine, AI, and the Future of Personalized Health Care. *Clin Transl Sci* 2021;14:86–93.

6. Story MD, Durante M. Radiogenomics. *Med Phys* 2018;45:e1111–e22.

7. Bodalal Z, Trebeschi S, Nguyen-Kim TDL, Schats W, Beets-Tan R. Radiogenomics: Bridging Imaging and Genomics. *Abdom Radiol* 2019;44:1960–84.

8. Shui L, Ren H, Yang X, et al. The Era of Radiogenomics in Precision Medicine: An Emerging Approach to Support Diagnosis, Treatment Decisions, and Prognostication in Oncology. *Front Oncol* 2020;10:570465.

9. Yogananda CGB, Shah BR, Yu FF, et al. A Novel Fully Automated MRI-based Deep-Learning Method for Classification of 1p/19q Co-deletion Status in Brain Gliomas. *Neurooncol Adv* 2020;2:vdaa066.

10. Robinet L, Siegfried A, Roques M, Berjaoui A, Cohen-Jonathan Moyal E, MRI-based deep learning tools for mgmt promoter methylation detection: a thorough evaluation. *Cancers* 2023;15:2253.

11. Hegi ME, Diserens AC, Gorlia T, et al. MGMT Gene Silencing and Benefit from Temozolomide in Glioblastoma. *N Engl J Med* 2005;352:997–1003.

12. Yan H, Parsons DW, Jin G, et al. IDH1 and IDH2 Mutations in Gliomas. *N Engl J Med* 2009;360:765–73.

13. Badic B, Hatt M, Durand S, et al. Radiogenomics-Based Cancer Prognosis in Colorectal Cancer. *Sci Rep* 2019;9:9743.

14. Bibault JE, Giraud P, Housset M, et al. Deep Learning and Radiomics Predict Complete Response after Neo-Adjuvant Chemoradiation for Locally Advanced Rectal Cancer. *Sci Rep* 2018;8:12611.

15. Trivizakis E, Manikis GC, Nikiforaki K, et al. Extending 2-D Convolutional Neural Networks to 3-D for Advancing Deep Learning Cancer Classification With Application to MRI Liver Tumor Differentiation. *IEEE J Biomed Health Inform* 2019;23:923–30.

16. Yeh AC, Li H, Zhu Y, et al. Radiogenomics of Breast Cancer Using Dynamic Contrast Enhanced MRI and Gene Expression Profiling. *Cancer Imaging* 2019;19:48.

17. Katabathina VS, Marji H, Khanna L, et al. Decoding Genes: Current Update on Radiogenomics of Select Abdominal Malignancies. *Radiographics* 2020;40:1600–26.

18. Gerdes H, Casado P, Dokal A, et al. Drug ranking Using Machine Learning Systematically Predicts the Efficacy of Anti-Cancer Drugs. *Nat Commun* 2021;12:1850.

19. Huang C, Clayton EA, Matyunina LV, et al. Machine Learning Predicts Individual Cancer Patient Responses to Therapeutic Drugs with High Accuracy. *Sci Rep* 2018;8:16444.

20. Trivizakis E, Papadakis GZ, Souglakos I, et al. Artificial Intelligence Radiogenomics for Advancing Precision and Effectiveness in Oncologic Care (Review). *Int J Oncol* 2020;57:43–53.

21. Hodneland E, Dybvik JA, Wagner-Larsen KS, et al. Automated Segmentation of Endometrial Cancer on MR Images Using Deep Learning. *Sci Rep* 2021;11:179.

22. Brugnara G, Isensee F, Neuberger U, et al. Automated Volumetric Assessment with Artificial Neural Networks Might Enable a More Accurate Assessment of Disease Burden in Patients with Multiple Sclerosis. *Eur Radiol* 2020;30:2356–64.

23. Sieren MM, Widmann C, Weiss N, et al. Automated Segmentation and Quantification of the Healthy and Diseased Aorta in CT Angiographies Using a Dedicated Deep Learning Approach. *Eur Radiol* 2022;32:690–701.

24. Soerensen SJC, Fan RE, Seetharaman A, et al. Deep Learning Improves Speed and Accuracy of Prostate Gland Segmentations on Magnetic Resonance Imaging for Targeted *Biopsy. J Urol* 2021;206:604–12.

25. Sheth KN, Mazurek MH, Yuen MM, et al. Assessment of Brain Injury Using Portable, Low-Field Magnetic Resonance Imaging at the Bedside of Critically Ill Patients. *JAMA Neurol* 2020;78(1):41–47.

26. Baribeau Y, Sharkey A, Chaudhary O, et al. Handheld Point-of-Care Ultrasound Probes: The New Generation of POCUS. *J Cardiothorac Vasc Anesth* 2020;34:3139–45.

27. Mazurek MH, Cahn BA, Yuen MM, et al. Portable, Bedside, Low-Field Magnetic Resonance Imaging for Evaluation of Intracerebral Hemorrhage. *Nat Commun* 2021;12:5119.

28. Koonjoo N, Zhu B, Bagnall GC, Bhutto D, Rosen MS. Boosting the Signal-to-Noise of Low-Field MRI with Deep Learning Image Reconstruction. *Sci Rep* 2021;11:8248.

29. Arnold TC, Baldassano SN, Litt B, Stein JM. Simulated Diagnostic Performance of Low-Field MRI: Harnessing Open-Access Datasets to Evaluate Novel Devices. *Magn Reson Imaging* 2021;87:67–76.

30. Banerjee I, Choi HH, Desser T, Rubin DL. A Scalable Machine Learning Approach for Inferring Probabilistic US-LI-RADS Categorization. *AMIA Annu Symp Proc* 2018;2018:215–24.

31. Pickhardt PJ, Graffy PM, Perez AA, Lubner MG, Elton DC, Summers RM. Opportunistic Screening at Abdominal CT: Use of Automated Body Composition Biomarkers for Added Cardiometabolic Value. *Radiographics* 2021;41:524–42.

32. Jang S, Graffy PM, Ziemlewicz TJ, Lee SJ, Summers RM, Pickhardt PJ. Opportunistic Osteoporosis Screening at Routine Abdominal and Thoracic CT: Normative L1 Trabecular Attenuation Values in More than 20 000 Adults. *Radiology* 2019;291:360–7.

33. Pickhardt PJ, Graffy PM, Zea R, et al. Utilizing Fully Automated Abdominal CT-Based Biomarkers for Opportunistic Screening for Metabolic Syndrome in Adults Without Symptoms. *AJR Am J Roentgenol* 2021;216:85–92.

34. Pickhardt PJ, Graffy PM, Zea R, et al. Automated CT Biomarkers for Opportunistic Prediction of Future Cardiovascular Events and Mortality in an Asymptomatic Screening Population: A Retrospective Cohort Study. *Lancet Digit Health* 2020;2:e192–e200.

35. Lenchik L, Boutin RD. Sarcopenia: Beyond Muscle Atrophy and into the New Frontiers of Opportunistic Imaging, Precision Medicine, and Machine Learning. *Semin Musculoskelet Radiol* 2018;22:307–22.

36. Pickhardt PJ, Lee SJ, Liu J, et al. Population-Based Opportunistic Osteoporosis Screening: Validation of a Fully Automated CT Tool for Assessing Longitudinal BMD Changes. *Br J Radiol* 2019;92:20180726.

37. Graffy PM, Liu J, O'Connor S, Summers RM, Pickhardt PJ. Automated Segmentation and Quantification of Aortic Calcification at Abdominal CT: Application of a Deep Learning-Based Algorithm to a Longitudinal Screening Cohort. *Abdom Radiol (NY)* 2019;44:2921–8.

38. Lee SJ, Liu J, Yao J, Kanarek A, Summers RM, Pickhardt PJ. Fully Automated Segmentation and Quantification of Visceral and Subcutaneous Fat at Abdominal CT: Application to a Longitudinal Adult Screening Cohort. *Br J Radiol* 2018;91:20170968.

39. Graffy PM, Liu J, Pickhardt PJ, Burns JE, Yao J, Summers RM. Deep Learning-based Muscle Segmentation and Quantification at Abdominal CT: Application to a Longitudinal Adult Screening Cohort for Sarcopenia Assessment. *Br J Radiol* 2019;92:20190327.

40. Anas EMA, Mousavi P, Abolmaesumi P. A Deep Learning Approach for Real Time Prostate Segmentation in Freehand Ultrasound Guided Biopsy. *Med Image Anal* 2018;48:107–16.

41. Unberath M, Gao C, Hu Y, et al. The Impact of Machine Learning on 2D/3D Registration for Image-Guided Interventions: A Systematic Review and Perspective. *Front Robot AI* 2021;8:716007.

42. Naito Y, Tsuneki M, Fukushima N, et al. A Deep Learning Model to Detect Pancreatic Ductal Adenocarcinoma on Endoscopic Ultrasound-Guided Fine-Needle Biopsy. *Sci Rep* 2021;11:8454.

43. Rana A, Lowe A, Lithgow M, et al. Use of Deep Learning to Develop and Analyze Computational Hematoxylin and Eosin Staining of Prostate Core Biopsy Images for Tumor Diagnosis. *JAMA Netw Open* 2020;3:e205111.

44. Shah BR, Lehman VT, Kaufmann TJ, et al. Advanced MRI Techniques for Transcranial High Intensity Focused Ultrasound Targeting. *Brain* 2020;143:2664–72.

45. Middlebrooks EH, Domingo RA, Vivas-Buitrago T, et al. Neuroimaging Advances in Deep Brain Stimulation: Review of Indications, Anatomy, and Brain Connectomics. *AJNR Am J Neuroradiol* 2020;41:1558–68.

46. Majdi MS, MKeerthivasan M, Rutt BK, Zahr NM, Rodrigues JJ, Saranathan M. Automated Thalamic Nuclei Segmentation Using Multi-Planar Cascaded Convolutional Neural Networks. arXiv:191207209 2021.

47. Liu Y, Zhang Q, Liu L, Li C, Zhang R, Liu G. The Effect of Deep Learning-Based QSM Magnetic Resonance Imaging on the Subthalamic Nucleus. *J Healthc Eng* 2021;2021:8554182.

48. Beliveau V, Norgaard M, Birkl C, Seppi K, Scherfler C. Automated Segmentation of Deep Brain Nuclei Using Convolutional Neural Networks and Susceptibility Weighted Imaging. *Hum Brain Mapp* 2021;42:4809–22.

49. Mainprize T, Lipsman N, Huang Y, et al. Blood-Brain Barrier Opening in Primary Brain Tumors with Non-invasive MR-Guided Focused Ultrasound: A Clinical Safety and Feasibility Study. *Sci Rep* 2019;9:321.

50. Curley CT, Mead BP, Negron K, et al. Augmentation of Brain Tumor Interstitial Flow Via Focused Ultrasound Promotes Brain-Penetrating Nanoparticle Dispersion and Transfection. *Sci Adv* 2020;6:eaay1344.

51. Brady SL, Trout AT, Somasundaram E, Anton CG, Li Y, Dillman JR. Improving Image Quality and Reducing Radiation Dose for Pediatric CT by Using Deep Learning Reconstruction. *Radiology* 2021;298:180–8.

52. Jensen CT, Gupta S, Saleh MM, et al. Reduced-Dose Deep Learning Reconstruction for Abdominal CT of Liver Metastases. *Radiology* 2022:211838.

53. Theruvath AJ, Siedek F, Yerneni K, et al. Validation of Deep Learning-based Augmentation for Reduced (18)F-FDG Dose for PET/MRI in Children and Young Adults with Lymphoma. *Radiol Artif Intell* 2021;3:e200232.

54. Johnson PM, Recht MP, Knoll F. Improving the Speed of MRI with Artificial Intelligence. *Semin Musculoskelet Radiol* 2020;24:12–20.

55. Rudie JD, Gleason T, Barkovich MJ, et al. Clinical Assessment of Deep Learning–based Super-Resolution for 3D Volumetric Brain MRI. *Radiol Artif Intell* 2022;4(2):e210059.

56. Knoll F, Zbontar J, Sriram A, et al. fastMRI: A Publicly Available Raw k-Space and DICOM Dataset of Knee Images for Accelerated MR Image Reconstruction Using Machine Learning. *Radiol Artif Intell* 2020;2:e190007.

57. Recht MP, Zbontar J, Sodickson DK, et al. Using Deep Learning to Accelerate Knee MRI at 3 T: Results of an Interchangeability Study. *AJR Am J Roentgenol* 2020;215:1421–9.

58. O'Neill TJ, Xi Y, Stehel E, et al. Active Reprioritization of the Reading Worklist Using Artificial Intelligence Has a Beneficial Effect on the Turnaround Time for Interpretation of Head CT with Intracranial Hemorrhage. *Radiol Artif Intell* 2021;3:e200024.

59. Baltruschat I, Steinmeister L, Nickisch H, et al. Smart Chest X-ray Worklist Prioritization Using Artificial Intelligence: A Clinical Workflow Simulation. *Eur Radiol* 2021;31:3837–45.

60. Heilbrun ME, Chapman BE, Narasimhan E, Patel N, Mowery D. Feasibility of Natural Language Processing-Assisted Auditing of Critical Findings in Chest Radiology. *J Am Coll Radiol* 2019;16:1299–304.

61. Mabotuwana T, Hall CS, Cross N. Framework for Extracting Critical Findings in Radiology Reports. *J Digit Imaging* 2020;33:988–95.

62. Zhou Y, Amundson PK, Yu F, Kessler MM, Benzinger TL, Wippold FJ. Automated Classification of Radiology Reports to Facilitate Retrospective Study in Radiology. *J Digit Imaging* 2014;27:730–6.

63. Becker AS, Erinjeri JP, Chaim J, et al. Automatic Forecasting of Radiology Examination Volume Trends for Optimal Resource Planning and Allocation. *J Digit Imaging* 2021:1–8.

64. Kalra A, Chakraborty A, Fine B, Reicher J. Machine Learning for Automation of Radiology Protocols for Quality and Efficiency Improvement. *J Am Coll Radiol* 2020;17:1149–58.

65. Lou R, Lalevic D, Chambers C, Zafar HM, Cook TS. Automated Detection of Radiology Reports that Require Follow-up Imaging Using Natural Language Processing Feature Engineering and Machine Learning Classification. *J Digit Imaging* 2020;33:131–6.

66. Carrodeguas E, Lacson R, Swanson W, Khorasani R. Use of Machine Learning to Identify Follow-Up Recommendations in Radiology Reports. *J Am Coll Radiol* 2019;16:336–43.

22 Technological View

Ruiwen Ding and William Hsu

INTRODUCTION

Artificial intelligence (AI)/machine learning (ML) has rapidly evolved, with new novel techniques constantly being proposed for extracting meaningful patterns from available biomedical data. The development of these algorithms is supported by a growing amount of data being generated as part of routine clinical care. These datasets are increasing in number and in complexity, spanning multiple modalities (e.g., clinical information, radiological imaging, digital pathology, labs), biological scales (e.g., cellular level, tissue level, organ level), and data streams (e.g., wearable sensors, physiological data).

In this chapter, we present a vision of AI's potential from integrating multimodal, multiscale, and multistream data to shed light on biological processes that influence the etiology and progression of diseases. Providing new biological details about what is observed on diagnostic images provides the basis for individualized and precision-targeted diagnosis and therapies. To envision this future, we assume the following:

- **Large, diverse datasets will be pervasive.** The amount of data being collected on patients is occurring at an unprecedented rate. Initiatives such as the National Institutes of Health's All of Us [1], the Veteran Administration's Million Veteran Program [2], and Verily's Project Baseline [3] are current examples of these efforts. Electronic health records (EHR) have also become ubiquitous and are increasingly becoming more interoperable across institutions. Data currently residing outside of the EHR, such as radiological images, digital pathology, sequencing, and mHealth data, will eventually be more closely integrated [4]. Individuals born in the last decade will have lifelong digital records. Vital signs are increasingly being measured and recorded in real-time on smart devices. A number of publications have identified specific genetic risk factors and targets for individualized treatments using datasets such as the Alzheimer's Disease Neuroimaging Initiative [5] and the UK Biobank [6]. At the David Geffen School of Medicine at UCLA, the Integrated Diagnostics (IDx) Shared Resource collects and curates high resolution, spatially, and longitudinally co-registered radiology and pathology data to improve early detection of cancer and precision diagnosis. By annotating and linking individual lesions characterized across multiple modalities (clinical encounters, radiological imaging, digital pathology) and streams (temporal observations) to outcomes of interest (histopathology, disease recurrence), IDx permits a variety of investigations to reveal detailed information about the significance of tumor microenvironments and how they evolve, characterized using quantitative findings extracted from diagnostic images, and their correlation with outcomes.

- **The types of data that a radiologist will routinely review will be increasingly diverse.** Radiologists are tasked with synthesizing large amounts of information to obtain case-relevant information before interpreting images and formulating a differential diagnosis [7]. The patient's medical history, prior imaging exams, and labs provide important context on potential findings observed in the images.

- **The role of the radiologist will be that of an information synthesizer.** The radiologist will play a key role in combining information—patient demographics, clinical history, labs—to provide more precise and confident interpretations of what is observed on diagnostic images. The radiologist will identify suspicious findings and identify the type of disease, its aggressiveness, and potential course of action. Shifting away from qualitative, subjective interpretations to more quantitative, objective assessments is necessary to achieve a more accurate and efficient diagnosis.

Given these assumptions, we discuss the role of AI/ML in enabling multimodal, multiscale, and multistream data fusion. We first discuss the vision of how the integration of clinical, imaging, and molecular data can enable precision diagnosis and targeted interventions. We highlight specific techniques that are being developed to perform data fusion, summarizing their strengths and limitations. We conclude by discussing open research areas to be addressed for realizing potential of multimodal data fusion

DOI: 10.1201/9781003095279-28

DERIVING ACTIONABLE INFORMATION FROM FUSED DATA

Traditionally, radiology scans obtained via different protocols (e.g., T1/T2-weighted, diffusion-weighted, dynamic contrast-enhanced) are presented to radiologists who inspect specific disease-related patterns on the images. With the current advances in image analysis, computers now can automatically extract a set of subvisual quantitative patterns (radiomic features) such as signal intensity distribution, which cannot be easily discerned by the naked eye. Such information has been shown to correlate with disease severity and treatment response. Nevertheless, our current understanding of how these pixel patterns relate to underlying biology is limited. Datasets (e.g., genomic, proteomic, metabolomic) that characterize diseases from multiple biological scales enables detailed investigation into the underlying biological processes and clinical characteristics.

For example, prostate cancer (PCa) is the second leading cause of cancer death among men in the United States and accounts for one in four new diagnoses of cancer in men [8]. Over the past decade, PCa diagnosis has become more precise due to the adoption of multiparametric magnetic resonance imaging (mpMRI) to detect and characterize suspicious tumors [9] and targeted image-guided biopsy to sample and histopathologically confirm malignant tumors [10]. PI-RADS (Prostate Imaging and Reporting and Data System) [11], which utilizes a combination of information from T2-weighted, diffusion, and dynamic contrast-enhanced MRI to derive a suspicion score for each tumor, has become standard practice. However, PI-RADS scoring has substantial reader variability and is limited to imaging characteristics easily interpreted by a radiologist [12]. Recent genomic studies have started to examine the relationship between mpMRI-derived findings and clinically significant PCa [13–15]. These studies, which link mpMRI-derived phenotypes to distinct underlying pathobiological mechanisms, are nascent [16]. Future multimodal, multiscale, multi-stream data fusion, illustrated in Figure 22.1, will elucidate 1) what characteristics of PCa that are characterizable on mpMRI are associated with aggressive molecular hallmarks and 2) whether joint clinical, radiologic, histopathologic, and genomic analysis yields robust imaging phenotypes that can improve the identification of aggressive tumors.

TECHNIQUES FOR DATA FUSION

Multimodal, multiscale, and multistream data fusion will permit the discovery of relationships between imaging phenotypes and underlying biological properties such as vascularity, tumor heterogeneity, and hypoxia [17]. Development and validation of these data fusion techniques are important in elucidating the pathophysiologic interpretation of radiomic features, discovering new visual patterns predictive of aggressive disease, and permitting models that incorporate quantitative image features to be interpretable and generalizable for clinical decision-making.

Data fusion techniques can be broadly categorized into two groups: early and late fusion. Different modalities are represented as a single joint feature space in early fusion, on which a downstream prediction task is performed. In late fusion, each modality separately produces one prediction, and the final decision is made by aggregating all the predictions. Early fusion captures the inherent relationships among different input modalities, and as such, we focus on early fusion techniques in this chapter. In subsequent sections, we highlight two classes of data fusion techniques (deep learning–based and dimensionality reduction–based), as illustrated in Figure 22.2.

Deep Learning–Based Approach and Potential Applications

Deep learning (DL), a subfield of ML, has been successful in computer vision, speech recognition, and natural language processing. Its advantage resides on the ability to automatically learn from high-dimensional input data without feature engineering and feature selection. Given its ability to learn from large-scale data, DL has been found to be promising for multimodal, multiscale, and multistream data fusion in applications including autonomous driving and emotion recognition in videos while its use in medicine is relatively understudied. Still, there are already several studies that leveraged this concept in neuroimaging with the aim of both understanding the disease and improving the diagnosis accuracy. Ning et al. [18] used a simple NN with two hidden layers to distinguish Alzheimer's disease (AD) from cognitively normality (CN) and predict progression from mild cognitive impairment (MCI) to AD using integrated magnetic resonance imaging (MRI) and single nucleotide polymorphism (SNP) data. The input consists of 16 AD-related imaging traits extracted from the original MRIs and 19 AD-related SNPs. A partial derivative method was used to quantify the contribution of each feature to the output and identify the most important ones. Several strong interactions between the important imaging and genetic features were found. In another study, NNs were used to classify AD from MCI and CN by integrating MRI, SNP, and

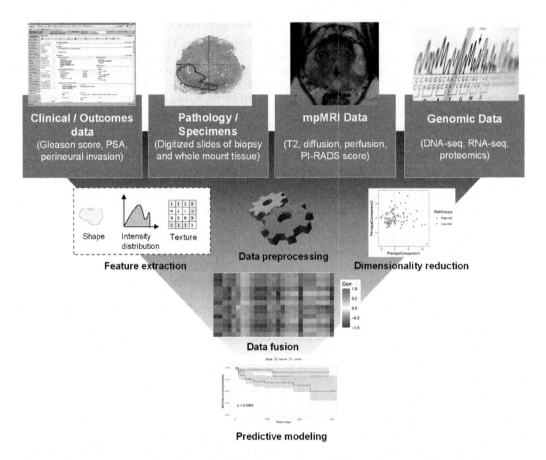

Figure 22.1 Multimodal, multiscale, and multistream modeling techniques. These help integrate data on prostate cancer patients from multiple sources, including clinical records, pathology/biospecimens, advanced diagnostic imaging, and gene sequencing. A variety of techniques can be applied for data preprocessing, feature extraction, and dimensionality reduction, yielding a set of relevant features from each source. Data fusion techniques attempt to uncover the correlations among features across data types with the goal of building better models for predicting cancer aggressiveness.

clinical data. The learned features from these three modalities were then concatenated into 1D feature vector ("intermediate features") on which the classification was performed. This study identified a set of top-performing features using drop in accuracy, where the features that lead to the largest reductions in accuracy are ranked higher. Most of the important features were found to be related to AD. In unsupervised clustering analysis, the intermediate features from the multimodal data generated better separation than the original features. This result indicates that the DL models were able to detect the subtle relationships in the intermediate features. Together, these two studies highlight that multimodal analysis can improve diagnosis accuracy and add value in understanding the disease.

Even though multimodal data fusion using DL has shown promise in current work, other models such as deep Boltzmann machine (DBM) can be explored. Connections between adjacent hidden units in DBMs are bidirectional, which allows the flow of information between modalities during representation learning. Suk et al. [19] leveraged DBMs to classify AD and MCI by jointly using MRI and positron emission tomography (PET) scans. However, this study did not include biological data and was limited in the interpretability of the resulting fused representation from DBM. Though studies discussed in this section improved the interpretability by investigating the important features, better techniques are needed to understand the fused representation from all DL models in the future. Also, most existing studies used cross-sectional data in their analysis.

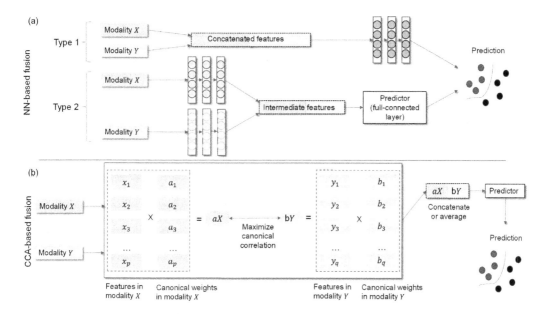

Figure 22.2   Visual comparison of (a) neural network–based fusion and (b) CCA-based fusion.

Future studies can predict disease progression based on longitudinal data using recurrent NNs to learn the temporal dynamics of the longitudinal data. The application of multimodal DL where imaging and non-imaging data are fused is currently limited in non-neuroimaging domains. As multimodal, multiscale, and multistream datasets become more abundant, DL analysis could easily be applied.

Dimensionality Reduction–Based Approach and Potential Applications

One major obstacle in multimodal, multiscale, and multistream data fusion is the inconsistent scale and dimensionality of each data type. Dimensionality reduction–based approaches aim to minimize the scale and dimensionality differences in data streams by projecting them into a homogeneous metaspace data [20]. Unlike DL-based approaches, dimensionality reduction typically does not require a large sample size and has better interpretability. However, feature engineering is typically required to extract features from imaging data before integrating it with non-imaging data, while in DL-based models, raw images can be directly used as input. One popular technique is canonical correlation analysis (CCA). CCA captures the complex relationship among multiple variables across data modalities. It tries to find a linear combination of variable sets from each data modality such that the correlation between these variable sets is maximized.

Given two modalities $X \in \mathbb{R}^{N \times p}$, $Y \in \mathbb{R}^{N \times q}$, where N is the sample size, p and q are the number of descriptors or features in X and Y respectively, the objective of CCA is to find a set of canonical weights, denoted as $a \in \mathbb{R}^{p \times 1}, b \in \mathbb{R}^{q \times 1}$, such that the correlation between the linear combination of variables in X and variables in Y is maximized:

$$a \in \mathbb{R}^{p \times 1}, b \in \mathbb{R}^{q \times 1} = argmax\ Corr(aX, bY) = \frac{a^T \left(X^T Y\right) b}{\sqrt{a^T (X^T X) a}\ \sqrt{b^T (Y^T Y) b}}$$

The weights a and b are learned through a mathematical process called singular value decomposition. aX and bY are called canonical components, and there are up to *min (p, q)* number of pairs of canonical components. Each pair of the canonical component has a canonical correlation coefficient that captures how strongly the components are associated with each other. Sparse CCA [21] is the most common variant where regularization terms are used in the learning objective. Doing so has several benefits, including reducing overfitting, facilitating interpretation of solutions, and applying CCA to high-dimensional datasets where the number of features exceeds the number of observations.

CCA has been broadly utilized in the neuroimaging domain. Fang et al. [22] proposed a sparse CCA framework to identify a set of SNP-functional MRI interaction modules for schizophrenia patients. They identified a set of SNP-voxel associations with both statistical and biological significance, where the SNPs are from genes related to the risk of the disease and brain development. Some studies have also added outcome prediction as another aspect in the joint analysis. For example, in Du et al. [23], sparse CCA was used together with regression to identify brain imaging-genomic associations relevant to diagnosis status. The aim was to learn the structural-MRI voxel canonical components and AD-related SNP canonical components predictive of the AD outcome status in a linear regression model. More comprehensive imaging–SNP associations were identified when incorporating the outcome status compared to when only using sparse CCA.

Despite its promising utility in neuroimaging, the use of CCA is nascent. Related to PCa, Lee et al. [24] used multiview CCA to predict biochemical recurrence using histologic and proteomic features with outcome status as the third view showed promising results. However, due to the small sample size of their study, their model was not able to predict outcome in a statistically significant manner in Kaplan–Meier analysis. Another study by Fan et al. [25] utilized CCA-fused features extract from T2-weighted MRI and dynamic contrast-enhanced MRI to predict histological grade in invasive ductal carcinoma. They showed that their support vector machine model utilizing CCA-fused inputs outperformed other approaches for data fusion (e.g., feature concatenation, classifier fusion).

OPEN QUESTIONS AND FUTURE DIRECTIONS

There are some potential barriers in fully extending the utility of multimodal, multiscale, and multistream data fusion. One issue is how to weigh different modalities and the accuracy of various data sources. Some modalities have higher noise and rates of missingness than others. Potential solutions include multiplicative multimodal modeling [26]. Spatial registration is also an important challenge. Current multimodal data fusion approaches fuse data sources at the feature level, avoiding the need for spatial registration. However, spatial correlation between radiological and digital pathology images may yield useful insights into how tissue microenvironments are observed under diagnostic imaging. Studies have developed methods for registration of radiology and pathology images in prostate cancer [27, 28]. Several barriers to accurate registration exist. Various artifacts due to motion and physics (e.g., beam hardening, partial volume effect) are introduced during image acquisition. Tissue deformation that occurs when tissue is removed from the body introduces variation between *in vivo* and *ex vivo* scans. Registration between lower resolution images (e.g., MRI acquired at millimeter resolution) and higher resolution images (e.g., digital slides scanned at micrometer resolution) is also a challenge. The IDx program at UCLA has explored the use of patient-specific 3D-printed molds generated from segmented diagnostic scans to align histopathology slices and key imaging slices and limit the amount of deformation that occurs during slide preparation. Nevertheless, these techniques work well with denser organs of relatively small sizes (e.g., prostate, kidney). The ability to do so for other organs such as the lung, liver, and breast remains an open area of investigation and may require other approaches such as introducing fiducial markers.

Even if we have large cohorts in the future, the majority will have been collected retrospectively. The selection biases introduced in retrospective cohorts can create confounders and lead to the discovery of false-positive signals. Statistically significant associations can be easily identified even when multiple comparison correction is applied, and the explanatory power of the data is small. Related to the UK Biobank, in Miller et al. [29], more than 100,000 subjects were included in the joint analysis of magnetic resonance imaging (MRIs), X-rays, genetic, lifestyle measures, and other biomedical data to enable the discovery of imaging biomarkers in various diseases. However, statistical significance was reached with only 1% of explained sample variance and a correlation of 0.1. Accounting for confounders will be an important part of future research since doing so can reduce the false positive signals discovered by the statistical tests.

CONCLUSION

This chapter focused on how multimodal, multiscale, and multistream data fusion opens a new potential for AI/ML applications in radiology. Different data sources and imaging modalities carry complementary information. Incorporating information across biological scales and modalities can lead to improved predictions of prognosis and treatment. In developing multimodal models, model developers should keep in mind desirable characteristics [30]: 1) similarity in the representation space implies similarity of related concepts; and 2) robustness to missing information enabling the ability to infer missing data sources given information about observed ones. In summary, the potential for these modeling techniques is vast and essential in providing radiologists with the tools needed to provide precise and actionable information for diagnosis and targeted treatment.

ACKNOWLEDGMENTS

The authors would like to gratefully acknowledge Dr. Akinyinka Omigbodun, Tianqing Li, and Yannan Lin for their insightful feedback on a draft of this chapter.

REFERENCES

1. All of Us Research Program I, Denny JC, Rutter JL, Goldstein DB, Philippakis A, Smoller JW, et al. The "All of Us" research program. *N Engl J Med.* 2019;381(7):668–76.

2. Gaziano JM, Concato J, Brophy M, Fiore L, Pyarajan S, Breeling J, et al. Million Veteran Program: A mega-biobank to study genetic influences on health and disease. *J Clin Epidemiol.* 2016;70:214–23.

3. Arges K, Assimes T, Bajaj V, Balu S, Bashir MR, Beskow L, et al. The Project Baseline Health Study: A step towards a broader mission to map human health. *NPJ Digit Med.* 2020;3(1):84.

4. Weber GM, Mandl KD, Kohane IS. Finding the missing link for big biomedical data. *JAMA.* 2014;311(24):2479–80.

5. Petersen RC, Aisen PS, Beckett LA, Donohue MC, Gamst AC, Harvey DJ, et al. Alzheimer's Disease Neuroimaging Initiative (ADNI): clinical characterization. *Neurology.* 2010;74(3):201–9.

6. Ollier W, Sprosen T, Peakman T. UK Biobank: From concept to reality. *Pharmacogenomics.* 2005;6(6):639–46.

7. Krupinski E, Bronkalla M, Folio L, Keller B, Mather R, Seltzer S, et al. Advancing the diagnostic cockpit of the future: An opportunity to improve diagnostic accuracy and efficiency. *Acad Radiol.* 2019;26(4):579–81.

8. Siegel RL, Miller KD, Fuchs HE, Jemal A. Cancer statistics, 2021. *CA Cancer J Clin.* 2021;71(1):7–33.

9. Furlan A, Borhani AA, Westphalen AC. Multiparametric MR imaging of the Prostate: Interpretation including prostate imaging reporting and data system version 2. *Radiol Clin North Am.* 2018;56(2):223–38.

10. Moore CM, Robertson NL, Arsanious N, Middleton T, Villers A, Klotz L, et al. Image-guided prostate biopsy using magnetic resonance imaging-derived targets: a systematic review. *Eur Urol.* 2013;63(1):125–40.

11. Panda A, Gulani V. *Multiparametric MRI of Prostate: Analysis and Recommendations of Prostate Imaging Reporting and Data System (PI-RADS) Version 2.1.* Reading MRI of the Prostate; 2020. 25–47.

12. Gupta RT, Mehta KA, Turkbey B, Verma S. PI-RADS: Past, present, and future. *J Magn Reson Imaging.* 2020;52(1):33–53.

13. Parry MA, Srivastava S, Ali A, Cannistraci A, Antonello J, Barros-Silva JD, et al. Genomic evaluation of multiparametric magnetic resonance imaging-visible and -nonvisible lesions in clinically localised prostate cancer. *Eur Urol Oncol.* 2019;2(1):1–11.

14. Jambor I, Falagario U, Ratnani P, Perez IM, Demir K, Merisaari H, et al. Prediction of biochemical recurrence in prostate cancer patients who underwent prostatectomy using routine clinical prostate multiparametric MRI and decipher genomic score. *J Magn Reson Imaging*. 2020;51(4):1075–85.

15. Houlahan KE, Salmasi A, Sadun TY, Pooli A, Felker ER, Livingstone J, et al. Molecular hallmarks of multiparametric magnetic resonance imaging visibility in prostate cancer. *Eur Urol*. 2019;76(1):18–23.

16. Emberton M. To see or not to see - what renders prostate cancer visible? *Nat Rev Urol*. 2019;16(5):274–5.

17. Tomaszewski MR, Gillies RJ. The biological meaning of radiomic features. *Radiology*. 2021;298(3):505–16.

18. Ning K, Chen B, Sun F, Hobel Z, Zhao L, Matloff W, et al. Classifying Alzheimer's disease with brain imaging and genetic data using a neural network framework. *Neurobiol Aging*. 2018;68:151–8.

19. Suk HI, Lee SW, Shen D, Alzheimer's disease neuroimaging I. Hierarchical feature representation and multimodal fusion with deep learning for AD/MCI diagnosis. *Neuroimage*. 2014;101:569–82.

20. Viswanath SE, Tiwari P, Lee G, Madabhushi A, Alzheimer's disease neuroimaging I. dimensionality reduction-based fusion approaches for imaging and non-imaging biomedical data: concepts, workflow, and use-cases. *BMC Med Imaging*. 2017;17(1):2.

21. Witten DM, Tibshirani R, Hastie T. A penalized matrix decomposition, with applications to sparse principal components and canonical correlation analysis. *Biostatistics*. 2009;10(3):515–34.

22. Fang J, Lin D, Schulz SC, Xu Z, Calhoun VD, Wang YP. Joint sparse canonical correlation analysis for detecting differential imaging genetics modules. *Bioinformatics*. 2016;32(22):3480–8.

23. Du L, Liu K, Yao X, Risacher SL, Guo L, Saykin AJ, et al. Diagnosis status guided brain imaging genetics via integrated regression and sparse canonical correlation analysis. 2019 IEEE 16th International Symposium on Biomedical Imaging (ISBI 2019)2019. 356–9.

24. Lee G, Singanamalli A, Wang H, Feldman MD, Master SR, Shih NN, et al. Supervised multiview canonical correlation analysis (sMVCCA): Integrating histologic and proteomic features for predicting recurrent prostate cancer. *IEEE Trans Med Imaging*. 2015;34(1):284–97.

25. Fan M, Liu Z, Xie S, Xu M, Wang S, Gao X, et al. Integration of dynamic contrast-enhanced magnetic resonance imaging and T2-weighted imaging radiomic features by a canonical correlation analysis-based feature fusion method to predict histological grade in ductal breast carcinoma. *Phys Med Biol*. 2019;64(21):215001.

26. Liu K, Li Y, Xu N, Natarajan P. Learn to combine modalities in multimodal deep learning. arXiv preprint arXiv:180511730. 2018.

27. Rusu M, Shao W, Kunder CA, Wang JB, Soerensen SJC, Teslovich NC, et al. Registration of presurgical MRI and histopathology images from radical prostatectomy via RAPSODI. *Med Phys*. 2020;47(9):4177–88.

28. Sood RR, Shao W, Kunder C, Teslovich NC, Wang JB, Soerensen SJC, et al. 3D Registration of pre-surgical prostate MRI and histopathology images via super-resolution volume reconstruction. *Med Image Anal*. 2021;69:101957.

29. Miller KL, Alfaro-Almagro F, Bangerter NK, Thomas DL, Yacoub E, Xu J, et al. Multimodal population brain imaging in the UK Biobank prospective epidemiological study. *Nat Neurosci*. 2016;19(11):1523–36.

30. Srivastava N, Salakhutdinov R. Multimodal learning with deep boltzmann machines. *J Mach Learn Res*. 2014;15(1):2949–80.

23 Societal View

Amy Patel

PROMISES OF AI IN IMAGING FOR SOCIETY

First, before a deeper dive is taken into the field of radiology, a broad overview of what AI in imaging means for society and humanity should be considered, as this is truly the foundation for which the house of medicine and doors of radiology are built. AI is transforming the way we solve complex problems, improve decision-making, and decipher data analytics. AI is touching many industries in addition to healthcare, including, but certainly not limited to, finance, transportation, criminal justice, and national security [1]. In healthcare, and particularly radiology, deep learning can be employed to solve complex medical issues such as distinguishing a normal versus abnormal lymph node on imaging, or to detect a subtle invasive cancer on mammography [1]. "Augmented intelligence," in which AI can assist humans in making improved decisions, is the overarching goal for the field, accelerating efficiency of tasks in so many arenas of society [2]. In fact, AI has great potential to aid in accelerated healthcare advancements such as those being seen with drug developments as well as expedited measures regarding information access [2].

AI imaging in society could also mean faster and cheaper diagnoses for patients, particularly in parts of the world where imaging access is scarce and expertise varies. A focus on education from a clinical standpoint, deployment of infrastructure, and a phased, methodical approach of AI introduction exclusive to this patient population could benefit and innovate global health radiology [3]. The promise is great, and the stakes are high when it comes to AI in imaging for society, as lives could literally depend on it for survival all over the world, regardless of geographic location or socioeconomic status.

AI'S POTENTIAL IMPLICATIONS ON THE FIELD OF RADIOLOGY

When it comes to AI imaging specifically to radiology, its potential is truly remarkable and will forever change the way in which we practice. However, currently only approximately one-third of radiologists are utilizing AI in their everyday practices and institutions [4]. AI in the field of radiology has the potential to enhance both interpretative and non-interpretive skills which affect our daily workflow and lives, and in an era of rising imaging volumes, a radiology workforce shortage, and burnout, AI applications could assist in mitigating time-intensive and laborious tasks. For example, AI complementing case interpretation to increase interpretation efficiency and diagnostic accuracy could be a gamechanger when one considers "diagnostic errors, intra- and interreader variability, and burnout" [4].

NON-INTERPRETIVE EFFICIENCY

AI algorithms improving non-interpretative tasks of the clinical day-to-day have the potential to greatly improve daily workflows. For example, improving workflows such as machine learning (ML) algorithms, creating hanging protocols that are unique to each radiologist's preferences, and include information such as clinical and ordering data and reviewing prior radiology images and reports specific to the patient, could provide a higher level of work efficiency than we have ever seen. There is also potential to improve image quality through deep learning (DL) by acting on the image that is processed or directly on raw imaging. This can result in reducing artifacts and optimizing contrast enhancement which, in turn, improves image quality. Additionally, a reduction in radiation dose is always at the top of the mind, and AI has the potential to assist with this by being trained with "high-quality and "noisy" images and to recognize what normal versus abnormal looks like at a lower dose compared to a higher dose [4]. Also, when it comes to "triaging," AI has the potential to handle tasks that would make time available for radiologists to spend with patients as well as on teaching and research. Thus, AI applications are being generated to put more critical finding exams at the top of the reading list, optimize scan time, and reduce patient

DOI: 10.1201/9781003095279-29

waiting times. Since AI applications can generate complex models and analyze many features at once, it has the potential to improve patient compliance of scheduled appointments. Currently in North America, approximately 23.5% of patients miss scheduled appointments and this percentage has not significantly wavered in over a decade [4].

INTERPRETATIVE EFFICIENCY AND ACCURACY

When it comes to the interpretative aspects of AI in imaging, studies are demonstrating an AI machine learning algorithm outperforming a radiologist regardless of variables that may be perceived by some as flawed methodology and conclusions, but, nevertheless, show its potential in elevating patient care. For example, a study was conducted where an AI deep learning–based system was trained using mammograms from women in the UK and United States and then was applied retrospectively to test sets resulting in reductions of 1.2% and 2.7% in the rates of false-positive and false-negative detection of biopsy-proven breast cancers, respectively, in the UK group, and 5.4% and 9.4%, respectively, in the US dataset [5]. Of course, in this study, demographics of the population were not clear which one cannot ensure broad applicability across all patient types, and monitoring of the AI system would need to be in place, by humans, to ensure performance did not deteriorate over time. However, more studies are demonstrating similar results, confirming the potential of AI utilization in mammography and the field of breast imaging. AI algorithms in mammography now encompass approximately 14.8% of the market as breast cancer is still the most common cancer in women throughout the world. These applications range from lesion characterization on mammography, ultrasound, and MRI, estimation of breast density, and improvement of image quality. When it comes to thoracic radiology, approximately 26.6% of AI algorithms encompass this market, as lung cancer has the highest death rates amongst all cancer types and most of these focus on the detection of lung nodules [4]. However, there are other chest AI algorithms that can detect pulmonary emboli, pneumothoraces, and pleural effusions. In the world of musculoskeletal radiology, AI algorithms for the detection on radiographs of fractures such as of the ankle, wrist, hand, and proximal humerus have great potential as well as detecting osteoarthritis and bone quality [4]. In neuroradiology, the misdiagnoses of ischemic strokes and subarachnoid hemorrhages is high (30–42% and 5–51%, respectively). AI algorithms have great potential to detect strokes and hemorrhages, as well as assess aneurysms and calcium scoring of brain anatomy [4]. Undoubtedly, AI has the potential to exponentially improve diagnostic accuracy for patients that will, in turn, save patient lives. As Pakdemirli poignantly states, "wherever you see yourself in the spectrum, nobody can stop the advancement, innovations, and implantation of AI in radiology".

CONCLUSION

Overall, stakeholders in AI feel that AI can improve the field of radiology and will not replace radiologists [4]. Its potential is limitless in elevating the way in which we practice, ultimately affecting society as a whole, which can result in a future of sound utilization, accuracy, and efficiency for the betterment of patient care.

REFERENCES

1. West D, Allen J. How Artificial Intelligence is Transforming the World. *Brookings* (2018). https://www.brookings.edu/articles/how-artificial-intelligence-is-transforming-the-world/ Accessed 12/30/23.

2. Forbes Expert Panel. 15 Tech Experts Share Potential Impacts of AI on Society. https://www.forbes.com/sites/forbestechcouncil/2020/06/26/15-tech-experts-share-potential-impacts-of-ai-on-society/?sh=5156b56c3714. Accessed 12/30/23.

3. Mollura D, Culp M, Pollack E, et al. Artificial Intelligence in Low- and Middle-Income Countries: Innovating Global Health Radiology. *Radiology* 297, no. 3 (2020).

4. Mello-Thoms C, Mello C. Clinical Applications of Artificial Intelligence in Radiology. *Br J Radiol* 96, no. 1150 (2023). 37099398.

5. Killock D. AI Outperforms Radiologists in Mammographic Screening. *Nat Rev Clin Oncol* 17, no. 3 (2020). 31965085.

24 Financial View

Sophia Mavrommatis, Madeleine Howard, Dina Belhasan,
Sinibaldo Rafael Romero Arocha, and David J.H. Wu

HEALTH OUTCOMES

It is anticipated that the use of AI in the field of radiology will contribute to decreased healthcare costs through several means: 1) reducing superfluous patient imaging, 2) providing information that will eliminate the need for secondary testing, 3) detecting pathologies earlier, and 4) reducing medical imaging interpretation errors.

First, the unnecessary imaging of patients has proved to be incredibly costly for the healthcare system. In fact, recent literature reports that unnecessary medical imaging costs the United States' healthcare system in the range of $200 billion to $250 billion annually [1]. The use of AI in radiology may help to reduce excess medical imaging and its associated cost. For instance, a recent study demonstrated that AI software used to detect lung cancer in patient X-rays was able to accurately determine from the X-rays whether or not follow-up imaging was needed and ultimately resulted in a 30% reduction in follow-up CT scans [2]. Reductions in superfluous imaging do not only reduce healthcare costs, but also avoid subjecting patients to unnecessary and potentially harmful radiation exposure [1].

Second, AI shows further promise in curtailing healthcare spending by minimizing the need for certain types of diagnostic testing. For example, the determination of MGMT methylation patterns in high-grade gliomas is a task that typically necessitates specialized laboratory testing performed in only a few locations across the United States. However, one study found that machine learning techniques can be used to determine the MGMT methylation status on low grade gliomas in MRIs with high accuracy [3], in turn eliminating the need for the aforementioned laboratory testing. As such, it is possible that with the implementation of AI in radiology, important patient health information that otherwise may have required additional costly testing will be made ascertainable from patient imaging.

Third, the use of AI in medical imaging analysis will allow for the earlier detection of pathologies in patients, potentially resulting in substantial healthcare cost reductions. Consider fragility fractures associated with osteoporosis. Annually, the United States sees an estimated 1.5 million fragility fractures with associated costs approximated at $17.9 billion [4]. Sectra OneScreen is a novel AI application of the traditionally used digital X-ray radiogrammetry, a diagnostic technique that can be used to diagnose bone-density disorders such as osteoporosis via a peripheral analysis of bone mineral density [5, 6]. If Sectra One Screen AI technology were implemented in primary care settings as a component of other routine screenings, the diagnosis and treatment of osteoporosis could potentially occur prior to the development of fractures and in turn prevent costs associated with the development of osteoporosis-related fractures [7]. It is clear that the AI-facilitated ability to diagnose conditions earlier, as opposed to when conditions become advanced and require more intense treatments, has massive cost-saving potential in healthcare.

Fourth, the increasing accuracy at which AI technology can diagnose and detect pathologies is particularly significant given the prevalence of missed diagnoses and misdiagnosis within the field of radiology; approximately 30% of abnormal results are missed while 4% of daily imaging interpretations contain inaccuracies [8]. These imaging mistakes may contribute to medical errors, which as a whole cost the United States upwards of $38 billion annually [8]. One way that AI can be used to reduce medical errors is through the use of computer-aided diagnosis (CAD). A recent article argues that CAD may be used as a supplement to the BI-RADS score—a score intended to standardize the measurement of breast health, but is often misread by healthcare providers—to help reduce false-positive results and associated costs [9]. As false-positive results are estimated to cost the US healthcare system approximately $2.18 billion per year [9, 10], it is clear that AI's potential to minimize the occurrence of these false-positive results has significant cost saving applications.

RADIOLOGIST BURNOUT

Burnout is a common phenomenon amongst all physicians, including radiologists, and is costly to healthcare systems. In fact, it is reported that physician burnout—measured through labor turnover and decreased clinical work hours—costs the United States approximately $4.6 billion per

DOI: 10.1201/9781003095279-30

year [11]. Luckily, the automation powers of AI can be harnessed to reduce radiologists' workload. An example of this includes the AI-mediated automation of transcription-related tasks, such as radiologists' dictated reports [11, 12]. It has been reported that the automation of this task could translate into one hour of saved time for radiologists per day. Thus, AI could be used by radiologists in administrative tasks as both a cost and time saving measure.

LABOR MARKET

As previously noted, the implementation of AI in the field of radiology has the potential to yield a plethora of benefits including, but not limited to, improved health outcomes and reduced healthcare costs. While considering the potentially beneficial impacts of AI in radiology, it is also necessary to consider the effects of AI on the labor market. Due to radiologists' relatively limited patient interaction as well as the digital nature of the patient data that they handle, some within the radiology community are anxious that AI will automate radiology-related tasks and in turn displace radiologists from their positions [12–14]. For instance, in resource-limited countries the AI-mediated detection of tuberculosis on radiographs can be automated to perform the task of what is generally considered the role of the radiologist [14]. Though anxiety surrounding radiologists' job security is understandable, the logistics of AI job automation in radiology is complex and its effects on the labor market remain uncertain.

Many argue that AI will not replace radiologists, but rather serve to supplement their work. AI may optimize radiologists' work by reducing their per scan analysis time [12–14]. The effect of increased image analysis capacity from AI on radiologists' employment depends on whether the demand for scans remains fixed or the increased efficiency results in a heightened demand for imaging. The former scenario suggests a decrease in the demand for radiologists as fewer would be needed to analyze the same number of scans. The latter circumstance requires a similar number of radiologists to be employed—when compared to today's workforce—to analyze the larger volume of scans [12]. It is unclear which situation will prevail. Further, it is hypothesized that AI may one day supersede the accuracy at which radiologists interpret scans, effectively replacing radiologists from performing this task [12].

The notion of the complete automation of radiologists' jobs by AI operates largely under the assumption that radiologists work exclusively with digital information. This is not true; radiologists also assist in many non-prediction related tasks, such as coordinating treatment plans, choosing exams, consulting on physician teams, and performing image-assisted procedures [12, 13]. However, the effects of AI on radiologists' employment will also depend on whether these non-imaging related tasks are handled by radiologists rather than others in the medical field.

Currently, AI is not in a place to fully automate the image interpretation roles of radiologists; AI technology is not yet sophisticated enough, and the necessary image repositories do not yet exist [12–14]. Further, many regulatory steps will first need to be taken to fully implement AI in radiology [13–15]. For these reasons, forecasts do not anticipate that employment in digitally centered facets of healthcare, such as radiology, to be significantly affected by AI for at least the next 20 years [13]. Ultimately, though the impacts of AI on the radiology job market remain unclear, it appears likely that for the foreseeable future AI will supplement, rather than replace, the work of radiologists.

REIMBURSEMENT

As AI is integrated into the field of radiology, it is important to consider how radiology services that employ AI technology will be reimbursed. Historically in the United States, medical services have been reimbursed through the fee-for-service model, which relies on the use of Current Procedure Terminology (CPT) codes [15, 16]. Based on a complex and multifaceted review process, medical services are assigned a CPT code which specifies the dollar value of the service provided. AI technologies in radiology are not expected to be reimbursed with this fee-for-service model. Rather, it is posited that value-based payment models such as the Merit-Based Incentive System (MIPS) and the Alternative Payment Models (APMs) will serve as the scaffolding for the development of reimbursement plans that will reward AI-mediated quality improvement in radiology [15, 16]. In these models, radiologists may be reimbursed for the use of AI in their practice, if the use of AI results in 1) improved health outcomes with no associated rise in costs, 2) a reduction in cost with no change in health outcomes, or 3) improved health outcomes in conjunction with decreased costs. The possible transition to a value-based reimbursement model for AI services in radiology could prove to be beneficial for radiologists given the previously discussed potential of AI to reduce costs, increase efficiency, and improve patient outcomes in radiology.

CONCLUSION

It is clear that AI will have a major influence on radiology's economic landscape. Some economic impacts of AI on the field of radiology appear more obvious; for instance, AI may contribute to a reduction in healthcare spending through improved medical image analysis accuracy, efficiency, and sensitivity which may also improve patient outcomes. On the other hand, there are many uncertain financial consequences of AI in radiology. In particular, it is unclear how AI will affect the radiology labor market and how AI radiology technologies will be reimbursed. Ultimately, one insight does remain clear—AI will continue to stake its claim in the field of radiology, and as it does, the financial norms of radiology and healthcare as a whole will unequivocally evolve.

REFERENCES

1. Hadian M, Jabbari A, Mazaheri E, Norouzi M. What is the impact of clinical guidelines on imaging costs? *J Educ Health Promot.* 2021;10(10). doi:10.4103/jehp.jehp_225_20

2. O'Connor M. *AI Software Helps Radiologists Cut Unnecessary Chest CT Scans by 30%.* AI & Emerging Technologies 2021.

3. Korfiatis P, Kline TL, Lachance DH, Parney IF, Buckner JC, Erickson BJ. Residual deep convolutional neural network predicts MGMT methylation status. *J Digit Imaging.* 2017;30(5):622–628. doi:10.1007/s10278-017-0009-z

4. Curtis EM, Moon RJ, Harvey NC, Cooper C. The impact of fragility fracture and approaches to osteoporosis risk assessment worldwide. *Bone.* 2017;104:29–38. doi:10.1016/j.bone.2017.01.024

5. Wilczek ML, Kälvesten J, Algulin J, Beiki O, Brismar TB. Digital X-ray radiogrammetry of hand or wrist radiographs can predict hip fracture risk - A study in 5,420 women and 2,837 men. *Eur Radiol.* 2013;23(5):1383–1391. doi:10.1007/s00330-012-2706-9

6. Dhainaut A, Rohde GE, Syversen U, Johnsen V, Haugeberg G. The ability of hand digital X-ray radiogrammetry to identify middle-aged and elderly women with reduced bone density, as assessed by femoral neck dual-energy X-ray absorptiometry. *J Clin Densitom.* 2010;13(4):418–425. doi:10.1016/j.jocd.2010.07.005

7. Rinaldi C, Bortoluzzi S, Airoldi C, et al. The early detection of osteoporosis in a cohort of healthcare workers: Is there room for a screening program? *Int J Environ Res Public Health.* 2021;18(3):1–7. doi:10.3390/ijerph18031368

8. Lee CS, Nagy PG, Weaver SJ, Newman-Toker DE. Cognitive and system factors contributing to diagnostic errors in radiology. *Am J Roentgenol.* 2013;201(3):611–617. doi:10.2214/AJR.12.10375

9. Akkus Z, Cai J, Boonrod A, et al. A survey of deep-learning applications in ultrasound: artificial intelligence–powered ultrasound for improving clinical workflow. *J Am Coll Radiol.* 2019;16(9):1318–1328. doi:10.1016/j.jacr.2019.06.004

10. Vlahiotis A, Griffin B, Stavros AT, Margolis J. Analysis of utilization patterns and associated costs of the breast imaging and diagnostic procedures after screening mammography. *Clin Outcomes Res.* 2018;10:157–167. doi:10.2147/CEOR.S150260

11. Simon AF, Holmes JH, Schwartz ES. Decreasing radiologist burnout through informatics-based solutions. *Clin Imaging.* 2020;59(2):167–171. doi:10.1016/j.clinimag.2019.10.014

12. Agrawal A, Gans JS, Goldfarb A. Artificial intelligence: The ambiguous labor market impact of automating prediction. *J Econ Perspect.* 2019;33(2):31–50. doi:10.1257/jep.33.2.31

13. Davenport T, Kalakota R. The potential for artificial intelligence in healthcare. *Futur Healthc J.* 2019;6(2):94–98. doi:10.7861/futurehosp.6-2-94

14. van Leeuwen KG, de Rooij M, Schalekamp S, van Ginneken B, Rutten MJCM. How does artificial intelligence in radiology improve efficiency and health outcomes? *Pediatr Radiol.* 2021;(level 6). doi:10.1007/s00247-021-05114-8

15. Chen MM, Golding LP, Nicola GN. Who will pay for AI? *Radiol Artif Intell.* 2021;3(3):e210030. doi:10.1148/ryai.2021210030

16. Golding LP, Nicola GN. A business case for artificial intelligence tools: The currency of improved quality and reduced cost. *J Am Coll Radiol.* 2019;16(9):1357–1361. doi:10.1016/j.jacr.2019.05.004

PART VII

EXPECTATIONS— RADIOLOGISTS' JOBS, JOB SATISFACTION, SALARY, AND ROLE IN SOCIETY

25 Clinical View

Amy Patel

SOCIETY'S PERCEPTION OF RADIOLOGISTS' FUTURE ROLE IN THE ERA OF AI APPLICATIONS

The majority of society's perception, including computer scientists and the general public, is that AI will improve the field of radiology and not replace radiologists [1]. So far, the reaction is, overall, positive and there is belief that the benefits exceed the risks in the utilization of AI by radiologists, but there are concerns of patients feeling that AI will replace doctors or, at the minimum, decrease interaction with them. Thus, feelings about the absence of the human factor due to AI are of concern to patients. However, patients currently have limited knowledge on the subject matter [2]. Certainly, as the utilization of AI continues to increase in society and therefore a certain level of understanding is achieved regardless of health literacy level, society will feel more comfortable with the utilization of AI in healthcare, radiology, and the fact that it is a tool to improve the everyday care provided by radiologists but certainly not replace them [3].

RADIOLOGISTS' EXPECTATIONS OF AI IN EVERYDAY PRACTICE

Radiologic imaging continues to increase in volume at a rapid pace, which in turn, is putting more pressure on the radiology workforce. This is exacerbated by a shortage of radiologists which is resulting in the burnout of those who are still in practice [4]. Therefore, there is hope that AI will assist in elevating patient care in terms of efficiency and diagnostic accuracy, allowing large amounts of imaging data to be digested in mere seconds. This can also help radiologists prioritize growing worklists and critical diagnoses [4]. Currently, approximately 30% of radiologists are using AI in clinical practice. Investing in a tool that potentially has inconsistent performance that could lead to decreased productivity are reasons why there are many who are in a "watchful waiting" period as well as a lack of reimbursement for most AI tools. However, approximately 25% of radiologists plan to incorporate AI in their clinical workflow in the next one to five years, acknowledging its value [5].

In my clinical practice, we have been utilizing AI breast ultrasound since 2019 as early adopters in private practice. Although we are still in the early stages, we are seeing remarkably positive changes. The radiologists who interpret breast imaging in my practice have over 100 years of breast-imaging interpretation experience combined and age ranges from 35 to 60. Regardless of training background or age, all feel that this tool improves practice efficiency and roughly 10–120 seconds saved per case due to the diagnostic confidence we feel with this clinical-decision support, second-opinion consult tool. Spending less time per case has allowed us to add two to three patients to the clinical schedule a day, and has reduced patient and referring provider result wait times. We have also seen a maintenance of cancer detection rate while reducing unnecessary biopsies, also in keeping with practices using similar technology [6]. For us, AI has been an enhancement to our practice in which we find *value*. However, not all feel the same.

There are skeptics in our profession, who believe that if AI replaces radiologists, there will be job cuts or there won't be a profession at all [7]. What about trainees? In a survey conducted involving radiology trainees, they were more likely to experience feelings of doubt regarding whether they would have pursued radiology if they had known the potential implications of AI on the specialty as a whole (P=0.0254). However, they were also more likely to make plans to learn about AI (P=0.0401) [7]. Also, we are seeing a great interest of AI in trainees, as evidenced by The American College of Radiology (ACR) Resident and Fellow Section AI Journal Clubs [8]. As AI is increasingly introduced and implemented in the field of radiology, more trust will be garnered, regardless of career level or practice type, and the utilization of it will be welcomed, regardless of the initial road being potentially rocky during the transition of adoption [7].

WHO WILL PAY FOR AI? A CURRENT AND FUTURE LOOK AT THE STATE OF AI VALUATION AND REIMBURSEMENT IN THE UNITED STATES

In an era of declining reimbursements, it can be difficult to justify radiology practices and institutions paying for the utilization of AI when not being reimbursed. Currently, only 30% of radiologists are using AI in practice, and lack of reimbursement is likely one of the reasons why AI utilization is still low in the field [5]. Therefore, the US healthcare payment system will need to

DOI: 10.1201/9781003095279-32

evolve to ensure that AI applications are valued and that we are being fairly reimbursed for the service(s) we are providing.

However, payment breakthroughs are on the horizon, as evidenced by CMS approving reimbursement for computer-assisted triage and patient notification with large-vessel stroke suspicion under the Inpatient Prospective Payment System (IPPS) and diagnosis of diabetic retinopathy under the Medicare Physician Fee Schedule (MPFS) [9, 10]. Economics experts Drs. Melissa Chen, Lauren Golding, and Gregory Nicola discuss in "Who Will Pay for AI" that "Payment for AI in the current fee-for-service environment may be challenging, and sustained adoption of AI may not occur within the framework of the IPPS and Physician Fee Schedule." They note that AI might have a bigger role in value-based systems including alternative payment models (APMs) a part of Medicare's Quality Payment Program [11].

Also promising, in 2022, two Category II Current Procedural Terminology (CPT) codes for nonelastographic quantitative analysis with ultrasound and one Category III CPT code for AI use in Computed Tomography (CT) became active. Although these are not Category I which typically are reimbursed by Medicare and other commercial payers, reimbursement is starting to be seen in certain states regardless of insurance carrier when it comes to AI breast ultrasound utilizing Category II CPT codes [12].

However, it will remain to be seen how valuation of AI services will affect the value of other imaging tools currently being reimbursed in radiology, particularly in an increasingly competitive healthcare environment where budget neutrality continues to be at the forefront. Therefore, one must ask the question: if radiologists are being reimbursed for utilizing AI making us better and potentially faster at what we do, will the value of other radiology services decrease or stay the same?

CONCLUSION

As Pakdemirli poignantly states, "wherever you see yourself in the spectrum, nobody can stop the advancement, innovations, and implantation of AI in radiology" [7]. Known in the house of medicine as the leaders of advanced technology and cutting-edge innovation, it is our duty in radiology, societal and beyond, to maintain an open mind and learn as much as we can when pertaining to AI. Hopefully, we can adopt as much as possible, regardless of geographic location, practice type, or payor mix, but continue to advocate for access to AI care for our patients and ensure we are being fairly reimbursed for the services we are providing.

REFERENCES

1. Yang L, Ene I, Belaghi R, et al. Stakeholders' perspectives on the future of artificial intelligence in radiology: A scoping review. *Eur Radiol* 32, no. 3 (2022). 1477–1495.

2. Gao S, He L, Chen Y, et al. Public perception of artificial intelligence in medical care: Content analysis of social media. *J Med Internet Res* 20, no. 7 (2020). 32673231.

3. Patel A. AI has the potential to elevate the way we practice. *ACR DSI Blog* (2019). https://www.acrdsi.org/DSIBlog/2019/09/10/00/36/AI-Has-the-Potential-to-Elevate-the-Way-We-Practice. Accessed 07/08/21.

4. Mello-Thoms C, Mello C. Clinical applications of artificial intelligence in radiology. *Br J Radiol* 96, no. 1150 (2023). 37099398.

5. Allen B. ACR data science institute artificial intelligence survey. *J Am Coll Radiol* 18, no. 8 (2021). 1153–1159.

6. Mango V.L, Sun M, Wynn R.T, Ha R. Should We Ignore, Follow, or Biopsy? Impact of artificial intelligence decision support on breast ultrasound lesion assessment. *Am J Roentgenol* 214, no. 6 (2020): 1445–1452. 32319794.

7. Pakdemirli E. Perception of artificial intelligence among radiologists. *Acta Radiol Open* 8, no.9 (2019). 31632696.

8. Participate in the RFS Journal Club. https://www.acr.org/Member-Resources/rfs/Journal -Club Accessed 07/08/21.

9. CMS Approves Reimbursement Opportunity for AI (2020). https://www.aidoc.com/blog/ cms-approves-reimbursement-ai/ Accessed 08/10/21.

10. HAP Radiology Billing and Coding Blog (2021). https://info.hapusa.com/blog-0/reimburse-ment-for-artificial-intelligence-in-radiology-is-more-than-just-billable-codes Accessed 08/10/21.

11. Chen M, Golding L, Nicola G. Who will pay for AI? *Radiology: Artificial Intelligence* 3, no.3 (2021). 34142090.

12. Smetherman D, Golding L, Moy L, et al. The economic impact of AI on breast imaging. *Journal of Breast Imaging* 4, no.3 (2022).

26 Technological View

Joshua D. Warner, Gian Marco Conte, Andrew J. Missert,
Panagiotis Korfiatis, and Timothy L. Kline

OVERVIEW

The general societal perception of AI is drastically different from reality [2]. Unfortunately, this widespread misperception significantly impacts and skews the expectations of radiologists. In the media, there exists an endless stream of stories showing how computers will replace radiologists [3–5]. In addition, debates at societal conferences constantly feature the radiologist's job as a sort of survival curve, with predictions of only a few years until the profession is deemed obsolete. Such discussions are not only sensationalist but also psychologically draining for practicing radiologists. One fallout from such exercises is that those who see these same stories and discussions often decide not to pursue radiology, thereby actively harming the recruitment of new candidates into the field.

Besides the psychological components around the role of AI in radiology, incorrect assumptions about what radiologists do and how valuable their contributions are may affect policies and reimbursements [6]. For example, relative value units (RVUs), a measure of value used in the Medicare reimbursement formula for physician services, could be misvalued due to incorrect assumptions about AI. This is because the RVUs are assigned by an American Medical Association (AMA) committee (the AMA Specialty Society Relative Value Scale Update Committee) based on the professional physician work required for a medical procedure. The primary components of the physician's work include the time it takes to perform the service, the level of technical skill required, and the mental effort and judgment necessary. This 32-person committee has a single radiologist member and has the power to devalue RVUs in radiology due to misperceptions about AI's real-world effects. Therefore, educating those setting policy and RVU rates about the capabilities of AI should be a priority.

Building on these prior misconceptions, there is also the element of radiologist salaries and the value radiologists provide to the medical system at large. The proliferation of mid-level providers often results in high utilization of imaging. Radiologists may be the first comprehensively trained clinician participating in the patient's care, especially in urgent and emergent settings. Radiologists are increasingly asked about management of the findings they report by such mid-level providers. Imaging utilization is increasing, but the quality and value are also growing as diagnostic yield and quality improves with the development of more powerful techniques, making radiologists essential partners with clinical and surgical referring providers. AI solutions can have a crucial role in enabling this transition to essential partners. In the early phases we experience today, AI can help automate tedious or repetitive noninterpretive tasks (protocoling, measurements, etc.), freeing radiologists' time to contribute in unique and valuable ways in the clinical workflow [7, 8].

TECHNICAL VIEW

AI has already had impacts in many areas of radiology [9, 10]. These include workflow enhancement, image reconstruction, image quality improvement, image synthesis, quantification, disease detection and classification, and report generation. In general, these algorithms require many curated examples to train a model to perform at the level required for a particular task. For example, an image segmentation algorithm may require hundreds if not thousands of examples, and a classification algorithm may require thousands if not millions of diverse samples.

Unfortunately, these approaches will likely never work for very rare diseases, patients with deformities, unique posttraumatic or postsurgical anatomy, non-standardized or degraded image acquisitions, and other situations commonly encountered in practice. However, society expects radiologists to perform just as well in the face of technical and patient variations. An important question to be asked is: What types of questions AI can help us answer? For image processing, questions like finding an organ, measuring morphological parameters, and identifying stroke are ripe for the utilization of AI. However, answering open-ended questions from referring providers such as discussing the best surgical approach, or the optimal follow-up approach (biopsy, additional imaging technique and modality), is a much bigger ask from AI and will likely not be fully realized for quite some time—possibly remaining out of reach until or unless AGI is developed.

DOI: 10.1201/9781003095279-33

Beyond these capabilities, the current outputs of an AI algorithm are not eloquent. For example, the output of a classification algorithm currently consists of the machine learning model's confidence in the presence of a disease, quite a departure from the current standard radiology report. Such AI approaches require substantial high-quality annotated training data. However, any time there is a change in clinical practice, such data will not be available for some time. In this sense, algorithmic approaches will always lag behind the cutting edge of clinical practice. For example, when a new publication proposes a new system to stage or classify disease, radiologists can read the paper and employ the approach immediately. In contrast, an AI system will need several years to achieve similar performance (in the best-case scenario), and it will also present practical challenges for its implementation [11–13], starting with the data preparation [14].

These realizations have led to more sobering conversations at recent conferences, in contrast to the ebullient enthusiasm of only a few years ago, in many cases focusing on how promising AI algorithms can fail in generalization and considering regulation of AI. Discussions of another AI winter are justified, with a significant factor being the inherent difficulty in generalization of the AI models. This issue is currently a major focus in the scientific community [15]. We have excellent algorithmic approaches for finding patterns and for making predictions. However, medical imaging is not currently standardized at the level required for the widespread adoption of an AI developed at one institution. Differences in images produced on scanners from different vendors, workstation software versions, reconstruction kernels, to name only a few, all impact model performance and degrade generalizability. Monitoring the algorithm and archiving the results in cases of AI used to reconstruct medical images are challenges that need to be addressed. Technological advances like federated learning will hopefully enable the training of algorithms in diverse datasets enabling more generalized algorithms [16–19].

Unfortunately, we are not archiving the data needed to fully realize the potential of these techniques, including raw k-space data (MR) and sinograms (CT). In all existing AI solutions, we are using reconstructed images. The sooner we move to a data archival and retrieval system (DACS) model instead of a picture archival and communication system (PACS), the better.

The current classes of available AI algorithms will not replace but instead support practicing radiologists. For example, automated search for specific high-impact findings like stroke and pulmonary embolism are already FDA-approved AI solutions available for clinical use. Tuning the sensitivity of these algorithms can facilitate their utility in worklist prioritization (i.e., review first those cases which have high probability of a significant finding). Also, natural language processing (NLP) can help collect train-of-thought dictation into a concise report. Automatic quality control and automated technologist feedback can also decrease radiologist interruptions.

Some perspective is also essential to realize the current state of AI in Radiology. For example, ECGs (which have had significant investment in AI-based approaches for interpretation) still require physician overreads for billing purposes. This problem is perhaps one of the most straightforward (i.e., a 1D digital signal processing problem, with only kilobytes of input data). On the other hand, medical imaging is a 2D–5D problem, with megabytes to gigabytes of input data. It is evident that radiologists will not be replaced any time soon.

Lastly, disruptive transformation is always possible in such a rapidly evolving field. Our viewpoint is limited by the state of the art at time of writing. But, to replace radiologists would require verifiable strong AI (i.e., AGI). At the time of writing, AGI does not exist, and no certain pathways to AGI are known [1]. However, if AGI were to be developed, the implications would be massive, with transformative effects felt throughout society, far beyond radiology.

SUMMARY

In this chapter, we highlighted the current state of AI in the Radiology practice. We focused on the technological viewpoint to portray the realities of what we can expect and anticipate from AI-based systems. The misperception of AI affects current radiologists' expectations and the recruitment of future practitioners. Unfortunately, this is a direct result of expectations and concerns being divergent from reality. In general, there is a sharp line between strong (i.e., artificial general intelligence or AGI) and weak, task-specific AI (i.e., artificial narrow intelligence). For AI in radiology, everything we currently have is weak AI. AGI does not currently exist, and humanity has no clear path to developing it. It is not clear that AGI represents more than science fiction. As such this viewpoint is focused on the current state of the art. While we can consider potential disruptive innovations such as strong AI which would indeed significantly impact the practice of radiology, such disruptions would also be transformative for all of society and the world. For the foreseeable future, the radiology practice will be dominated by human reasoning, with light integration of AI in the loop.

ABBREVIATIONS

AGI: artificial general intelligence
AI: artificial intelligence
AMA: American Medical Association
RVU: Relative value units

REFERENCES

1. Fjelland R. Why general artificial intelligence will not be realized. Humanities and Social Sciences Communications. *Palgrave*. 2020;7(1):1–9.

2. Langlotz CP. Will artificial intelligence replace radiologists? Radiology: Artificial intelligence. *RSNA*. 2019;1(3):e190058.

3. Obermeyer Z, Emanuel EJ. Predicting the future - big data, machine learning, and clinical medicine. *N Engl J Med. ncbi.nlm.nih.gov.* 2016;375(13):1216–1219.

4. Susskind R, Susskind D. Technology will replace many doctors, lawyers, and other professionals. *Harv Bus Rev.* 2016;11.

5. Morgenstern M. Automation and anxiety: Will smarter machines cause mass unemployment. *The Economist June.* 2016;25.

6. Chen MM, Golding LP, Nicola GN. Who Will Pay for AI? Radiology: Artificial intelligence. *RSNA*. 2021;3(3):e210030.

7. Pesapane F, Codari M, Sardanelli F. Artificial intelligence in medical imaging: threat or opportunity? Radiologists again at the forefront of innovation in medicine. *Eur Radiol Exp.* 2018;2(1):35.

8. Hricak H. 2016 New horizons lecture: Beyond imaging-radiology of tomorrow. *Radiology.* 2018;286(3):764–775.

9. Montagnon E, Cerny M, Cadrin-Chênevert A, et al. Deep learning workflow in radiology: A primer. *Insights Imaging.* 2020;11(1):22.

10. Cheng PM, Montagnon E, Yamashita R, et al. Deep learning: An update for radiologists. *Radiographics.* 2021;41(5):1427–1445.

11. He J, Baxter SL, Xu J, Xu J, Zhou X, Zhang K. The practical implementation of artificial intelligence technologies in medicine. *Nat Med. Springer Science and Business Media LLC.* 2019;25(1):30–36.

12. Recht MP, Dewey M, Dreyer K, et al. Integrating artificial intelligence into the clinical practice of radiology: challenges and recommendations. *Eur Radiol.* 2020;30(6):3576–3584.

13. Juluru K, Shih H-H, Keshava Murthy KN, et al. Integrating Al algorithms into the clinical workflow. Radiology: Artificial intelligence. *RSNA*. 2021;3(6):e210013.

14. Willemink MJ, Koszek WA, Hardell C, et al. Preparing medical imaging data for machine learning. *Radiology.* 2020;295(1):4–15.

15. Eche T, Schwartz LH, Mokrane F-Z, Dercle L. Toward generalizability in the deployment of artificial intelligence in radiology: Role of computation stress testing to overcome underspecification. *Radiol Artif Intell. RSNA.* 2021;3(6):e210097.

16. Sheller MJ, Edwards B, Reina GA, et al. Federated learning in medicine: facilitating multi-institutional collaborations without sharing patient data. *Sci Rep.* 2020;10(1):12598.

17. Kaissis GA, Makowski MR, Rückert D, Braren RF. Secure, privacy-preserving and federated machine learning in medical imaging. *Nat Mach Intell. Nature Publishing Group.* 2020;2(6):305–311.

18. Dayan I, Roth HR, Zhong A, et al. Federated learning for predicting clinical outcomes in patients with COVID-19. *Nat Med.* 2021. http://dx.doi.org/10.1038/s41591-021-01506-3.

19. Rieke N, Hancox J, Li W, et al. The future of digital health with federated learning. npj Digital Medicine. *Nat Res.* 2020;3(1). http://dx.doi.org/10.1038/s41746-020-00323-1.

27 Societal View

Benard Ohene Botwe and Raphael Nicholas Mayeden

INTRODUCTION

Following the discovery of X-rays in 1895 by Wilhelm Rontgen, the practice and workflow of radiology staff has continued to expand to the present day. Radiologists' role in society has become even more diverse and crucial with the dramatic technological improvements of highly sophisticated equipment and systems in medical imaging [1]. More recent technological advancements have found new focus on increasing integration of artificial intelligence (AI) algorithms into medical imaging modalities [2–7]. AI can mimic the cognitive functions of humans, such as in learning and problem-solving, and perform more mundane tasks faster and more consistently than humans [8–11]. Currently, there are more than 150 AI products for radiology in the USA and European markets which have been approved by the Food and Drug Administration (FDA) or have European Conformity (CE) [12]. Radiologists and other members of radiology departments continue to deal with the influx of sophisticated equipment and systems in medical imaging that affect their core duties and their role in society. This chapter presents a discussion on the expectations of some of the radiologists' role in society as implementation of AI advances. Figure 27.1 is an infographic depicting aspects of expected impact of AI on radiologists' role in society.

Expectations of AI Impact on Radiologists' Image Interpretation

The public and medical societies have long known and expect radiologists to be the main health professionals with responsibility for providing image interpretations for diagnosing and treating medical conditions and injuries. The high expertise of these professionals provides referring clinicians with the confidence to request radiological reports on radiological examinations. However, many parts of the world experience shortages of radiologists. The shortage of radiologists and the technological advancement in computation have, among other things, led to the application of artificial intelligence (AI) in image interpretations. AI tools are now being used to perform tasks that are traditionally performed by humans and are associated with human intelligence [13]. One key area where AI tools have dominated and have been mostly employed in medical imaging is image interpretation and reporting, among others [14–16]. Computer algorithms are being trained to interpret images, in similar ways to how human brains process images, thereby helping in triaging abnormal images from normal, enhancing efficiency and speed of workflow, reducing reporting turnaround time, and improving patient management [17, 18]. As a result of the demonstrated remarkable progress in image-interpretation tasks, some images which ordinarily would have been reported by a radiologist are sent to referring clinicians without radiological reports, hence taking up the radiologist's role to some extent [17]. For instance, in Ghana, the national government has worked with an equipment vendor to install a commercially developed AI tool—Computer-Aided Detection for Tuberculosis (CAD4TB)—to help with prompt diagnosis and treatment of tuberculosis from chest radiographs. This was introduced to improve TB detection within highly affected regions of the country, and the system automatically predicts TB by generating heatmap scores ranging between 0 and 100 to indicate the extent of abnormalities [19]. This process then provides opportunity for the disease to be classified into low or high risk, with good or poor prognosis, thereby allowing radiographic technicians to send heatmap reports to referring clinicians for interim management without waiting for the radiological report. This is one example of AI-assisted imaging evaluation which is taking place in some developing countries [19].

In some circles, radiologists are perceived as "invisible image readers," and this perception has been heightened further with the advent of teleradiology, where radiologists report more and more complex imaging studies remotely with virtually no direct contact with referring clinicians or patients. The lack of clinical visibility with the advancement of AI fed into the belief that radiologists' role as image interpreters would be taken over by "machines." In the study by Coppola et al. [20], which surveyed 1032 radiologists, some (18.9%) respondents feared that computers will replace radiologists for reporting of imaging examinations. In a survey of radiographers in Ghana, some 53.6% of respondents expect that AI would take over the image interpretations jobs of the radiologist [6]. On the contrary, in a traditionally human-centered discipline such as healthcare, new technologies are introduced not necessarily to compete with the human workforce over image interpretation, but to supplement and support human intelligence efficiently, and to improve

DOI: 10.1201/9781003095279-34

Figure 27.1 Expectations of AI impact of radiologists and their roles

clinical decision-making. This is understood by many radiologists, as Coppola et al. [20] reported that a majority of radiologists surveyed recognized AI as a means to aid daily working practice (66.2%), lower diagnostic error rate (73.0%), optimize radiologists' work (67.9%), enrich radiological reports with quantitative data (70.6%), use the superior capabilities of computers to detect molecular markers, and to perform such tasks as detection and characterization of pulmonary nodules (61.7%), etc.

While AI-enabled technologies may be superior in interpreting sensory perceptual information in a bottom-up fashion, such algorithms currently remain deficient in contextual knowledge and the complex rational associations needed in processing clinical information. AI is therefore not likely to take over more complex image interpretation tasks that are currently performed by radiologists or replace radiologists in the foreseeable future [21]. A large portion of patients value the important human interactions they receive in the radiology department and are skeptical that computers would replace radiologists [21]. AI technologies thrive on the availability of data for curation and the development of multilayered algorithms for machine learning. The majority of illnesses are caused by common disorders for which data abounds. But for rare and evolving diseases for which there is paucity of data for building AI imaging systems, the interpreting expertise of radiologists applying their clinical knowledge to disease remain crucial; although as more complex technologies are developed, the job description of a radiologist is expected to change. The influx of more sophisticated AI systems being developed for use in clinical practice in medical imaging is expected to increasingly revolutionize the practice of radiology in image interpretation in the coming decades [17, 18, 22]. Recent surveys have also suggested that the radiology community itself expects alteration to the duties of the workforce [20]. In order to remain relevant, however, radiologists would need to improve upon their clinical visibility, such as by participating in multidisciplinary team (MDT) clinical activities and interacting more with patients.

Expectations of AI Impact on Residency Training/Education

Radiological residency training is crucial in creating future radiologists, and is not expected to be discontinued, despite sentiments to the contrary. However, as it is now anticipated that AI will likely have a profound impact on the future practice of radiology and other medical subspecialties, concerns have arisen about reduced learning opportunities for radiologists, thus leading to decreased interest to pursue radiology by some future doctors and society [20, 23, 24]. For example, in a study conducted by Sit et al. [24], among 484 medical students in 19 UK medical schools, 49% reported that they were less likely to consider a career in radiology due to the integration of AI in the field. Similarly, a study in Canada also reported that a significant proportion (one sixth) of Canadian medical students were less likely to consider radiology as a career due to fear of replacement by the perceived success of AI. The general perception is that AI would replace the job of image interpretation, hence making their future profession less lucrative. The Canadian medical students, in particular, according to Gong et al. [23] were more concerned about job displacement, rather than replacement, of radiologists which could lead to reduced workforce demands. With so much uncertainty about the impact of AI in radiology, some authors such as Andrew Ng and Geoffrey Hinton have also expressed their fears about the impact of AI in radiology training [11, 25–27]. Hinton, a legendary computer scientist, in particular, asserted in 2016 that "people should stop training radiologists now. It's just completely obvious that within 5 years, deep learning is going to do better than radiologists" [11, 25–27]. However, evidence also suggests that it very unlikely that AI would entirely take over or displace radiologists of their work [28]. This is because although there has been recent major progress in AI tools, "the progress in performance has not yet materialized in every aspect of radiology as … predicted. Rather, specific 'narrow' applications have proven successful; and generalized superhuman performance remains elusive" [25, p. 2]. Thus, until extraordinary technologies that substantially surpass radiologists' abilities are invented that would wipe out radiologists' duties, radiologists' work would exist and likely even improve, and residency training will continue, with the incorporation of AI training. Therefore, as the field of healthcare AI continues to expand, it is increasingly apparent that AI education for clinicians and medical students will be needed [24]. This education will help to dispel some of the myths surrounding AI in radiology so as to encourage physicians-in-training to take up radiology as a career.

In order to make the discipline of radiology and radiology residency training more attractive to medical students, radiologists should be actively involved in medical student education and also participate in general clinical rounds with other clinical colleagues. Another expected role is the provision of teaching services to radiology residents, junior doctors in non-radiology disciplines, as well as radiographical technicians, nurses, and other paramedical staff. Radiologists should guide and become more integral to the development of curricula and guidelines in these other healthcare related disciplines.

Expectations of AI Impact on Radiologist Professional Outlook/Salaries

AI tools are expected to impact radiologists' professional outlook and salaries. Whilst some radiologists are uncertain of their future careers with AI, a section of radiologists perceive that AI in radiology would lead to a diminished professional reputation of radiologists compared with non-radiologists. These reasons are compounded by the fear of losing visibility and professional authoritativeness as AI takes over the radiologist's traditional role, with attendant uncertainties over recruitment opportunities [20, 29]. Correspondingly, the European Society of Radiology (ESR) has raised concerns over the strong dependence of future radiologists' computer-assisted diagnostic systems with "potentially deleterious consequences" [20]. Consequently, there have been calls for policies to be enacted to regulate the utilization of AI in radiology practice to protect radiologists' profession [20].

The counter view to this assertion is that radiologists do more than just image interpretation. Radiologists are clinicians, consulting with colleague physicians and making inputs in clinical management decisions, as well as providing therapeutic and interventional services, such as tailoring the technical parameters of imaging examinations to be performed on specific patients, correlating imaging findings with clinical history and other ancillary test results, and discussing procedures and results with referring clinicians and patients. Therefore, AI technologies do not threaten the job security and professional outlook of radiologists, but rather may help to improve diagnostic accuracy, increase efficiency, and lower imaging costs in a win–win for radiologists, their clinical colleagues, and patients [30].

Expectations of AI Impact on Radiologists to Take Up Role Extension

As artificial intelligence continues to integrate into radiology, enhancing radiologists with additional skills, fellow clinicians and society may also expect radiologists to extend their roles in healthcare. AI capabilities not only influence traditional radiological activities, but expand to additional areas including radiomics, imaging biobanks, clinical decision support systems, and workflow management [11, 31]. Studying the trajectory of AI's progresses and creating new niches for the radiological professional would be well advised. This is important for radiology's future, as, like it or not, AI is here to stay, and AI's substantial medical and productivity benefits will irrevocably change the field. Consequently, radiologists can be expected to start re-training in the areas of information technology, computer science, and software usage while collaborating with software developers and computer engineers for appropriate integration of AI in radiology for patients' interests. Although AI would be largely owned and controlled by developers, companies, and governments, radiologists would need to be in the forefront of protocol development and standardization. Assumption of managerial roles by radiologists in clinical environments would promote advocacy and involvement in critical decision-making, and the drafting of legal and regulatory frameworks guiding the development, acquisition, and utilization of AI imaging technologies.

EXPECTATIONS OF AI IMPACT ON VETTING OF RADIOLOGY EXAMINATIONS

Currently radiologists as practitioners play vital roles in evaluating and deciding on the appropriateness of requested radiological procedures. However, this role is also expected to be impacted by AI. This is because AI software systems have the capabilities to accurately evaluate radiological referrals much faster and could eventually be used to save costs and free up radiologists' time that could be better used to interact with clinicians or patients. With more advances in AI, it is further expected that the technology would help with automated vetting of referrals, verifying appropriate clinical indications, and protocoling the appropriate imaging modality and techniques to be employed via interactions with the electronic health record [32].

Expectations of AI Impact on Radiologists' Research Activities

Several studies have shown that AI tools would accelerate and transform radiological research [5–8, 11, 20, 25]. These new technologies are envisaged to improve data collection, process workflow and analysis, and thereby improving the research activities of radiologists and allied professionals. The ESR [11, p. 6], posits that

> a vast amount of AI research is ongoing; image interpretation is an attractive target for researchers, given that the tasks involved (at least in part) involve analysis of large amounts of data to produce an output. Radiologists cannot, and should not, wish this research away, but rather embrace it and integrate it as much as possible into the daily work, guiding AI research directions to ensure the maximum clinical benefit to patients from new developments.

Improved Value of Radiology Services

Considering the context of value-based health care, AI tools are also expected to create value by either reducing costs or decreasing required time of radiology services rendered by radiologists, which may in turn help to improve health outcomes [9, 33]. AI can increase workflow efficiency and aid in performing time-consuming and iterative tasks, thereby allowing radiologists to complete more complex duties, possibly with reduced resources [34]. Studies using AI tools have shown reduction in turnaround time for reporting critical findings on chest radiographs from 80 minutes to 30–35 minutes [35]. In particular, AI could improve radiology departmental efficiency, such as by scheduling and planning of radiology procedures including imaging guided biopsies, and better predict "no-shows," hence helping to reduce the lost opportunity costs of unfulfilled appointments [9]. As more sophisticated diagnostic equipment with improved sensitivities emerges, AI tools trained to characterize incidental findings can play vital roles in guiding the early detection of potentially dangerous lesions while saving time and cost of unnecessary subsequent investigations and interventions [9]. For instance, AI has been used to detect and characterize lung nodules in computed tomography (CT) studies, faster and with higher diagnostic accuracy [36, 37].

EXPECTATIONS OF AI IMPACT ON RADIOLOGISTS' RADIATION DOSE OPTIMIZATION

Due to potentially harmful side effects of radiation exposure associated with certain radiologic modalities, the public and the medical profession have relied on radiologists, radiologic technicians, medical physicists, biomedical engineers, and all radiation protection stakeholders to ensure radiation protection of patients. Radiologists play critical roles in balancing the risks-versus-reward calculations in obtaining radiologic studies, and to some extent, optimization of radiation exposure to patients and minimizing exposure for staff. More recently and with increasing frequency, AI tools such as deep learning are being used to improve and accelerate image reconstruction and post-processing [37, 38]. Such AI technologies facilitate diagnostic image quality with lower radiation doses [9]. Hence, more effective dose optimization is another area of expected impact of radiologists' role in society from implementation of AI in medical imaging/radiology continues.

CONCLUSION

The field and practice of radiology continue to evolve rapidly. Like it or not, AI is here with us and is increasing in power and reach, driven by increasing abundance of data and rapid technical improvements, particularly in deep learning. Computer-assisted quantification, triage, and diagnostic tools are already being used in radiology. Over the next decade, it is expected that more models will become widely available and adopted, with radiologists increasingly utilizing algorithmic support regardless of their area of subspecialty practice. AI-based models will be better able to detect and identify rapidly declining disease states, perform quantitative evaluation of lesions, and predict morbidity and mortality. While AI will not replace radiologists for the foreseeable future, AI will affect the way radiology is practiced. As a corollary to this, radiologists should not sit aloof, nor feel threatened, but rather position themselves for the gains that AI offers. With effective implementation of AI, radiologists can practice at the peak of their profession, and those who adopt early will have a head start.

REFERENCES

1. Boland GW. Stakeholder expectations for radiologists: Obstacles or opportunities? *J Am Coll Radiol*. 2006 Mar; 3(3):156–163. https://doi.org/10.1016/j.jacr.2005.10.008

2. Oren O, Gersh BJ, Bhatt DL. Artificial intelligence in medical imaging: Switching from radiographic pathological data to clinically meaningful endpoints. *Lancet Digit Health* 2020; 2(9):e486–e488. https://doi.org/10.1016/S2589-7500(20)30160-62

3. Wagner JB. Artificial intelligence in medical imaging. *Radiol Technol*. 2019; 90(5):489–501.

4. Pearce C, McLeod A, Reinhart N, Whyte R, Deveny E, Shearer M. Artificial Intelligence and the clinical world: a view from the front line. *Med J Aust*. 2019; 210(6):S38–40.

5. Botwe BO, Akudjedu TN, Antwi WK, Rockson P, Mkoloma SS, Balogun EO, Elshami W, Bwambale J, Barare C, Mdletshe S, Yao B, Arkoh S. The integration of artificial intelligence in medical imaging practice: Perspectives of African radiographers. *Radiography (Lond)*. 2021 Aug; 27(3):861–866. https://doi.org/10.1016/j.radi.2021.01.008

6. Botwe BO, Antwi WK, Arkoh S, Akudjedu TN. Radiographers' perspectives on the emerging integration of artificial intelligence into diagnostic imaging: The Ghana study. *J Med Radiat Sci*. 2021 Feb 14. https://doi.org/10.1002/jmrs.460

7. Antwi WK, Akudjedu TN, Botwe BO. Artificial intelligence in medical imaging practice in Africa: a qualitative content analysis study of radiographers' perspectives. *Insights Imaging*. 2021 Jun 16; 12(1):80. https://doi.org/10.1186/s13244-021-01028-z

8. Abuzaid MM, Elshami W, Tekin H, Issa B. Assessment of the willingness of radiologists and radiographers to accept the integration of artificial intelligence into radiology practice. *Acad Radiol*. 2020 Oct 29:S1076-6332(20)30553-5. https://doi.org/10.1016/j.acra.2020.09.014

9. van Leeuwen KG, de Rooij M, Schalekamp S. et al. How does artificial intelligence in radiology improve efficiency and health outcomes?. *Pediatr Radiol* (2021). https://doi.org/10.1007/s00247-021-05114-8.

10. Thwaites D, Moses D, Haworth A, Barton M, Holloway L. Artificial intelligence in medical imaging and radiation oncology: Opportunities and challenges. *J Med Imaging Radiat Oncol.* 2021 Aug;65(5):481–485. https://doi.org/10.1111/1754-9485.13275

11. European Society of Radiology (ESR). What the radiologist should know about artificial intelligence – an ESR white paper. *Insights Imaging.* 2019;10:44. https://doi.org/10.1186/s13244-019-0738-2.

12. Diagnostic Imaging Analysis Group. *AI for Radiology. Products.* Radboud University Medical Center; 2020. https://www.aiforradiology.com. Accessed 25 July 2021

13. Ranschaert ER, Duerinckx AJ, Algra P, Kotter E, Kortman H, Morozov S. Advantages, challenges, and risks of artificial intelligence for radiologists. In: Ranschaert ER, Morozov S, Algra PR, editors. *Artificial Intelligence in Medical Imaging: Opportunities, Applications and Risks.* Cham: Springer; 2019. p. 329e46.

14. Morozov S, Ranschaert ER, Introduction Algra PR. Game changers in radiology. In: Ranschaert ER, Morozov S Algra PR, editors. *Artificial Intelligence in Medical Imaging: Opportunities, Applications and Risks.* Cham: Springer, 2019. p. 3–5.

15. Ranschaert ER, Duerinckx AJ, Algra P, Kotter E, Kortman H, Morozov S. Advantages, challenges, and risks of artificial intelligence for radiologists. In: Ranschaert ER,Morozov S Algra PR, editors. *Artificial Intelligence in Medical Imaging: Opportunities.* Cham: Springer, Applications and Risks, 2019. p. 329–346.

16. Murphy A, Liszewski B. Artificial intelligence and the medical radiation profession: how our advocacy must inform future practice. *J Med Imaging Radiat Sci.* 2019; 50 (4S2): S15–19. https://doi.org/10.1016/j.jmir.2019.09.001

17. Hosny A, Parmar C, Quackenbush J, Schwartz LH, Aerts HJWL. Artificial intelligence in radiology. *Nat Rev Cancer.* 2018;18(8):500–510. https://doi.org/10.1038/s41568-018-0016-5.

18. Thrall JH, Li X, Li Q et al. Artificial intelligence and machine learning in radiology: opportunities, challenges, pitfalls, and criteria for success. *J Am Coll Radiol.* 2018; 15(3 Pt B):504–508.

19. Wuni A.-R, Botwe BO, Akudjedu TN. Impact of artificial intelligence on clinical radiography practice: Futuristic prospects in a low resource setting. https://doi.org/10.1016/j.radi.2021.07.021.

20. Coppola F, Faggioni L, Regge D, Giovagnoni A, Golfieri R, Bibbolino C, Miele V, Neri E, Grassi R. Artificial intelligence: Radiologists' expectations and opinions gleaned from a nationwide online survey. *Radiol Med.* 2021 Jan;126(1):63–71. https://doi.org/10.1007/s11547-020-01205-y

21. Ongena, Y.P., Haan, M., Yakar, D. et al. Patients' views on the implementation of artificial intelligence in radiology: Development and validation of a standardized questionnaire. *Eur Radiol.* 2020; 30:1033–1040. https://doi.org/10.1007/s00330-019-06486-0.

22. LeCun Y, Bengio Y, Hinton G. Deep learning. *Nature* 2015; 521:436–444.

23. Gong B, Nugent JP, Guest W et al. Influence of artificial intelligence on Canadian medical students' preference for radiology specialty: A national survey study. *Acad Radiol.* 2019; 26:566–577.

24. Sit, C., Srinivasan, R., Amlani, A. et al. Attitudes and perceptions of UK medical students towards artificial intelligence and radiology: A multicentre survey. *Insights Imaging*. 2020; 11:14. https://doi.org/10.1186/s13244-019-0830-7.

25. Kelly B, Judge C, Bollard SM, Clifford SM, Healy GM, Yeom KW, Lawlor A, Killeen RP. Radiology artificial intelligence, a systematic evaluation of methods (RAISE): A systematic review protocol. *Insights Imaging*. 2020 Dec 9;11(1):133. https://doi.org/10.1186/s13244-020-00929-9

26. Morgenstern M. Automation and anxiety. *The Economist*. 2017. https://www.economist.com/special-report/2016/06/25/automation-and-anxiety. Accessed 7 Aug 2017.

27. Mukherjee S. A.I Versus M.D. *New Yorker*. 2017. http://www.newyorker.com/magazine/2017/04/03/ai-versus-md. Accessed 7 Aug 2017.

28. Langlotz CP. Will artificial intelligence replace radiologist? *Radiology: Artificial Intelligence*. 2019; 1:e190058.

29. Eltorai AEM, Bratt AK, Guo HH. Thoracic radiologists' versus computer scientists' perspectives on the future of artifcial intelligence in radiology. *J Thorac Imaging*. 2019. https://doi.org/10.1097/RTI.0000000000000453.

30. Davenport TH, Dreyer KJDO. AI will change radiology, but it won't replace radiologists. 2018. https://hbr.org/2018/03/ai-will-change-radiology-but-it-wont-replace-radiologists. Accessed 20 Aug 2021

31. Nilsson NJ. *Artificial Intelligence: A New Synthesis*. Morgan Kaufmann Publishers, Inc; 1998.

32. Hardy M, Harvey H. Artificial intelligence in diagnostic imaging: Impact on the radiography profession. *Br J Radiol*. 2020;93(1108):20190840. https://doi.org/10.1259/bjr.20190840.

33. Chong LR, Tsai KT, Lee LL et al. Artificial intelligence redictive analytics in the management of outpatient MRI appointment no-shows. *AJR Am J Roentgenol*. 2020; 215:1155–1162.

34. Gore JC. Artificial intelligence in medical imaging. *Magn Reson Imaging*. 2020 May; 68:A1–A4. https://doi.org/10.1016/j.mri.2019.12.006

35. Baltruschat I, Steinmeister L, Nickisch H et al. Smart chest X-ray worklist prioritization using artificial intelligence: A clinical workflow simulation. *Eur Radiol*. 2021; 31:3837–3845. https://doi.org/10.1007/s00330-020-07480-7.

36. Liu B, Chi W, Li X, et al. Evolving the pulmonary nodules diagnosis from classical approaches to deep learning-aided decision support: Three decades' development course and future prospect. *J Cancer Res Clin Oncol*. 2020; 146:153–185. https://doi.org/10.1007/s00432-019-03098-5.

37. Hsieh J, Liu E, Nett B et al. *A New Era of Image Reconstruction: True Fidelity™ Technical White Paper on Deep Learning Image Reconstruction*. GE Healthcare Online Document; 2019. https://www.gehealthcare.com/-/jssmedia/040dd213fa89463287155151fdb01922.pdf. Accessed Aug 24, 2021.

38. Willemink MJ, Noël PB. The evolution of image reconstruction for CT — from filtered back projection to artificial intelligence. *Eur Radiol*. 2019; 29:2185–2195.

28 Financial View

Mohammad Mirza-Aghazadeh-Attari and Armin Zarrintan

INTRODUCTION

Artificial intelligence is considered to be the centerpiece of the fourth industrial revolution, following in the path of innovations such as steam power, electricity, and computers. Like its predecessors, the wide-scale application of artificial intelligence could have significant impacts on the labor market of any sector of the economy [1]. Notably, it is now speculated that AI will be an even greater force in shaping the future of economies, as it will not only be a simple tool that increases productivity in the hands of low- and medium-skilled workers, but instead will be a substitute to these workers, and could even threaten the jobs of high-skilled ones [2].

Much of this speculation has been caused by lack of empiric evidence and the nature of the matter at hand, which is complicated to study, primarily owing to the fact that true economic impacts of innovations as profound as AI (or electricity, computers, and steam power) may take decades to unravel [3].

From a purely economic perspective, utilizing AI systems will only levy marginal operating costs, and employers will skip paying salaries, healthcare costs, and other marginal wage costs. Empirically, the existing evidence suggests that introduction of intelligent machines results in reduced wages and reduced employment at local levels in the short term, with studies lacking consensus on the national economic effects [4, 5]. Some retrospective investigations have failed to show the positive impact of artificial intelligence on gross domestic product, productivity, and the labor market, whereas others predict an increase in employment, labor productivity, and business opportunities [4, 6].

Studies focusing on the previous revolutions in sectors such as agriculture and industry have found that reduced marginal costs were coupled with increased productivity and demand. In contrast, the healthcare market has been traditionally different, as historically increased demand has led to changes in the labor market (not vice versa) and compelled providers to reduce marginal costs [6, 7]. One example of this has been the delegation of manual tasks to non-physician healthcare workers, such as obtaining blood samples or certain injections. An ever-increasing demand has increased the opportunity cost of physicians, which has led providers to consider less-educated workers to perform the same tasks [8]. The same drive is currently persuading an increasing number of health sector managers to consider AI solutions for old problems. It is estimated that the AI-associated healthcare market grew rapidly in recent years (with an annual growth rate of 40%) to reach a size of $6.6Bn USD in 2021 and is projected to grow at similar rates for the next half-decade to reach more than $45Bn USD in 2026 [9]. Interestingly, observers of international health markets have found that as time has passed, AI solutions have been increasingly used in different applications ranging from patient data management and risk-analysis solutions for health maintenance organizations to wearables used by patients [10].

In the medical profession, there is a consensus that due to the nature of radiology, it will be the first field to be dramatically impacted by artificial intelligence. This notion is further supported by the vast array of available evidence that hints at the potential efficacy of automated computer-assisted diagnostic (CAD) systems [11]. Fryback and Thornbury proposed a hierarchical model to assess the effectiveness of AI software in the process of diagnostic imaging [12], where higher-level products have more significant impacts compared to lower-level ones. Currently, most commercially available software belongs to lower levels, meaning they are either meant to increase sensitivity or aid in clinical decision-making, or, simply put, they are tools that are not meant to substitute radiologists [13].

However, the application of AI in radiology and its effect on the labor market could be much more diverse, as radiologists do not function alone and are situated as the cornerstone of a very complex task force that includes individuals from different backgrounds. Boland et al. provided a constructed view of the services radiology departments offered for patients and other clinicians and subdivided the process into incremental tasks performed consecutively to ensure that timely service is provided (Figure 28.1) [14].

Focusing on the "imaging value chain" results in the realization that our current estimations and expectations of how AI may shape the future of the labor market in the field of radiology may not be entirely based on holistic objective observations, as most of the evidence existent in

DOI: 10.1201/9781003095279-35

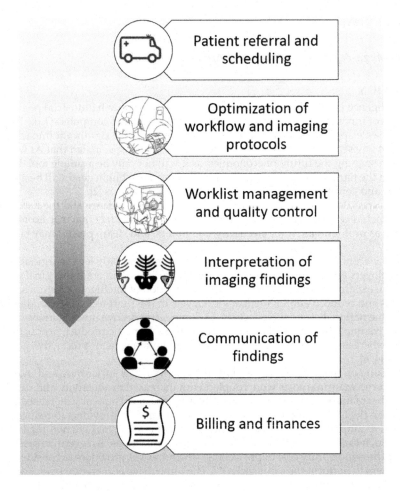

Figure 28.1 Imaging chain of value: different increments of the imaging process

this regard only focuses on the middle part of the value chain, which is the evaluation of medical images by radiologists, and does not take into considerations the steps before and after this task which are as crucial in ensuring service delivery [15]. The application of AI systems and intelligent machines may completely change how radiology departments function and will possibly have unexpected effects on the labor market. For one, the use of AI algorithms may increase the productivity of single radiology units by reducing missed imaging sessions and lost revenue, which could be coupled with increased pay for radiologists and other working staff, or could mean more investment in more diagnostic appliances which would translate to more job opportunities [16]. AI algorithms could also reduce the cost of performing necessary diagnostic tasks by aiding imaging protocol optimization or dedicating more expensive instruments for more appropriate tasks. AI systems could also help radiology departments to utilize resources better, find better providers, choose better reimbursement options and formulate better business strategy plans, and aid in business analytics [17]. Radiologists, data scientists, and information technology experts largely neglect the aforementioned and likewise alterations caused by AI application in radiology departments, which may contribute to the grim outlook some of the previously mentioned specialists have [18].

HOW IS AI AFFECTING RADIOLOGIST'S JOB OPPORTUNITIES?

As mentioned, advances in the field of artificial intelligence have led to the utilization of AI in many imaging-related tasks such as patients' appointment ordering and scheduling, modality selecting, image acquisition, image reporting and interpretation, and billing [19]. AI can calculate the time for each imaging session, and by arranging the patients' appointment reduce wasted time and increase patient satisfaction. AI may also have an essential role in integrating and interpreting patients' medical records and selecting the best and most helpful imaging modality for more precise evaluation and diagnostic workup. Some AI systems have been shown to optimize image acquisition by using imaging protocols precisely, minimizing image artifacts, and reducing radiation dosage [20–22].

Many studies have demonstrated that AI can help radiologists using deep learning methods with lesions measurement, volumetric calculation, and image segmentation. Interestingly, studies have shown that AI can also detect and diagnose pathologies based on large data inputs. These studies revealed that AI could help diagnose breast cancer, prostate cancer, and pulmonary nodules with accuracies resembling expert radiologists [19]. Such studies have caused some specialists, including radiologists, to believe that AI will make radiologists obsolete and impose drastic changes in the job market.

However, monitoring the current trends in AI applications in health industries does not show any signs of such changes happening any time soon.

First of all, validating AI software as a reliable method for analyzing and interpreting images needs a very complicated, expensive, and time-consuming approval process, which needs to be redone any time the protocols are changed or expanded [23]. Furthermore, the industry shows no signs of institutions risking the potential legal challenges of solely relying on AI solutions, and developers of AI solutions do not seem to be implying that their algorithms will shoulder the immense responsibility of handling patients [24, 25]. Based on proposals submitted for approval of regulatory agencies, it seems that almost all solutions are being considered as tools to assist radiologists [13, 26].

Based on these trends, AI solutions will further enable radiologists to detect diseases accurately and participate in treating patients. It will also allow health systems to adapt to new trends in the health market, such as increased focus on patient autonomy and communication [27]. Allowing AI solutions to undertake some of the more repetitive tasks will enable radiologists to better focus on high-value services, such as dedicating more time for patients and patient education and increasing time spent on continuous medical education and medical research [28, 29].

To assert the aforementioned does not mean that radiologists should expect that job markets will be static. Based on a survey performed on radiologists, more and more radiologists are taking courses to enable them to work with AI algorithms, and more job opportunities are being presented to those with such qualifications [19]. In response to this increased demand for such radiologists, some residency programs in Western countries have started to integrate principles of artificial intelligence and machine learning into educational curricula, and health provider organizations have begun requesting their radiologists to take part in CME courses focused on the application of AI in radiology [30].

WILL AI AFFECT THE INCOME OF RADIOLOGISTS?

To understand the effects of AI on radiology's income, we must consider factors that will alter the input and output of radiology departments. It is evident that utilization of AI solutions will reduce waste and overhead costs and increase the efficiency of radiology departments as a whole which could have positive effects on the income of radiologists. However, the prospect of task delegation and reduction of overall imaging needs may negatively affect radiologist salaries [4, 31]. One possible effect of AI on radiologists' income that is not yet extensively studied is the effect of increased education and technical capabilities of radiologists. As AI becomes an even more dominant paradigm, more and more radiologists will be educated to better understand, utilize, and even develop such tools, which could significantly alter salaries and benefit packages. Importantly, better communication among clinical units due to AI solutions may alternate payment methods and increase performance-based schemes (instead of simple fee-for-service methods), which could increase salaries for subgroups of radiologists and shift the duties of radiologists to those which generate more value [32, 33].

Importantly, external factors may be more detrimental than AI itself on radiologist salaries and reimbursement rates. Considering the recent trends, there has been a push to reduce the number of unnecessary medical procedures, such as diagnostic imaging studies, and a reluctance to adopt novel imaging modalities in fear of increasing health care costs [34, 35]. An example is an initiative promoted by the American College of Physicians, which aims to decrease the unnecessary utilization of 13 services directly performed by radiologists [36, 37]. Another example would be insurance companies' reluctance to cover CT colonography amid the evidence of its benefit in detecting gastrointestinal cancers [38]. It seems that the interplay between health-service purchasers, clinicians, and radiologists will determine if AI solutions will be viewed in the same manner, as an accessory tool, or otherwise considered as a necessary integration to the practice.

The effect of AI on radiologist salaries will also be determined by the response that active radiologists, academics, and service providers will have to the changing marketplace [39]. Will academic and university radiologists stop accepting new residents and fellows because of the effects of AI on the market? Will radiologists work longer hours to compensate for the possible lost reimbursement? Will radiologists be outsourced to teleradiology companies?

Overall, there are no clear indications that a catastrophe looms for the radiology job market. However, like any significant economic force, different individuals will be differently impacted based on various factors such as institutional matters (private vs. public, large vs. small, degree of sophistication, etc.), work experience, personal attributes (readiness to undertake new tasks, learn new concepts and accept new roles), and other macroeconomic and microeconomic trends.

RADIOLOGIST CURRENT PERSPECTIVE TOWARD AI

Focus on Job Opportunities and Salaries

Many studies have explored the attitude and view of radiologists toward AI and its effect on the daily life and practice of radiologists [23]. Table 28.1 briefly shows the main points of some of these studies.

Although there are limited large-scale studies, the consensus seems to be that most radiologists believe that the application of AI will overall benefit the practice of radiology, increase diagnostic accuracy, and reduce medical errors. However, most radiologists do not seem to think that these changes will be brought out sooner than ten years, and a considerable number have doubts regarding the broad scale of integration of AI in radiology practice [40–42]. Furthermore, radiologists from different backgrounds seem to think specific modalities such as mammography and CT will be impacted more profoundly than ultrasound and interventional techniques [43].

When surveying finances and other related matters, the consensus among different studies still holds true. Most radiologists still do not believe that AI poses a significant threat to their occupations, and most expect to have AI solutions integrated into their professions. Most radiologists also think that their peers should take the lead in implementing AI in diagnostic imaging, and more than three-quarters are willing to spend time learning associated skills [40, 43, 44]. Interestingly, as radiologists become more educated on the principles of AI, the more they become confident that AI will only not pose a challenge but will instead contribute to the quality of life of radiologists [40, 41].

Although the economic effects of AI on the reimbursement of radiologists are not studied sufficiently, the existing evidence shows that most radiologists are not concerned in this regard but rather think that introduction of AI may have unintended consequences on the reputation and authority of physicians [41].

Studies have shown that as the knowledge and information of radiologists regarding AI increases, their fear of AI about radiology careers decreases, and they have a more positive perspective toward it [45].

CONCLUSION

In conclusion, it seems that most radiologists have a reasonable outlook towards the future of radiology and the possible impacts of AI. There are limited resources available for assessing the financial impact of AI on radiologist's salaries. However, if precedent holds, the addition of AI to radiology practice will create new opportunities for those prepared to learn and adapt and will also possibly negatively affect some accepts, though there is no evidence to show that the overall effect will be negative. Based on current visible trends it seems that radiology will not drastically impact radiology job opportunities and salaries in the short term (5–10 years), and radiologists, both as individuals and as members of the house of medicine, will be able to adapt to these changes. This is also reflected in the fact that most radiologists see themselves being educated regarding AI and acknowledge its significant contribution to the accurate diagnosis of diseases.

Table 28.1 List of Surveys Assessing the Opinions and Views of Radiologists Regarding the Implementation of AI in Their Practice

Author	Country of Origin	Year Study Was Conducted	Target Group	Number of Paticipants	Significant Findings in the Survey	Ref
Waymel et al.	France	2019	Radiologists and radiology residents	270	• 81.1% of radiologists think there will be fewer errors in image interpretation. • 60% of radiologists think radiologists should be the sole bearer of responsibility for patients in cases of error. • Radiologists believe interventional radiology and antenatal imaging will be the least affected by AI	(43)
Huisman et al.	The Netherlands (the study include subjects from 54 countries)	2021	Radiologists and radiology residents	1041	• 75% of radiologists have no coding skills but are willing to learn necessary skills, and 79% are ready to take the lead in feature implementation of AI in radiology. • Only 13% of radiologists think the radiology job market may be in jeopardy.	(40)
Coppola et al.	Italy	2020	Radiologists	1032	• 60.3% of radiologists fear that AI implementation will lead to a worse professional reputation for radiologists among other physicians. • 18.9% think AI implementation will lead to reduced hiring opportunities. • 77.9% of respondents had no ethical concerns regarding AI. • 77% of respondents had a favorable view of AI implementation in radiology.	(41)
Jungmann et al.	Germany	2020	Radiologists, information technologists, and industry representative	123	• Only 14% of respondents believe medicine will become more humane with the implementation of AI. • There is a disagreement among physicians and IT specialists regarding whether medical societies should be involved in validating AI software. • The AI industry representatives are significantly more likely to believe that AI will make the radiology profession more accurate.	(46)

(Continued)

Table 28.1 (Continued)   List of Surveys Assessing the Opinions and Views of Radiologists Regarding the Implementation of AI in Their Practice

Author	Country of Origin	Year Study Was Conducted	Target Group	Number of Paticipants	Significant Findings in the Survey	Ref
Tajaldeen and Alghamdi	Saudi Arabia	2020	Radiologists	93	• Only 17% of radiologists in Saudi Arabia use any form of AI solution in their daily practice. • Only 25% of radiologists had read more than four articles regarding the use of AI in radiology.	(44)
Eltorai et al.	USA	2019	Thoracic radiologists and computer scientists	95	• 15.6% of computer scientists believe radiology jobs will be obsolete in 10-20 years, compared to 3.2% of radiologists. • 89.5% of radiologists and 86.7% of computer scientists believe job satisfaction of radiologists will increase due to AI solutions. • It is more probable for computer scientists to have had an AI article compared to radiologists.	(47)
Ooi et al.	Singapore	2021	Radiologists and radiology residents	125	• Only 1.6% of radiologists had used AI software for automated diagnosis of conditions in the workspace. • Only 7% of radiologists feel discouraged by the application of AI in their workspace. • Only 12% of radiologists view AI solutions as ever replacing radiologists.	(48)
Abuzaid et al.	UAE	2020	Radiologists and radiographers	153	• Only 14.4% of radiologists are comfortable using AI software on a daily basis. • 7.2% of respondents are worried about the impact of AI on radiology. • 39.9% of radiologists with knowledge regarding AI are self-taught.	(45)
Alelyani et al.	Saudi Arabia	2021	Radiologists	714	• 89% of respondents think AI will never replace radiologists. • 86% of respondents think AI is an essential part of radiology practice in the future. • 82% of respondents think it's necessary to include AI in residency programs.	(49)

REFERENCES

1. Syam N, Sharma A. Waiting for a sales renaissance in the fourth industrial revolution: Machine learning and artificial intelligence in sales research and practice. *Industrial Marketing Management.* 2018;69:135–46.

2. Agrawal A, Gans JS, Goldfarb A. Artificial intelligence: The ambiguous labor market impact of automating prediction. *Journal of Economic Perspectives.* 2019;33(2):31–50.

3. Brynjolfsson E, Rock D, Syverson C. *1. Artificial Intelligence and the Modern Productivity Paradox: A Clash of Expectations and Statistics.* University of Chicago Press; 2019.

4. Lu Y, Zhou Y. A review on the economics of artificial intelligence. *Journal of Economic Surveys.* 2021;35(4):1045–72.

5. Acemoglu D, Autor D, Hazell J, Restrepo P. *AI and Jobs: Evidence from Online Vacancies.* National Bureau of Economic Research; 2020.

6. Jha S. The economics of automation. *Academic Radiology.* 2020;27(1):3–5.

7. Burda D. Layoffs rise as pace of cost-cutting accelerates. *Modern Healthcare.* 1994;24(50):33–4, 6, 8.

8. Beyer A, Rehner L, Hoffmann W, van den Berg N. Task-sharing between pediatricians and non-physician healthcare professionals in outpatient child health care in Germany: Assessment of need and acceptance for concept development. *Inquiry.* 2020;57:46958020969299.

9. Artificial Intelligence in Healthcare Market with Covid-19 Impact Analysis by Offering (Hardware, Software, Services), Technology (Machine Learning, NLP, Context-Aware Computing, Computer Vision), End-Use Application, End User and Region - Global Forecast to 2026 MARKETSANDMARKETS; 2020. Available from: https://www.marketsandmarkets .com/Market-Reports/artificial-intelligence-healthcare-market-54679303.html.

10. Park CW, Seo SW, Kang N, Ko B, Choi BW, Park CM, et al. Artificial intelligence in health care: Current applications and issues. *Journal of Korean Medical Science.* 2020;35(42):e379.

11. Hosny A, Parmar C, Quackenbush J, Schwartz LH, Aerts H. Artificial intelligence in radiology. *Nature Reviews Cancer.* 2018;18(8):500–10.

12. Fryback DG, Thornbury JR. The efficacy of diagnostic imaging. *Medical Decision Making.* 1991;11(2):88–94.

13. van Leeuwen KG, Schalekamp S, Rutten M, van Ginneken B, de Rooij M. Artificial intelligence in radiology: 100 commercially available products and their scientific evidence. *European Radiology.* 2021;31(6):3797-804.

14. Boland GW, Duszak R, Jr., McGinty G, Allen B, Jr. Delivery of appropriateness, quality, safety, efficiency and patient satisfaction. *Journal of the American College of Radiology: JACR.* 2014;11(1):7–11.

15. Enzmann DR. Radiology's value chain. *Radiology.* 2012;263(1):243–52.

16. Martin-Carreras T, Chen PH. From data to value: How artificial intelligence augments the radiology business to create value. *Seminars in Musculoskeletal Radiology.* 2020;24(1):65–73.

17. Lee D, Yoon SN. Application of artificial intelligence-based technologies in the healthcare industry: Opportunities and challenges. *International Journal of Environmental Research and Public Health.* 2021;18(1).

18. Gurgitano M, Angileri SA, Rodà GM, Liguori A, Pandolfi M, Ierardi AM, et al. Interventional Radiology ex-machina: Impact of Artificial Intelligence on practice. *La Radiologia Medica.* 2021;126(7):998–1006.

19. Gampala S, Vankeshwaram V, Gadula SSP. Is artificial intelligence the new friend for radiologists? A review article. *Cureus.* 2020;12(10):e11137.

20. Curtis C, Liu C, Bollerman TJ, Pianykh OS. Machine learning for predicting patient wait times and appointment delays. *Journal of the American College of Radiology: JACR.* 2018;15(9):1310–6.

21. Sachs PB, Gassert G, Cain M, Rubinstein D, Davey M, Decoteau D. Imaging study protocol selection in the electronic medical record. *Journal of the American College of Radiology: JACR.* 2013;10(3):220–2.

22. Zhu B, Liu JZ, Cauley SF, Rosen BR, Rosen MS. Image reconstruction by domain-transform manifold learning. *Nature.* 2018;555(7697):487–92.

23. Park SH, Choi J, Byeon JS. Key principles of clinical validation, device approval, and insurance coverage decisions of artificial intelligence. *Korean Journal of Radiology.* 2021;22(3):442–53.

24. Jaremko JL, Azar M, Bromwich R, Lum A, Alicia Cheong LH, Gibert M, et al. Canadian association of radiologists white paper on ethical and legal issues related to artificial intelligence in radiology. *Canadian Association of Radiologists Journal = Journal l'Association canadienne des radiologistes.* 2019;70(2):107–18.

25. Brady AP, Neri E. Artificial intelligence in radiology-ethical considerations. *Diagnostics (Basel).* 2020;10(4):231.

26. Benjamens S, Dhunnoo P, Meskó B. The state of artificial intelligence-based FDA-approved medical devices and algorithms: An online database. *NPJ Digit Med.* 2020;3:118.

27. Vogenberg FR, Santilli J. Key trends in healthcare for 2020 and beyond. *American Health & Drug Benefits.* 2019;12(7):348–50.

28. Ahuja AS. The impact of artificial intelligence in medicine on the future role of the physician. *PeerJ.* 2019;7:e7702.

29. Hazarika I. Artificial intelligence: Opportunities and implications for the health workforce. *International Health.* 2020;12(4):241–5.

30. Richardson ML, Adams SJ, Agarwal A, Auffermann WF, Bhattacharya AK, Consul N, et al. Review of artificial intelligence training tools and courses for radiologists. *Academic Radiology.* 2021;28(91):238–52.

31. Chockley K, Emanuel E. The end of radiology? Three threats to the future practice of radiology. *Journal of the American College of Radiology: JACR.* 2016;13(12 Pt A):1415–20.

32. van Leeuwen KG, de Rooij M, Schalekamp S, van Ginneken B, Rutten M. How does artificial intelligence in radiology improve efficiency and health outcomes? *Pediatric Radiology.* 2021:1–7.

33. Frank MR, Autor D, Bessen JE, Brynjolfsson E, Cebrian M, Deming DJ, et al. Toward understanding the impact of artificial intelligence on labor. *Proceedings of the National Academy of Sciences of the United States of America.* 2019;116(14):6531–9.

34. Levin DC, Rao VM. The declining radiology job market: How should radiologists respond? *Journal of the American College of Radiology: JACR.* 2013;10(4):231–3.

35. Prabhakar AM, Oklu R, Harvey HB, Harisinghani MG, Rosman DA. The radiology job market: Analysis of the ACR jobs board. *Journal of the American College of Radiology: JACR*. 2014;11(5):507–11.

36. Brownlee S, Chalkidou K, Doust J, Elshaug AG, Glasziou P, Heath I, et al. Evidence for overuse of medical services around the world. *Lancet*. 2017;390(10090):156–68.

37. Lyu H, Xu T, Brotman D, Mayer-Blackwell B, Cooper M, Daniel M, et al. Overtreatment in the United States. *PLoS One*. 2017;12(9):e0181970.

38. Smith MA, Weiss JM, Potvien A, Schumacher JR, Gangnon RE, Kim DH, et al. Insurance coverage for CT colonography screening: impact on overall colorectal cancer screening rates. *Radiology*. 2017;284(3):717–24.

39. Forman HP, Larson DB, Kaye AD, Kazerooni EA, Norbash A, Crowe JK, et al. Masters of radiology panel discussion: The future of the radiology job market. *AJR American Journal of Roentgenology*. 2012;199(1):127–32.

40. Huisman M, Ranschaert E, Parker W, Mastrodicasa D, Koci M, Pinto de Santos D, et al. An international survey on AI in radiology in 1,041 radiologists and radiology residents part 1: fear of replacement, knowledge, and attitude. *European Radiology*. 2021;31(9):7058–66.

41. Coppola F, Faggioni L, Regge D, Giovagnoni A, Golfieri R, Bibbolino C, et al. Artificial intelligence: Radiologists' expectations and opinions gleaned from a nationwide online survey. *La Radiologia Medica*. 2021;126(1):63–71.

42. Chiwome L, Okojie OM, Rahman A, Javed F, Hamid P. Artificial intelligence: Is it armageddon for breast radiologists? *Cureus*. 2020;12(6):e8923.

43. Waymel Q, Badr S, Demondion X, Cotten A, Jacques T. Impact of the rise of artificial intelligence in radiology: What do radiologists think? *Diagnostic and Interventional Imaging*. 2019;100(6):327–36.

44. Tajaldeen A, Alghamdi S. Evaluation of radiologist's knowledge about the Artificial Intelligence in diagnostic radiology: a survey-based study. *Acta Radiologica Open*. 2020;9(7):2058460120945320.

45. Abuzaid MM, Elshami W, Tekin H, Issa B. Assessment of the willingness of radiologists and radiographers to accept the integration of artificial intelligence into radiology practice. *Academic Radiology*. 2020;29(1):87–94.

46. Jungmann F, Jorg T, Hahn F, Dos Santos DP, Jungmann SM, Düber C, et al. Attitudes toward artificial intelligence among radiologists, IT specialists, and industry. *Academic Radiology*. 2021;28(6):834–40.

47. Eltorai AEM, Bratt AK, Guo HH. Thoracic radiologists' versus computer scientists' perspectives on the future of artificial intelligence in radiology. *Journal of Thoracic Imaging*. 2020;35(4):255–9.

48. Ooi SKG, Makmur A, Soon AYQ, Fook-Chong S, Liew C, Sia SY, et al. Attitudes toward artificial intelligence in radiology with learner needs assessment within radiology residency programmes: A national multi-programme survey. *Singapore Medical Journal*. 2021;62(3):126–34.

49. Alelyani M, Alamri S, Alqahtani MS, Musa A, Almater H, Alqahtani N, et al. Radiology community attitude in Saudi Arabia about the applications of artificial intelligence in radiology. *Healthcare (Basel, Switzerland)*. 2021;9(7):834.

PART VIII

ATTITUDES— IMPLEMENTATION FEASIBILITY

29 Clinical View

Christina Malamateniou

INTRODUCTION

Artificial intelligence (AI) principles have been applied for at least eight decades thanks to pioneers like Alan Turing. AI is certainly not a new concept, with its theoretical underpinnings tracing further back [1]. For many years, AI use remained the privilege of the few and totally inaccessible to the uninitiated, such as "non-computer scientists." There were many hurdles to overcome before AI jumped from theory to practice and from computer labs to patients' bedsides. Lack of adequate computational power, and therefore speed, made AI algorithms too slow to be clinically useful. Lack of well-curated big data prevented optimal training and validation efforts for the new algorithms. Inadequate understanding of neuroscience and on how the human brain learns, limited applications of principles now widely used in machine and deep learning. The big data revolution in medical imaging, acceleration in computational speed and efficiency, and advancements in neuroscience enabled AI to finally (and gradually) reach clinical practice in the last decade [2].

Artificial intelligence is now increasingly being used in different applications of medical imaging. Different AI-enabled tools are not only streamlining operational tasks such as improving image acquisition, post-processing, and workflow optimization, but also for medical image interpretation and reporting.

Every AI tool lifecycle involves the following five stages:

1. Conceptualization: the initial phase of development during which a clinical need is identified and an AI tool or device is conceptualized to address this need

2. Development: the phase during which the AI tool is built, trained, and internally tested to ensure it does what it promises to do

3. Validation: the phase during which the AI tool is subjected to a series of assessments on clinical, economic, and ethical performance prior to deployment

4. Implementation: the phase during which the AI tool is finally translated into the intended clinical setting and context

5. Monitoring: this might include regular quality control and quality assurance during the product's use and also its monitoring after deployment, including the process of decommissioning

There is great interest by clinical practitioners, academics, and researchers alike in this translation from the computer labs to the bedside. The stakes are high, as AI holds promise of optimizing the management of workflows for regular imaging and screening practices by predicting, addressing, and minimizing non-shows; streamlining medical imaging data acquisition, postprocessing, segmentation, and registration for optimal image quality and diagnosis; standardizing processes such as examination vetting, identity checks, medical implant MRI safety procedures, patient positioning; facilitating interoperability across different scanners and clinical settings; and minimizing the time spent on mechanistic or repetitive tasks and allowing, instead, more time with the patient to reflect on and addressing patients' needs and preferences. Ultimately, the aim and wish is that AI can truly enable the provision of precision medicine and patient-centered radiology, whereby all medical data and examination results will be integrated for a particular patient, allowing optimal, holistic decision-making for individualized diagnosis and treatment pathways [3–5].

While AI has been met with great enthusiasm, there has also been proportional skepticism by clinicians, researchers, and patients alike. One would not be able to say that the radiology AI ecosystem has immediately embraced this change. There have been a lot of mixed feelings, controversy, denial, fear and hope, anxiety of being replaced by mere machines, all of which came down to unrealistic expectations, lack of knowledge, and lack of trust, accentuated by inadequate or sub-optimal (often retrospective) research and evidence base, to ascertain what AI could really achieve. As a new technology, AI raises many ethical issues for proper evaluation, such as validity, clinical usefulness, ethical use, safety and ensuring no harm to patients, credibility and acceptability, regular monitoring, and contextualization in local practice. The seemingly limitless promise of AI can often elicit an often-unrealistic wish list from clinical teams, who hope AI will

DOI: 10.1201/9781003095279-37

correct many healthcare shortcomings from the past, such as inequality, lack of transparency, lack of clear accountability frameworks, challenges with explainable healthcare provision, issues with data-sharing and accessibility, and sub-optimal or non-existent service-user or patient involvement [6–8].

Therefore, alongside the wish lists and promises of AI to revolutionize the way medicine is practiced, there are still unfortunately many obstacles and controversies that impact, delay, or even hinder AI's clinical implementation and threaten the transition to personalized care for all patients. Recently different publications have attempted to discuss the challenges to AI adoption by clinical practitioners and to propose solutions that are essential to harness the potential benefits of AI implementation into clinical practice [9–11].

This chapter aims to discuss these priorities for AI implementation into radiology practice.

Educating Medical Imaging Professionals for a Digital Future

One of the first steps to successfully integrate any new technological innovation is providing adequate training on the new technology for involved parties [12]. Such training will benefit practitioners, academics, researchers, patients, and service users alike.

Radiology and radiography (also referred to as medical radiation technology) are on the frontline of this innovation and radiology practitioners have become accustomed to the use of constantly new and improved medical imaging technologies, both software and hardware, that provide the best diagnosis and treatment for patients. However, AI is a truly disruptive technology that learns by feeding on newly provided data, similar to the human brain, such that AI algorithms can grow dynamically while being exposed to new experiences and clinical scenarios. The quality of the data we provide AI algorithms with and the scientific rigor by which we assess the algorithms' output determines the added value and reliability of each AI application (or, like computer scientists playfully like to say, this is the infallible principle of "garbage in, garbage out").

Education of human users is an intervention that builds trust in a bottom-up approach. Having the necessary knowledge, skills, and attitude to interact with AI means that clinical practitioners can harness benefits and minimize risks by acting as informed gatekeepers. Training will enable practitioners to learn the language of AI so they can train other healthcare professionals, contribute to the evidence base as clinical researchers, query the algorithms in case of malfunction, erroneous use, or poor performance, and, very importantly, be able to explain AI in layperson's terms to the patients, and if required, allay patients' fears [13–16].

Investing in practitioner training can crucially help minimize automation bias or over-reliance on AI technology. New skills will be needed for practitioners to navigate the digital future safely [10, 11, 13–16]; a key professional responsibility in the care of their patients.

The drive for AI innovation is not currently matched by proportionate AI educational provision offerings. Many radiologists still rely on personal study to build their knowledge. Fortunately, there is currently a small but increasing availability of formal AI programs led by higher education institutions [10, 13, 17–20]. Many of the radiologists and radiographers' national AI surveys reveal a lack of knowledge, skills, and confidence in AI [14, 18–25]. New knowledge on AI helps reframe AI as an opportunity to embrace and not a threat to fear. As Professor Curt Langlotz has elegantly explained: "AI will not replace radiologists; radiologists with AI knowledge will replace those without AI knowledge" [26].

AI Governance as a Necessary Sense-Checking for AI Innovation

Thousands of new AI tools flooded the market in the last ten years, and the recent COVID-19 pandemic accelerated this trend. To facilitate AI adoption in a safe and effective way and to offer standardization, different governance initiatives are currently being developed [27–32]. These include regulatory frameworks within which every AI tool classified as a medical device needs to meet safe use standards; whereas with AI products that are not categorized as medical devices and are designed, for example, to improve operational efficiency rather than support specific clinical or care-related decision-making, manufacturers should develop their technology in line with ISO 82304-1 for healthcare software [27]. The Medicines and Healthcare products Regulatory Agency (MHRA) and British Standards Institute (BSI) are leading in the UK the race to provide sufficient regulations that fine-tune the clinical applications of AI in medical imaging, the EU council with its recently published Artificial Intelligence Act [28] and Medical Device Regulation (MDR) [29] offers the baseline in the EU area, whereas the US Food and Drug Administration (FDA) holds this role for the USA and other regions [30]. All of this guidance is co-constructed with the mediation of the respective professional bodies locally.

In the last two to three years there has been a concerted effort to develop robust regulatory frameworks [27–30] to help regulate the AI market. A BSI publication on auditable standards of safe and ethical use of AI tools is imminent in winter 2023 and is eagerly anticipated. Furthermore, different professional bodies have attempted to provide guidance on AI use [12, 30–33], and specific expert consulting services have emerged to facilitate this transition and support developers and clinicians.

Some people have expressed concerns that regulation might curb creativity and innovation. However, investing in robust AI regulation, standardization, and governance is vital to reinforce the trust required for any clinical practitioner–service user relationship, and also to ensure AI developers and innovators have better-defined frameworks of due-diligence and practice. Furthermore, AI regulation and accreditation ensure that procurement agencies within radiology have the tools and reference to assess available AI technologies and decide on those that are more closely aligned with particular clinical contexts and needs.

AI Ethics

AI needs large datasets for training, validation, and evaluation. Unfortunately, due to some recent examples of misuse of patient data by large AI companies without the appropriate ethical approval and consents in place [34–38], the use of AI has gained some negative publicity. This has inevitably damaged the reputation of AI in the eyes of end-users such as healthcare practitioners and patients. These scenarios clearly demonstrate that, despite AI's huge potential, use of AI must respect the established clinical governance and research ethics frameworks to build a robust and ethical evidence base and to secure trust. Research ethics processes might often seem unnecessarily complex and to be bureaucratic tick-box exercises, but these processes are in place to protect everyone: patients, healthcare professionals, and AI developers. Federated learning may provide excellent opportunities for optimal efficiency and data accessibility while maintaining patient privacy and ethical data use [39, 40].

Accountability after erroneous results using AI-enabled tools to reach a decision is still quite a gray area. Early work proposed that the responsibility should be with the end users, but this may cause healthcare workers to become more reluctant to adopt AI, and therefore delay implementation. More recent suggestions propose that accountability is shared by all key agencies involved in the AI tool life cycle: the AI companies, healthcare providers, and practitioners [41–43].

Validate Your AI

Before an AI tool can be used clinically, internal validation has to occur, to ensure AI tools work as expected. After internal validation, external (clinical) validation on unseen data is required to ensure the same tool works in the respective clinical context. This type of evaluation needs to consider patient demographics for external clinical settings, local equipment, and incidence of pathologies or regional clinical priorities. Furthermore, regular quality control and monitoring are necessary to ensure continually optimal AI performance [31, 32, 44, 45].

External validation and evaluation are far from standardized processes. There are different measures of AI tool evaluation pertaining to sensitivity and specificity in pathology detection, reproducibility under different conditions, or interoperability when used with different medical imaging hardware. Different institutions use different tools, and some have created their own well-curated databases of clinical data to provide platforms for others to validate their AI (services that often require financial compensation). There are also increasingly standardized evaluation checklists stemming from newly established ethics, governance, and regulatory frameworks [32, 44, 46]. These are vital steps toward AI adoption that will develop rapidly in the next decade, with both more standardization needed but also customization to address local contexts.

AI Black-Box No More

The fine detail and intricacies of AI technology, underlined by many lines of coding, can often be fully comprehended only by the AI developers and computer scientists, like a well-kept secret. There is no necessary requirement for this to change in the future. However, for practitioners to trust and adopt such algorithms before their use on patients, the details of clinical application of any AI tool, its affordances, and its potential risks need to be fully grasped by clinical practitioners. Understanding of how the AI algorithm has reached its decision, such as through the use of heat maps or displaying of comparable examples, would help build end-user trust. Transparency about the methods of training, validation, evaluation, and quality assurance will be helpful to ensure clinical practitioners feel confident in algorithmic output and know how to better interpret the outcomes of any form of assessment in evaluation of AI performance [47–49]. Such knowledge also will enable practitioners to improve the functionality of AI algorithms.

With Great Power Comes Great Responsibility: AI's Potential for Fairer Healthcare

Current healthcare systems have inherited a wealth of technological know-how from the collective efforts of many researchers and clinical scientists in the past, but they have also inherited an onerous legacy of healthcare inequalities and disparities, often producing poor health outcomes, poor quality of life, with higher morbidity and mortality for specific patient populations. Such disparities might relate to social status, gender, race, intellectual ability, capacity for verbal communication, presence of disability, and geographical location of clinical setting [15]. It is the hope of many AI enthusiasts that AI tech can bring about a fairer healthcare system that works for everyone.

This, of course, relies on acquiring and curating high-quality data that is representative of the target population, on robust clinical governance and ethics frameworks to prevent conscious and unconscious biases, and on streamlined validation and evaluation processes to ensure the clinical added value of any AI tool [32, 37, 38, 44, 45]. Done wrong, AI can also unfortunately amplify these disparities and inequalities on a large scale, with far-reaching consequences (see big data, big troubles). There is no better (or worse) evidence than the health disparities and varying health outcomes resulting from the delayed diagnosis of COVID-19 infection on Black, Asian, Minority Ethnic (BAME) populations in the UK and other countries [50, 51]. There is great responsibility not only for AI developers but also for healthcare providers, clinical practitioners, researchers, patient advocates, and policymakers to ensure implementation processes are fair and minimize potential biases [52–56].

The Importance of Research in Building the Evidence Base, Trust, and Key Stakeholder Acceptability

Health tech is a nascent discipline. Practitioners are not an easy audience to convince, particularly when it comes to clinical value and patient safety. It is the "do no harm" principle that governs, and often delays, much of the technology adoption in healthcare. For practitioners to be convinced, well designed, ethical, prospective research studies are needed to build the evidence base required [57–60]. Patients are also key stakeholders as the end-users. So, they need to be reassured that the results produced by AI tools are valid and meaningful [57, 60]. Clear communication of the AI research available, using simple language, offering options, involving them as co-producers in AI development and testing, addressing their queries, are all means by which patients can trust the process and the outcome and feel better informed and involved [15].

Patient and Person-Centeredness: A Case for Meaningful, User-Friendly AI Tools

While clinical practitioners are experts by training and nurture, patients are experts by experience and nature. Therefore, patients' suggestions to improve development pipelines and products are insightful and invaluable as suggestions come from the lived experience of a disease [61–62]. These suggestions often relate to how things feel; they might relate to sensory adaptations, environment improvements, optimize communication, format, and timing of healthcare service delivery, all things that matter to patient-centeredness.

Every decision in healthcare should be made for the patients, with the patients. Therefore, patients and clinical practitioners should be included throughout the AI tool lifecycle such as conceptualization, design, and evaluation [63–65]. Otherwise, AI developers risk creating a product that is not clinically relevant, functional, or user-friendly enough. The cost of the afterthought can be huge for AI companies, so patient and public involvement initiatives are necessary ingredients to successful AI adoption [15].

The AI Ecosystem

A new AI ecosystem is being developed to support AI adoption, with new roles, new responsibilities, new opportunities, and new challenges [66]. AI ambassadors can further facilitate AI implementation [4]. These are knowledgeable AI practitioners who are well connected and great communicators, and who are the groups that safeguard and uphold AI ethics, integrity, and trust. AI ambassadors could troubleshoot on the spot, or also escalate concerns to an extensive fellow network of AI experts. They could be leads of AI quality assurance and governance. They can act as AI allies, "translators" for clinical practitioners and patients and liaisons, as a "bridge" between technology and end-users to offer a holistic management approach. Radiologists and radiographers, who are accustomed to working at the interface between technology and patients, could be ideal for this position. But anyone with a robust technical skillset and excellent communication skills can assume such a central role to facilitate AI adoption. After all, AI needs the whole ecosystem to collaborate for proactive and not reactive solutions for optimal outcomes [67]. Different medical imaging disciplines need not compete with each other, given so much need for different expertise. Healthcare practitioners' efforts should continue to be invested in improving healthcare outcomes and patient experiences. The priorities reported above must be addressed to facilitate a smooth and uncomplicated transition for the benefit of patients. After all, implementation is not owned by just one group of professionals or experts, but would be the outcome of collective intelligence within the diverse AI ecosystem that will grasp the opportunities and tackle the challenges that implementation of AI will bring to radiology [9, 10, 15, 66].

REFERENCES

1. https://sitn.hms.harvard.edu/flash/2017/history-artificial-intelligence/ (accessed April 20, 2022)

2. https://online.stanford.edu/programs/artificial-intelligence-professional-program (accessed June 10, 2022)

3. Lewis SJ, Gandomkar Z, Brennan PC. Artificial Intelligence in medical imaging practice: Looking to the future. *J Med Radiat Sci.* 2019;66:292–295.

4. Hardy M, Harvey H. Artificial intelligence in diagnostic imaging: Impact on the radiography profession. *Br J Radiol.* 2020 Apr;93(1108):20190840.

5. Malamateniou C, Knapp KM, Pergola M, Woznitza N, Hardy M. Artificial intelligence in radiography: Where are we now and what does the future hold? *Radiography (Lond).* 2021 Oct;27 Suppl 1:S58–S62.

6. Kim B, Koopmanschap I, Mehrizi MHR, Huysman M, Ranschaert E. How does the radiology community discuss the benefits and limitations of artificial intelligence for their work? A systematic discourse analysis. *Eur J Radiol.* 2021 Mar;136:109566.

7. Scheek D, Rezazade Mehrizi MH, Ranschaert E. Radiologists in the loop: the roles of radiologists in the development of AI applications. *Eur Radiol.* 2021 Oct;31(10):7960–7968.

8. Huisman M, Ranschaert E, Parker W, Mastrodicasa D, Koci M, Pinto de Santos D, Coppola F, Morozov S, Zins M, Bohyn C, Koç U, Wu J, Veean S, Fleischmann D, Leiner T, Willemink MJ. An international survey on AI in radiology in 1,041 radiologists and radiology residents part 1: Fear of replacement, knowledge, and attitude. *Eur Radiol.* 2021 Sep;31(9):7058–7066.

9. Strohm L, Hehakaya C, Ranschaert ER, Boon WPC, Moors EHM. Implementation of artificial intelligence (AI) applications in radiology: Hindering and facilitating factors. *Eur Radiol.* 2020;30:5525–5532. https://doi.org/10.1007/s00330-020-06946-y

10. Huisman M, Ranschaert E, Parker W, Mastrodicasa D, Koci M, Pinto de Santos D, Coppola F, Morozov S, Zins M, Bohyn C, Koç U, Wu J, Veean S, Fleischmann D, Leiner T, Willemink MJ. An international survey on AI in radiology in 1,041 radiologists and radiology residents part 2: Expectations, hurdles to implementation and education. *Eur Radiol.* 2021. https://doi.org/10.1007/s00330-021-07782-4

11. Kotter E, Ranschaert E. Challenges and solutions for introducing artificial intelligence (AI) in daily clinical workflow. *Eur Radiol.* 2021 Jan;31(1):5–7.

12. Topol, E. *The Topol Review: Preparing the Healthcare Workforce to Deliver the Digital Future.* The NHS Constitution; 2019.

13. Malamateniou C, McFadden S, McQuinlan Y, England A, Woznitza N, Goldsworthy S, Currie C, Skelton E, Chu KY, Alware N, Matthews P, Hawkesford R, Tucker R, Town W, Matthew J, Kalinka C, O'Regan T. Artificial Intelligence: Guidance for clinical imaging and therapeutic radiography professionals, a summary by the Society of Radiographers AI working group. *Radiography (Lond)*. 2021 Aug 19:S1078–8174.

14. Rainey C, O'Regan T, Matthew J, Skelton E, Woznitza N, Chu KY, Goodman S, McConnell J, Hughes C, Bond R, McFadden S, Malamateniou C. Beauty Is in the AI of the beholder: Are we ready for the clinical integration of artificial intelligence in radiography? An exploratory analysis of perceived AI knowledge, skills, confidence, and education perspectives of UK radiographers. *Front Digit Health*. 2021 Nov 11;3:739327. doi: 10.3389/fdgth.2021.739327. PMID: 34859245; PMCID: PMC8631824.

15. Malamateniou C, McEntee M, Integration of AI in radiography practice: ten priorities for implementation. *Rad Magazine*. 2022 Aug;48:567.

16. Rainey C, O'Regan T, Matthew J, Skelton E, Woznitza N, Chu KY, Goodman S, McConnell J, Hughes C, Bond R, Malamateniou C, McFadden S. An insight into the current perceptions of UK radiographers on the future impact of AI on the profession: A cross-sectional survey. *J Med Imaging Radiat Sci*. 2022 Jun 14:S1939–8654(22).

17. Schuur F, Rezazade Mehrizi MH, Ranschaert E. Training opportunities of artificial intelligence (AI) in radiology: a systematic review. *Eur Radiol*. 2021 Aug;31(8):6021–6029.

18. European Society of Radiology (ESR). What the radiologist should know about artificial intelligence – an ESR white paper. *Insights Imaging*. 2019;10:44.

19. Wiggins WF, Caton MT, Magudia K, Glomski SA, George E, Rosenthal MH, Gaviola GC, Andriole KP. Preparing radiologists to lead in the era of artificial intelligence: Designing and implementing a focused data science pathway for senior radiology residents. *Radiol Artif Intell*. 2020 Nov 4;2(6):e200057.

20. Wiggins WF, Caton MT Jr, Magudia K, Rosenthal MH, Andriole KP. A conference-friendly, hands-on introduction to deep learning for radiology trainees. *J Digit Imaging*. 2021 Aug;34(4):1026–1033.

21. HCIAC Corporate Roundtable Subcommittee on Artificial Intelligence. *White Paper on the Artificial Intelligence Era: The Role of Radiologic Technologists and Radiation Therapists*. HCIAC; 2019.

22. Antwi WK, Akudjedu TN, Botwe BO. Artificial intelligence in medical imaging practice in Africa: A qualitative content analysis study of radiographers' perspectives. *Insights Imaging*. 2021 Jun 16;12(1):80. doi: 10.1186/s13244-021-01028-z. PMID: 34132910; PMCID: PMC8206887.

23. Botwe BO, Antwi WK, Arkoh S, Akudjedu TN. Radiographers' perspectives on the emerging integration of artificial intelligence into diagnostic imaging: The Ghana study. *J Med Radiat Sci*. 2021 Sep;68(3):260–268. doi: 10.1002/jmrs.460. Epub 2021 Feb 14. PMID: 33586361; PMCID: PMC8424310.

24. Ryan ML, O'Donovan T, McNulty JP. Artificial intelligence: The opinions of radiographers and radiation therapists in Ireland. *Radiography (Lond)*. 2021 Oct;27 Suppl 1:S74–S82. doi: 10.1016/j.radi.2021.07.022. Epub 2021 Aug 25. PMID: 34454835.

25. Abuzaid MM, Tekin HO, Reza M, Elhag IR, Elshami W. Assessment of MRI technologists in acceptance and willingness to integrate artificial intelligence into practice. *Radiography (Lond)*. 2021 Oct;27 Suppl 1:S83–S87. doi: 10.1016/j.radi.2021.07.007. Epub 2021 Aug 4. PMID: 34364784.

26. Reardon S. Rise of Robot Radiologists: Deep-learning algorithms are peering into MRIs and x-rays with unmatched vision, but who is to blame when they make a mistake? *Nature* 2019;S58(576):19–26.

27. NHSx. *A Buyer's Guide to AI in Health and Care*. NHSx; 2020.

28. EU artificial Intelligence Act. Regulation of the European parliament and of the council laying down harmonised rules on artificial intelligence (Artificial intelligence act and amending certain union legislative acts) (published Apr 21, 2021).

29. https://www.gov.uk/guidance/medical-devices-eu-regulations-for-mdr-and-ivdr (accessed Sept 2021).

30. FDA. *Artificial Intelligence/Machine Learning (AI/ML)-Based Software as a Medical Device (SaMD): Action Plan*. FDA; 2021.

31. Larson DB, Harvey H, Rubin DL, Irani N, Tse JR, Langlotz CP. Regulatory frameworks for development and evaluation of artificial intelligence-based diagnostic imaging algorithms: Summary and recommendations. *J Am Coll Radiol*. 2021 Mar;18(3 Pt A):413–424.

32. Omoumi P, Ducarouge A, Tournier A, Harvey H, Kahn CE Jr, Louvet-de Verchère F, Pinto Dos Santos D, Kober T, Richiardi J. To buy or not to buy-evaluating commercial AI solutions in radiology (the ECLAIR guidelines). *Eur Radiol*. 2021 Jun;31(6):3786–3796.

33. Gillan C, Hodges B, Wiljer D, Dobrow M, Health care professional association agency in preparing for artificial intelligence: Protocol for a multi-case study. *JMIR Res Protoc* 2021;10(5):e27340.

34. Wong SH, Al-Hasani H, Alam Z. et al. Artificial intelligence in radiology: How will we be affected?. *Eur Radiol*. 2019;29:141–143.

35. Royal College of radiologists (RCR). *Integrating Artificial Intelligence with the Radiology Reporting Workflows (RIS and PACS)*. RCR; 2021.

36. RANZCR and ASMIRT. *Ethical Principles for AI in Medicine*. RANZCR and ASMIRT; 2019.

37. Geis JR, Brady AP, Wu CC, Spencer J, Ranschaert E, Jaremko JL, Langer SG, Borondy Kitts A, Birch J, Shields WF, van den Hoven van Genderen R, Kotter E, Wawira Gichoya J, Cook TS, Morgan MB, Tang A, Safdar NM, Kohli M. Ethics of artificial intelligence in radiology: Summary of the joint European and North American multisociety statement. *Radiology*. 2019 Nov;293(2):436–440.

38. Brady AP, Neri E. Artificial intelligence in radiology-ethical considerations. *Diagnostics (Basel)*. 2020 Apr 17;10(4):231.

39. Darzidehkalani E, Ghasemi-Rad M, van Ooijen PMA. Federated learning in medical imaging: Part I: Toward multicentral health care ecosystems. *J Am Coll Radiol*. 2022 Apr 25:S1546-1440(22)00280-0.

40. Darzidehkalani E, Ghasemi-Rad M, van Ooijen PMA. Federated learning in medical imaging: Part II: Methods, challenges, and considerations. *J Am Coll Radiol*. 2022 Apr 25:S1546-1440(22)00281-2.

41. European Society of Radiology (ESR). What the radiologist should know about artificial intelligence - an ESR white paper. *Insights Imaging*. 2019 Apr 4;10(1):44.

42. Naik N, Hameed BMZ, Shetty DK, Swain D, Shah M, Paul R, Aggarwal K, Ibrahim S, Patil V, Smriti K, Shetty S, Rai BP, Chlosta P, Somani BK. Legal and ethical consideration in artificial intelligence in healthcare: Who takes responsibility? *Front Surg.* 2022 Mar 14;9:862322.

43. Čartolovni A, Tomičić A, Lazić Mosler E. Ethical, legal, and social considerations of AI-based medical decision-support tools: A scoping review. *Int J Med Inform.* 2022 May;161:104738.

44. Bizzo BC, Ebrahimian S, Walters ME, Michalski MH, Andriole KP, Dreyer KJ, Kalra MK, Alkasab T, Digumarthy SR. Validation pipeline for machine learning algorithm assessment for multiple vendors. *PLoS One.* 2022 Apr 29;17(4):e0267213.

45. Tulk Jesso S, Kelliher A, Sanghavi H, Martin T, Henrickson Parker S. Inclusion of clinicians in the development and evaluation of clinical artificial intelligence tools: A systematic literature review. *Front Psychol.* 2022 Apr 7;13:830345.

46. Haller S, Van Cauter S, Federau C, Hedderich DM, Edjlali M. The R-AI-DIOLOGY checklist: a practical checklist for evaluation of artificial intelligence tools in clinical neuroradiology. *Neuroradiology.* 2022 May;64(5):851–864; Zerilli J, Bhatt U, Weller A. How transparency modulates trust in artificial intelligence. *Patterns (N Y).* 2022 Feb 24;3(4):100455.

47. Zeineldin RA, Karar ME, Elshaer Z, Coburger J, Wirtz CR, Burgert O, Mathis-Ullrich F. Explainability of deep neural networks for MRI analysis of brain tumors. *Int J Comput Assist Radiol Surg.* 2022 Apr 23. doi: 10.1007/s11548-022-02619-x. Epub ahead of print. PMID: 35460019.

48. Salahuddin Z, Woodruff HC, Chatterjee A, Lambin P. Transparency of deep neural networks for medical image analysis: A review of interpretability methods. *Comput Biol Med.* 2021 Dec 4;140:105111.

49. Fuhrman JD, Gorre N, Hu Q, Li H, El Naqa I, Giger ML. A review of explainable and interpretable AI with applications in COVID-19 imaging. *Med Phys.* 2022 Jan;49(1):1–14.

50. Paremoer L, Nandi S, Serag H, Baum F. Covid-19 pandemic and the social determinants of health. *BMJ* 2021;372 :n129 doi:10.1136/bmj.n129.

51. Public Health England. Disparities in the risk and outcomes of COVID-19. 2020. https://assets.publishing.service.gov.uk/government/uploads/system/uploads/attachment_data/file/908434/Disparities_in_the_risk_and_outcomes_of_COVID_August_2020_update.pdf (accessed Apr 12, 2022).

52. Lovejoy CA, Arora A, Buch V, Dayan I. Key considerations for the use of artificial intelligence in healthcare and clinical research. *Future Healthc J.* 2022 Mar;9(1):75–78.

53. Puyol-Antón E, Ruijsink B, Mariscal Harana J, Piechnik SK, Neubauer S, Petersen SE, Razavi R, Chowienczyk P, King AP. Fairness in cardiac magnetic resonance imaging: Assessing sex and racial bias in deep learning-based segmentation. *Front Cardiovasc Med.* 2022 Apr 7;9:859310.

54. Straw I, Wu H. Investigating for bias in healthcare algorithms: A sex-stratified analysis of supervised machine learning models in liver disease prediction. *BMJ Health Care Inform.* 2022 Apr;29(1):e100457.

55. Kelly CJ, Karthikesalingam A, Suleyman M, Corrado G, King D. Key challenges for delivering clinical impact with artificial intelligence. *BMC Med.* 2019 Oct 29;17(1):195.

56. Wilson A, Saeed H, Pringle C. Artificial intelligence projects in healthcare: 10 practical tips for success in a clinical environment. *BMJ Health & Care Informatics* 2021;28:e100323.

57. Knitza J, Muehlensiepen F, Ignatyev Y, Fuchs F, Mohn J, Simon D, Kleyer A, Fagni F, Boeltz S, Morf H, Bergmann C, Labinsky H, Vorbrüggen W, Ramming A, Distler JHW, Bartz-Bazzanella P, Vuillerme N, Schett G, Welcker M, Hueber AJ. Patient's perception of digital symptom assessment technologies in rheumatology: Results from a multicentre study. *Front Public Health*. 2022 Feb 22;10:844669.

58. Calisto FM, Santiago C, Nunes N, Nascimento JC. BreastScreening-AI: Evaluating medical intelligent agents for human-AI interactions. *Artif Intell Med*. 2022 May;127:102285.

59. Šendelj R, Ognjanović I, Roganović M, Krikščiūnienė D. AI enhanced person-centred care services for monitoring stroke outpatient rehabilitation. *Stud Health Technol Inform*. 2022 Jan 14;289:22–24.

60. Young AT, Amara D, Bhattacharya A, Wei ML. Patient and general public attitudes towards clinical artificial intelligence: a mixed methods systematic review. *Lancet Digit Health*. 2021 Sep;3(9):e599–e611.

61. Fancott C, Baker GR, Judd M, Humphrey A, Morin A. Supporting patient and family engagement for healthcare improvement: Reflections on "engagement-capable environments" in Pan-Canadian learning collaboratives. *Healthc Q*. 2018 Dec;21(SP):12–30.

62. Bergerum C, Wolmesjö M, Thor J. Organising and managing patient and public involvement to enhance quality improvement-Comparing a Swedish and a Dutch hospital. *Health Policy*. 2022 Apr 9:S0168-8510(22)00084-7.

63. Usher S, Denis JL. Exploring expectations and assumptions in the public and patient engagement literature: A meta-narrative review. *Patient Educ Couns*. 2022 Apr 14:S0738-3991(22)00147-1.

64. McGill B, Corbett L, Grunseit AC, Irving M, O'Hara BJ. Co-Produce, Co-Design, Co-create, or Co-construct-who does it and how is it done in chronic disease prevention? A scoping review. *Healthcare (Basel)*. 2022 Mar 30;10(4):647.

65. Smith H, Budworth L, Grindey C, Hague I, Hamer N, Kislov R, van der Graaf P, Langley J. Co-production practice and future research priorities in United Kingdom-funded applied health research: A scoping review. *Health Res Policy Syst*. 2022 Apr 2;20(1):36.

66. NHSx, MAAS, (2021) https://www.nhsx.nhs.uk/ai-lab/ai-lab-programmes/regulating-the-ai-ecosystem/ (accessed Apr 20, 2022).

67. Valtiner D, Reidl C. On change management in the age of artificial intelligence: a sustainable approach to overcome problems in adapting to a disruptive, technological transformation. *J Adv Manag Sci*. 2021 Sept;9(3):53–58.

30 Technological View

Ranveer Vasdev, Ibrahim Abdalla, Amy Song, Tyler Gathman,
Adrianna M. Rivera-León, and Thomas Kane

PATIENTS AND PROVIDERS

Healthcare providers' attitudes toward AI will have a heavy influence on clinical implementation of AI. Several studies have demonstrated lack of awareness and knowledge of AI among clinicians. One study examined provider familiarity of AI through a survey, revealing that more than half had never came across AI in the workplace [1]. When asked about machine learning versus deep learning, only half knew at least one of these terms and almost 90% of participants could not differentiate between the two [1]. Furthermore, 40% of providers in the survey believed that AI was extremely dangerous, comparing it to nuclear weapons. Though there were fears and a lack of knowledge regarding AI, almost 80% of providers believed that AI had beneficial applications [1]. Similarly, a French study found that almost three quarters of radiologists felt that they had received inadequate education on AI. Interestingly, one specific study found that the level of AI knowledge correlated with radiologists' attitudes toward AI. Those with only a basic understanding of AI exhibited fear towards AI while those that had more advanced knowledge of AI showed a positive attitude towards AI [2]. This is not surprising as people are often fearful of what they do not understand. Currently, physicians do want to learn more about AI. One study showed that 70% of providers were agreeable to attending educational classes on AI.[3] These studies collectively highlight that physician education can facilitate a favorable perspective of AI in medicine. By bridging the knowledge gap that exists between many practicing providers and their understanding of AI, the implementation of AI could proceed in a more expedited fashion.

Education can be a means to facilitate the implementation of AI, not only at the provider level but also for medical students. Unfortunately, medical student education is currently lacking in AI. The implementation of AI would benefit if healthcare students received basic knowledge of AI in medical education. Student learners are already advocating for basic AI training in their curriculum [4]. Though many are proponents of AI in medical education, there remains a lack of implementation of AI into the medical curriculum. A big factor that seems to play into AI implementation in medical curriculum is a lack of understanding of AI among faculty, further complicating its integration [5].

Another factor that can facilitate clinical AI implementation is an advocate of AI in the workplace. An ideal advocate is likely a radiologist with a strong understanding of AI. This advocate can combat negative perceptions colleagues have by providing information and education on AI while also allowing providers to experiment with AI applications to gain familiarity with its function [6]. This instills more trust amongst other radiologists to use AI in their daily practice.

A hindering factor in the implementation of AI amongst providers is that many users of AI find that it is inconsistent with readings and how this affects the workflow of radiologists. Many providers question the validity of AI algorithms [6]. Specifically, false positive or false negative readings from AI could negatively affect the safety of patients [7]. These inaccurate readings could also disrupt the workflow of radiologists by requiring them to take additional time to correct mistakes.

Although healthcare providers are important to consider when implementing AI in healthcare, so too are the patients AI will serve. Positive attitudes toward AI from the public will likely facilitate the implementation of AI in medicine. Though there have been mixed opinions that the public has on AI in medicine, ranging from distrust of AI to trusting AI over providers, it appears this varies by medical subspecialties. One study looked at the patient perception of AI in three specialties (dermatology, radiology, and surgery) and found that the public was more trusting of AI in the field of dermatology over radiology and surgery [8]. The authors suggested that this heterogeneity of public attitudes may be partially explained by educational background. There is evidence that higher distrust of AI is strongly correlated with lower educational attainment. Conversely, a higher public level of trust in AI was found more often in higher-educated respondents [8]. This parallels the studies previously mentioned when comparing provider's attitudes toward AI. A knowledge gap exists between both patients and providers regarding AI. This barrier to implementation can be reduced if more education is projected to the general public and providers alike.

Not surprisingly, another survey study found that AI implementation in medicine depended on people's trust in AI. However, this study also found that distrust towards and feelings of neglect

DOI: 10.1201/9781003095279-38

from providers increased patients' willingness to implement AI in medicine [9]. Furthermore, people who felt a lack of social belonging tended to favor AI implementation in medicine. In the future, it will be important to increase people's trust in AI but not through the default of individuals becoming discontent with physicians or feeling a general lack of community.

REGULATORY BODIES

An important topic regarding the use of AI in medical practice, and specifically radiology, is the regulation of the technology [10]. As innovations in the field continue to make the use of AI more feasible, a framework must exist to ensure software quality as well as patient safety [11].

Many forms of AI in current use in radiology have been approved by the FDA as medical devices in accordance with Section 510(k) of the Food, Drug and Cosmetic Act [11]. The job performed by these programs are generally low risk and aim to reduce the number of time radiologists spend conducting repetitive and time-consuming tasks [12]. Though many have suggested that radiologists will soon be replaced by AI, the devices in use and development are marketed as tools of medical professionals, not their rivals [13]. Here we will attempt to define and explain some of the regulatory bodies overseeing the development of these new technologies as well as the proposed policies for implementing these new systems into actual practice.

Although AI systems such as computer-assisted detection (CADe), computer-assisted triage (CADt), or computer-assisted diagnosis (CADx) have found their way into clinical practice, their designation as medical devices come with certain guidelines for introducing updates or new features [10]. For instance, as is the case for all FDA-approved static AI software, a 510(k) must be submitted when any of the following is made [14] :

1. "A change that introduces a new risk or modifies an existing risk that could result in significant harm"

2. "A change to risk controls to prevent significant harm"

3. "A change that significantly affects clinical functionality or performance specifications of the Device"

These principles are meant to ensure changes to existing programs avoid introducing previously averted risks to patients. These guidelines are useful but do not apply broadly to all forms of AI. As was mentioned by Wichmann et al., for technologies that employ self-learning mechanisms, categorization as a traditional medical device may be inappropriate. This is due to the inherent nature of these systems which are founded on making continuous changes and adjustments to their output. In response, the FDA has made an avenue for technologies employing self-learning to be marketed to the public *de novo* or through the pre-market approved application pathway [14]. Another change made by the FDA was to implement a total product life-cycle approach to the regulation of these machine-learning products. A benefit of this approach is that it allows producers to submit their devices as well as the changes they predict the software will make at the same time they submit their application for pre-market approval [10]. This reduces the need to fill out multiple 510(k) applications which could take up to 90 days to be approved [14].

Attaining FDA approval may be the first regulatory hurdle that must be overcome but it certainly is not the last. With the use of AI-assisted technology, a host of potential security and privacy issues arise and must be addressed. The use of AI in radiology requires enormous amounts of data: data that often may not be purged of patient identifiers. A proposed solution to mitigate this issue was to advocate for laws concerning patient health information to be changed [15].

The proposed changes would allow some patient information to be shared between developers and institutions but protected from other third parties. This may lead to pushback, however, because attempts such as in 2016 when startup DeepMind partnered with the Royal Free London NHS Foundation Trust, concerns were raised about the use and transfer of patient information that did not comply with data protection law [16, 17]. There are many financial implications of developing AI that achieves even modest implementation. Among them is the ethical concern of balancing the interests of the company and those of patients [18.]

INDUSTRY

Industry stakeholders can be well-positioned to facilitate the implementation of AI into the everyday practice of radiologists. However, this requires targeted clinical and systems-based product design as well as consideration of systemic problems with data sharing.

Beginning with initial product design, radiology AI products have largely focused on chest CT and radiograph interpretation as these are routinely utilized in medical scenarios. However, less frequent imaging modalities such as extremity MR or retrograde pyelograms are an untapped source for additional product design. This is all to say that the implementation of AI into radiology should first begin with routinely captured imaging modalities; however, it is not too early to begin with less frequently obtained but still clinically meaningful images [19]. The process of product design can require several process development cycles including product planning, risk management, documentation, configuration, outsourcing, and verification of the clinical relevance [20]. Implementation of AI models during the development can be designed with extensive hardware to support product processing power, which can limit scalability and transportation [21]. In the context of medicine, model transparency is critical for implementation into existing workflows as physicians and staff members are more likely to "buy into" AI if they can understand how the algorithm works. There is a risk that more transparent models and subsequent simplification may reduce performance [22]. At the same time, transparency that demonstrates how behavioral or clinical risk factors correlate to a disease process, may facilitate research into novel therapies and improved patient outcomes [23].

Product design must be compatible with existing infrastructures. For any imaging modality that translates patient anatomy to diagnosis, there are several layers of data manipulation including protocolling, study prioritization, feature analysis, and report generation that can serve as consideration for design or an opportunity for innovation. The complexity and interdependence of these layers are exponentially higher when you consider the input of both autonomous and human contributions, varying types of imaging (i.e., CT or MRI), and institutional or regional differences in clinical practice. As a result, data standardization must be utilized for not only product design but also scalability. At the same time, flexible or adaptive programs designed for multiple or heterogeneous data formats may offer an advantage for some AI platforms and aid in implementation [24].

Even taking into consideration these factors, integration and implementation of AI into a routine clinical setting will not be achieved without a demonstration of non-inferiority. This can be demonstrated through randomized clinical trials that can quantify not only non-inferiority or benefit, but also cost–benefit ratios, time until diagnosis, and whether specific patient population subgroups would directly benefit from AI products. Mammography for breast cancer screening is one such example where there is early evidence demonstrating non-inferiority which warrants additional large-scale prospective studies [21]. As a result, it is in the best interest of AI developers to align themselves with clinical researchers, even at the level of product design and testing, to facilitate the long-term vision of clinical integration.

REFERENCES

1. Castagno S, Khalifa M. Perceptions of artificial intelligence among healthcare staff: A qualitative survey study. *Front Artif Intell*. 2020;3:578983. doi:10.3389/frai.2020.578983

2. Huisman M, Ranschaert E, Parker W, et al. An international survey on AI in radiology in 1,041 radiologists and radiology residents part 1: fear of replacement, knowledge, and attitude. *Eur Radiol*. 2021;31(9):7058–7066. doi:10.1007/s00330-021-07781-5

3. Waymel Q, Badr S, Demondion X, Cotten A, Jacques T. Impact of the rise of artificial intelligence in radiology: What do radiologists think? *Diagn Interv Imaging*. 2019;100(6):327–336. doi:10.1016/j.diii.2019.03.015

4. Teng M, Singla R, Yau O, et al. Health care students' perspectives on artificial intelligence: Countrywide survey in Canada. *JMIR Med Educ*. 2022;8(1):e33390. doi:10.2196/33390

5. Grunhut J, Wyatt AT, Marques O. Educating future physicians in artificial intelligence (AI): An integrative review and proposed changes. *J Med Educ Curric Dev*. 2021;8:23821205211036836. doi:10.1177/23821205211036836

6. Xiang Y, Zhao L, Liu Z, et al. Implementation of artificial intelligence in medicine: Status analysis and development suggestions. *Artif Intell Med*. 2020;102:101780. doi:10.1016/j.artmed.2019.101780

7. Strohm L, Hehakaya C, Ranschaert ER, Boon WPC, Moors EHM. Implementation of artificial intelligence (AI) applications in radiology: hindering and facilitating factors. *Eur Radiol.* 2020;30(10):5525–5532. doi:10.1007/s00330-020-06946-y

8. Yakar D, Ongena YP, Kwee TC, Haan M. Do people favor artificial intelligence over physicians? A survey among the general population and their view on artificial intelligence in medicine. *Value in Health.* 2022;25(3):374–381. doi:10.1016/j.jval.2021.09.004

9. Frank DA, Elbæk CT, Børsting CK, Mitkidis P, Otterbring T, Borau S. Drivers and social implications of artificial intelligence adoption in healthcare during the COVID-19 pandemic. *PLoS One.* 2021;16(11):e0259928. doi:10.1371/journal.pone.0259928

10. Pianykh OS, Langs G, Dewey M, et al. Continuous learning AI in radiology: Implementation principles and early applications. *Radiology.* 2020;297(1):6–14. doi:10.1148/radiol.2020200038

11. Wichmann JL, Willemink MJ, De Cecco CN. Artificial intelligence and machine learning in radiology: Current state and considerations for routine clinical implementation. *Invest Radiol.* 2020;55(9):619–627. doi:10.1097/RLI.0000000000000673

12. Recht MP, Dewey M, Dreyer K, et al. Integrating artificial intelligence into the clinical practice of radiology: Challenges and recommendations. *Eur Radiol.* 2020;30(6):3576–3584. doi:10.1007/s00330-020-06672-5

13. European Society of Radiology (ESR). What the radiologist should know about artificial intelligence - an ESR white paper. *Insights Imaging.* 2019;10(1):44. doi:10.1186/s13244-019-0738-2

14. Health C for D and R. Artificial intelligence and machine learning in software as a medical device. *FDA*, September 22, 2021. Accessed May 8, 2022. https://www.fda.gov/medical-devices/software-medical-device-samd/artificial-intelligence-and-machine-learning-software-medical-device

15. Liew C. The future of radiology augmented with artificial intelligence: A strategy for success. *Eur J Radiol.* 2018;102:152–156. doi:10.1016/j.ejrad.2018.03.019

16. Murdoch B. Privacy and artificial intelligence: Challenges for protecting health information in a new era. *BMC Medical Ethics.* 2021;22(1):122. doi:10.1186/s12910-021-00687-3

17. Kharpal A. Google DeepMind patient data deal with UK health service illegal, watchdog says. *CNBC*, July 3, 2017. Accessed May 8, 2022. https://www.cnbc.com/2017/07/03/google-deepmind-nhs-deal-health-data-illegal-ico-says.html

18. Currie G, Hawk KE. Ethical and Legal challenges of artificial intelligence in nuclear medicine. *Semin Nucl Med.* 2021;51(2):120–125. doi:10.1053/j.semnuclmed.2020.08.001

19. Tadavarthi Y, Vey B, Krupinski E, et al. The state of radiology AI: Considerations for purchase decisions and current market offerings. *Radiol Artif Intell.* 2020;2(6):e200004. doi:10.1148/ryai.2020200004

20. Larson DB, Harvey H, Rubin DL, Irani N, Tse JR, Langlotz CP. Regulatory frameworks for development and evaluation of artificial intelligence–based diagnostic imaging algorithms: Summary and recommendations. *J Am Coll Radiol.* 2021;18(3, Part A):413–424. doi:10.1016/j.jacr.2020.09.060

21. Sechopoulos I, Mann RM. Stand-alone artificial intelligence - The future of breast cancer screening? *The Breast.* 2020;49:254–260. doi:10.1016/j.breast.2019.12.014

22. Patrzyk PM, Link D, Marewski JN. Human-like machines: Transparency and comprehensibility. *Behav Brain Sci.* 2017;40:e276. doi:10.1017/S0140525X17000255

23. Plan S. The national artificial intelligence research and development strategic plan. *National Science and Technology Council, Networking and Information Technology Research and Development Subcommittee.* 2016.

24. Tang A, Tam R, Cadrin-Chênevert A, et al. Canadian Association of Radiologists White Paper on Artificial Intelligence in Radiology. *Can Assoc Radiol J.* 2018;69(2):120–135. doi:10.1016/j.carj.2018.02.002

31 Societal View

Sara J. Fardin and Christopher G. Filippi

INTRODUCTION

The growing impact of artificial intelligence (AI) in radiology is no longer in question given the rapid, exponential increase in AI-related publications in radiology journals with neuroimaging leading the way, accounting for almost one-third of that output [1]. As AI makes its way into routine clinical workflow, from image acquisition, to diagnosis, to treatment management, radiologists are appropriately concerned about AI's ultimate impact on the future of the imaging landscape and the profession of radiology. Recent consensus statements and reviews have raised alarms about issues of fairness and bias with respect to AI deployment in healthcare that center around issues of positive or negative impacts in favor of specific groups, places, or timescales [2–3]. It will be incumbent on radiologists to be responsible leaders in the developing AI ecosystem to establish best practices that will minimize harm and maximize benefit. In this chapter, we will briefly discuss AI in radiology from the perspective of fairness and bias.

DEFINITIONS

What is meant by ethics in AI? According to the Merriam-Webster dictionary, ethics can be defined as a "set of moral principles or a theory or system of moral values," or ethics may refer to "principles of conduct governing an individual or group" [4]. More simply, ethics can be thought of as a guiding philosophy. The terms ethics and morals are used interchangeably, but there are subtle differences between the two. Morals invoke an individual's concepts of right versus wrong, which is subjective. Ethics deals with questions of fairness and the social responsibility of actions, which is more universal. Bias in scientific research refers to a systematic error that undermines the data obtained and the conclusions which can be derived from this data. In AI algorithmic development, bias refers to a disproportionate weight in favor of or against certain individuals or groups of people that may be both unfair and cause harm. Radiologists need to develop the guiding principles around what constitutes fair, unbiased development and deployment of AI in clinical imaging that is used for medical diagnosis, treatment, patient management, and prognostication.

ETHICS AND FAIRNESS OF ALGORITHMS

What Could Go Wrong?

A Massachusetts Institute of Technology Media Lab Researcher, Joy Buolamwini, is a founder of the Algorithmic Justice League, which seeks accountability in AI algorithmic development based on her research on facial recognition [5]. She tested facial recognition software being developed by IBM, Microsoft, and Megvi, a Chinese company, in which the Microsoft software had an error rate for darker-skinned women of 21% and IBM and Megvi softwares exhibited error rates of 35%, but all of the programs had error rates of < 1% for light-skinned males [5]. In a similar story, the *New York Times* reported on the inaccuracies of facial recognition software in which up to 35% of darker-skinned females were gender misidentified in a set of 271 photos using facial recognition software [6]. In another well-known example, Amazon's face recognition software falsely matched 28 members of Congress to felons, in which 20% of members of Congress are people of color but 39% of false matches on Amazon's facial recognition were people of color [7], likely an unintended consequence of this algorithm. This explains why 26 expert AI researchers including Yoshua Bengio, a 2018 Turing Prize winner, and a coalition of nearly 70 civil rights groups, over 400 members of the academic community, and more than 150,000 members of the public, all called upon Amazon to cease its selling of its Rekognition AI service (facial recognition) to police departments [7]. These algorithmic failures may reflect poor input data that does not include enough under-represented minorities (i.e., people of color) and women. These tools, if deployed, could cause real harm to these groups.

Examples already exist in which AI technology used by companies produced unintended consequences of bias against certain groups. Amazon scrapped its hiring algorithm for recruitment after it was shown to be biased against women [8]. In a pivotal study, banking-sector AI algorithms used to determine financial fitness for mortgages have demonstrated racial discrimination against Black and Latino applicants with higher rejections rates (61% to 48%) [9]. Even the justice system

DOI: 10.1201/9781003095279-39

and law enforcement agencies have used AI algorithms developed by private companies that have inaccurately assessed the recidivism risks of criminal defendants [10]. A Pro Publica report has shown that nearly 45% of African American defendants who were labeled higher risk did not re-offend, but 48% of Caucasian defendants who were labeled lower risk did re-offend [10]. All of these examples underscore an urgent need to develop regulatory and legal frameworks to ensure that individual rights, particularly from underrepresented groups, are not transgressed by rapidly developing AI technology.

This is perhaps most important when it comes to a person's health and wellness. In medicine, the guiding principle of *primum no nocere* (first, do no harm) must be coupled to AI algorithmic fairness to inform healthcare management decisions given similar examples in healthcare of AI deployment leading to unintended consequences of bias and unfairness. IBM created a "Watson for Oncology" program that used AI to recommend cancer treatment strategies; however, it often made unsafe and incorrect treatment decisions, which largely were attributable to training problems such as including a small number of hypothetical cancer patients rather than real patient data and an overreliance on the expertise of a few specialists at a particular cancer center rather than society guidelines (standard practice) or scientific evidence [11]. In another example, an AI algorithm widely used by US hospitals to improve the care for sick patients with complex medical problems showed significant bias, in that it was far less likely to refer Black patients than White patients for treatment who were equally sick. Correction for such biases would have improved referrals from roughly 18% to 47% [12]. In that study, healthcare costs appeared to be an effective proxy for health by some measures of predictive accuracy, but large racial biases occurred, which suggests that well-intentioned proxies for ground truth can lead to bias in AI algorithms [12].

TRUST IN AI DECISION-MAKING IN HEALTHCARE

The code architecture of AI algorithms is created by individuals and organizations whose values, beliefs, and preferences may intentionally or unintentionally inform this process. Software programs running on such trained algorithms based on the entered data and its dataset features, may propagate such hidden bias [3, 13]. A study by Shankar et al. [14] used two of the largest publicly available datasets, ImageNet [15] and Open Images [16], and analyzed classifiers trained on them and found representation bias because most images came from North America and Europe. The authors concluded that there cannot be accurate or reliable classification without representation so there is a need for geographic diversity considerations if constructed datasets are to be deployed to the developing world [14]. Bias may occur in any data based on gender, socioeconomic, environmental, geographic, cultural, and ethnic factors; and additional biases in medicine may occur from comorbidities of disease [3]. In imaging, there are many additional sources of potential bias from technical factors such as heterogeneity of scanner parameters for imaging to the post-processing of images.

Research and education are needed to elevate awareness of these risks and to develop ways to minimize bias in data given how elemental data is to the development of AI tools in radiology. Trust in AI depends on data quality, data collection, its management, and its evaluation. There needs to be transparency to allay privacy concerns and to be compliant with the Health Insurance Portability and Accountability Act (HIPAA). Both radiologists and patients need to understand and hold to account the decision-making processes of AI algorithms given that the workings of AI are often invisible and poorly understood [17]. Given that large amounts of data are typically needed for AI development, there are unresolved issues, including the need for informed consent, ownership, privacy and data protection, transparency, objectivity, gaps in access to data, and moral rights [18]. A new institutional review board (IRB) process may be needed for the digital age. Can patients' images be automatically used, if anonymized for data to build algorithms without explicit consent, meaning that they are assumed to "opt in" unless they indicate otherwise? Or do patients need to give consent to their data being used? How would such consent be documented in real time? If patient data is used to develop an AI algorithm, should the patients be financially compensated? These questions underscore how best practices around data management in AI are still evolving, with no clear consensus.

FRAMEWORKS TO PUT ETHICS INTO AI IN HEALTHCARE

Beneficence, maleficence, autonomy, and justice are the core existing bioethical principles [19], which can build a foundational framework to anchor ethics and fairness into the development of AI in radiology. Beneficence (do only good) and non-maleficence (do no harm) seem logically equivalent but represent distinct principles [17]. Autonomy preservation in the era of AI will mandate compromises between decision-making powers that radiologists retain, and how much autonomy is ceded to the AI algorithm. Floridi et al. [17] argue that

> autonomy of humans should be promoted and that the autonomy of machines should be restricted and made intrinsically reversible, should human autonomy need to be protected and re-established (consider the case of a pilot able to turn off the automatic pilot and regain full control of the plane).

The concept of justice in AI algorithmic use in radiology practice includes concepts of fairness that should seek to eliminate bias or discrimination in healthcare delivery, particularly given the risks of inherent bias in the datasets used to train and validate algorithms that could then potentially harm different groups of patients [17].

For AI, there needs to be a new bioethical concept in addition to beneficence, non-maleficence, justice and autonomy, specifically a fifth one termed "explicability" as described by Floridi and Cowls that includes intelligibility (how does it work?) with accountability (who is responsible for how it works?) [17, 19, 20]. Knowledge of how AI decisions are made is needed to understand what good or harm it may be doing in society and in which ways both beneficence and non-maleficence can be preserved [17, 21]. For example, the use of a visual saliency map that delineates on images where the AI algorithm focused its attention to arrive at a prediction could be a useful way to drive acceptance of AI-based decisions for both healthcare professionals and patients [22]. For AI to be just, there must be a process to address real or perceived grievances for any unintended harm wrought by its use [17]. For AI to preserve autonomy, a patient-centered approach is needed so that decisions about the use of AI should be informed by the knowledge of how that AI would work in ways that patients comprehend [17, 21]. Misplaced concerns, ignorance, and fear around AI and its uses in radiology practice could thwart the development for all the wrong reasons [21].

Transparency is needed for explicability, but unchecked transparency could violate privacy by revealing personal, protected information in data sets. The EU General Data Protection Regulation [23] requires transparency in data processing, although it is unclear to what degree the technical detail needs to be transparent. Radiologists, as the primary drivers of AI use in clinical practice, should be able to explain in lay language how data will be used to build an AI tool so that a lay person can understand why a model made a particular decision and comprehend conditions in which it succeeds or fails [21, 24]. Transparency is challenging considering that deep learning convolution neural networks have many hidden layers and are proverbial "black boxes," such that detection of potential biases may be difficult without reverse engineering to understand how such algorithms arrived at their decisions and predictions. On the other hand, if the transparency is too translucent, intellectual property may be in jeopardy, which may diminish investment and may risk integrity of the model [3].

ACTION PLANS FOR RADIOLOGISTS IN THE AI SPACE

Makers of algorithms and their users should be able to explain an algorithm's decision-making process, the reliability and predictability of the algorithm, and how they protect the AI tools and patients against privacy breaches and potential cyberattacks before the algorithms are routinely used in clinical practice [3, 17]. Guidelines for the customization, development, and implementation of ethical codes have been published [17], and relevant stakeholders including radiologists, patient advocates, institutions, and regulatory bodies will need to be a part of these discussions around intelligibility and explicability of AI algorithms [17, 21]. Radiologists need to determine the required level of training for implementation of AI, which should be informed by current radiology experts in AI. Curricula to train radiology residents, fellows, and board-certified radiologists in the basics of AI are needed so that there is a general understanding of how algorithms work, and such tools are intelligible to the entire radiology team. There needs to be rigorous quality assurance (QA) and quality improvement (QI) processes, including for the discovery of potential biases and unfairness in algorithmic deployment, to have a redress process for any perceived unfairness, and to quantify patient outcomes where AI is deployed [25].

Radiologists should lead discussions on best practices within the radiology community and the greater medical field, as radiologists historically are physicians who are first to utilize novel technology. Leaders in imaging societies such as the American College of Radiology (ACR) and Radiology Society of North America (RSNA) should strengthen working partnerships with federal stakeholders, including with the National Institutes of Health, the National Institute of Standards and Technology, and the US Food and Drug Administration.

SUMMARY

As well-summarized in a joint North American and European consortium white paper by Geis et al. [3], AI in radiology should "promote any use that helps individuals such as patients and providers and should block the use of radiology data and AI algorithms for irresponsible financial gains." In keeping with the principle *primum no nocere*, AI algorithms in radiology must have benefits that outweigh risks to minimize harm, promote good outcomes, and be as equitable as possible. Radiologists have critically important roles in patient care and must adapt to these technological changes by acquiring new skills to do their best for patients in this rapidly evolving AI ecosystem. It is up to radiologists to champion best practices for ethical, non-biased, and fair deployment of AI tools in routine radiology practice.

REFERENCES

1. Pesapane F, Codari M, Sardanelli F. "Artificial Intelligence in Medical Imaging: Threat or Opportunity." *Eur Exp Radiol* 2018; 2 (1): 35.

2. Lui YW, Chang PD, Zaharchuk G, et al. "Artificial Intelligence in Neuroradiology: Current Status and Future Directions." *AJNR Am J Neuroradiol* 2020; 51 (9): e227–231.

3. Geis JR, Brady AP, Wu CC, et al. "Ethics of Artificial Intelligence in Radiology. Summary Statement of the Joint European and North American Multisociety Statement." *Radiology* 2019; 293: 436–440.

4. https://www.merriam-webster/dictionary/fairness

5. Buolamwini J, Gebru T. "Gender Shades: Intersectional Accuracy Disparities in Commercial Gender Classification." Proceedings of Machine Learning Research; Conference of Fairness, Accountability, and Transparency, New York, NY, 2018, vol. 81, pp. 1–15.

6. Lohr S. "Facial Recognition Software is Accurate if You're a White Guy." *New York Times*, February 9, 2018.

7. Snow J. "Amazon's Face Recognition Falsely Matched 28 Members of Congress with Mugshots." July 26, 2018. https://www.aclu.org/blog/privacy-technology/surveillance-technologies/amazons-face-recognition-falsely-matched-28

8. Dastin J. "Amazon Scraps Secret AI Recruiting Tool That Showed Bias Against Women." *Reuters News*, October 10, 2018. https://www.reuters.com/article/us-amazon-com-jobs-automation-insight/amazon-scraps-secret-ai-recruiting-tool-that-showed-bias-against-women-idUSKCN1MK08G

9. Bartlett R, Morse A, Stanton R, Wallace N. *Consumer Lending Discrimination in the FinTech Era.* National Bureau of Economic Research; 2019. https://www.nber.org/papers/w25943

10. Angwin J, Larson J, Matta S, Kirchner L. "Machine Bias: There's Software Used Across the Country to Predict Future Criminals. And It's Biased Against Blacks." *ProPublica*, May 23, 2016. https://www.propublica.org/article/machine-bias-risk-assessments-in-criminal-sentencing

11. Ross C, Aguilar R. "Inside the Fall of Watson Health: How IBM's Audacious Plan to 'Change the Face of Health Care' With AI Fell Apart." *Stat News*, March 8, 2021. https://www.statnews.com/2021/03/08/ibm-watson-health-sale

12. Obermeyer Z, Powers B, Vogeli C, Mullainathan S. "Dissecting Racial Bias in an AI Algorithm Used to Manage the Healthcare of Populations." *Science* 2019; 366 (6464): 447–453.

13. Shahriari K, Shahriari M. "IEEE Standard Review. Ethically Aligned Design: A Vision for Prioritizing Human Wellbeing with Artificial Intelligence and Autonomous Systems." IEEE Canada International Humanitarian Conference (IHTC). 2017. https://doi.org/10.1109/ihtc .2017.8058187.

14. Shankar S, Halpern Y, Breck E, Atwood J, Wilson J, Sculley D. "No Classification without Representation: Assessing Geodiversity Issues in Open Data Sets for the Developing World." Proceedings of 31st Conference on Neural Information Processing Systems, NIPS 2017, Long Beach, CA. November 22, 2017 in arXiv:1711:08536

15. Russakovsky O, Deng J, Su H, et al. "ImageNet Large Scale Visual Recognition Challenge." *Int J Computer Vision* 2015; 115 (3): 211–252.

16. Krasin I, Duerig T, Alldrin N, et al. "A Public Dataset for Large-scale Multi-label and Multi-class Image Classification." *Dataset*, 2017. https://github.com/openimages.

17. Floridi L, Cowls J. "A Unified Framework of Five Principles for AI in Society." *Harvard Data Science Review*, 2019. DOI: 10.1162/99608f92.8cd550dl. https://hdsr.mitpress.mit.edu/pub/ l0jsh9d1/release/7.

18. Middlestadt BD, Floridi L. "The Ethics of Big Data: Current and Foreseeable Issues in Biomedical Contexts." *Sci Eng Ethics* 2016; 22 (2): 303–341.

19. Varkey B. "Principles of Clinical Ethics and Their Application to Practice." *Med Princ Prac* 2021; 30 (1): 17–28.

20. Floridi L, Taddeo M. "What is Data Ethics." *Philos Trans A Math Phys Eng Sci* 2016; 372 (2083): 20160360. DOI: 10.1098/rsta.2016.0360.

21. Floridi L, Cowls J, Beltrametti M, et al. "AI4People—An Ethical Framework for a Good AI Society: Opportunities, Risks, Principles, and Recommendations." *Minds and Machines* 2018; 28: 689–707.

22. Choy G, Khalilzadeh O, Michalski M, et al. "Current Applications and Future Impact of Machine Learning in Radiology." *Radiology* 2018; 288: 318–328.

23. https://gdpr.eu/what-is-gdpr/

24. Gilpin LH, Bau D, Yuan BZ, Bajwa A, Specter M, Kagal L. "Explaining Explanations: An Overview of Interpretability of Machine Learning". 2018 IEEE 5th International Conference of Data Science and Advanced Analytics (DSAA). https://doi.org/10.1109/dsaa.2018.00018.

25. Kingston JKC. Research and Development in Intelligent Systems XXXIII. 2016. arXiv:1802.07782. https://arxiv.org/abs/1802.07782

32 Financial View

Alexander E. Jacobs, Tyler Gathman, Aaron Schumacher, Ranveer Vasdev, and David J. H. Wu

THEMES IN HEALTH INFORMATION TECHNOLOGY

Health information technology investments are some of the largest administrative costs for providers and hospitals in the United States. In the case of electronic health record (EHR) implementation projects at large hospitals and hospital groups, costs are often measured in the hundreds of millions or billions of dollars [3]. Once projects are implemented, they require a team of technical support staff, software developers, and managers to keep critical health information technologies running without interruption.

EHR systems are used throughout hospitals, in nearly every patient encounter. AI algorithms in radiology would naturally affect a smaller percentage of the overall patient pool. The technologies operate on different financial scales, yet the case of EHR systems demonstrates that implementation of AI technologies will require sustained investment in data and personnel management [1].

Given the experimental nature of AI in radiology thus far, the financial costs of implementation are largely speculative. Projections regarding the financial impact of implementing AI in clinical radiology practice, both in the United States and abroad, must be treated as useful estimates rather than precise predictions. Such financial projections are made challenging by the complexity of the US multi-payer system, the trend toward ever higher administrative costs, and the lack of clarity regarding Center for Medicare & Medicaid Services (CMS) reimbursement practices for AI-derived care. Indeed, in one comprehensive review of the economic impact of AI in healthcare, the reviewers concluded that, "existing impact assessments show methodological deficits and … upcoming evaluations require more comprehensive economic analyses to enable economic decisions for or against implementing AI technology in health care" [4].

Despite the difficulty of predicting the future, previous technological transformations within and without healthcare provide a useful framework for assessing the financial costs associated with the implementation of AI in radiology and healthcare in general.

CASE STUDY

AI Detection of Diabetic Retinopathy

One of the first AI imaging technologies to be FDA-approved is the IDx-DR system, which uses a proprietary fundus camera and machine learning algorithm to detect diabetic retinopathy (DR) and macular degeneration [5]. For observers of AI in radiology, the promise of IDx-DR is familiar: accurate point-of-care imaging that can be deployed in many practice settings, thus increasing access to essential care, especially for patients with diabetes. Digital Diagnostics, the firm that developed the IDx-DR system, explicitly guarantees that its customers will have "no need for specialist overread or telemedicine call backs" [5].

As of 2020, the IDx-DR system is available to customers for a single purchase of approximately $13,000 [6]. It can also be acquired via lease or through a plan involving no upfront investment but requiring a minimum number of examinations per camera. In each of these three scenarios, Digital Diagnostics assesses a $25 fee to the provider per patient encounter. The original American Medical Association Current Procedural Terminology (CPT) reimbursement code for the IDx-DR system is 92250-TC, which is reimbursed at an average of $23.82. Whether any given test will be reimbursed can be unclear to providers. Reimbursement policies vary between Medicare Administrative Contractor payers, with some reimbursing only positive-result DR tests and others reimbursing all tests, regardless of result [6].

For the scenario in which only positive results are reimbursed, one analysis found an average reimbursement of approximately $80 would be required before the IDx-DR system allows providers to achieve a break-even point of positive marginal revenue [6]. In the scenario in which all results are reimbursed, positive or negative, the average reimbursement rate of $23.82 results in a marginal revenue of −$1.18 per exam. As the authors report, "The benefits of AI on health outcomes and costs are concurrent in part to its scalability, ease of use, facilitation of early detection and treatment, as well as efficiency," yet their results "indicate that at existing reimbursement rates and patterns, practices using IDx-DR would operate at a deficit with a negative marginal revenue for every patient evaluated." The authors further note that they may be underreporting this effect, given that their calculations do not include costs such as staffing or purchase fees [6].

DOI: 10.1201/9781003095279-40

The authors note that new AI-specific billing codes "must incorporate device and other operating costs in reimbursement determination." Failing to do so may result in providers lacking "financial incentive to implement this technology," even if it has demonstrated clinical benefit [6]. As of January 1, 2021, a new CPT code (9225X) for IDx-DR has been put into effect [6]. There is not yet any published information on the average reimbursement rates under the new procedural code, though projections anticipate a significant increase over the previous CPT code [6].

None of this is to say that the IDx-DR system is ineffective—rather, the reimbursement ecosystem is not yet able to reliably account for the implementation of clinical AI. IDx-DR is currently approved for the detection of two medical conditions: diabetic retinopathy and macular degeneration. In coming years, similar systems will proliferate for a host of different conditions. When presented individually, the cost of IDx-DR is dwarfed by overhead costs like payroll and EHR and may at first seem like a financially sound investment. Nevertheless, it is easy to see how a multitude of such AI systems in radiology and other specialties can quickly add up into a tangle of overlapping financial obligations, with hospitals and providers managing contracts with multiple smaller private firms. In assessing the financial consequences of implementing such technologies, it is imperative to understand whether these technologies offer a reasonable return on investment.

In short, stakeholders must carefully assess the lag between which one branch of the federal government—the FDA—approves AI-powered technologies and another branch of the federal government—CMS—makes it financially feasible to use them.

AI AND CMS

While economic considerations are not the sole incentive considered in the adoption of new healthcare technologies, they cannot be discounted. The adoption of AI in radiology may be slowed by AI firms that set prices too high or by payers offering reimbursement rates unacceptable to providers. As Allen et al. note,

> a key ingredient in moving [AI] algorithms for healthcare into routine clinical practice will be ensuring our healthcare system supports the fair compensation for the development of these algorithms…but developing a process for how that will happen may not be as simple as it might seem. [7]

Currently, CMS reimbursement protocols for FDA-approved AI radiology are split between the Medicare Physician Fee Schedule (MPFS) and the Inpatient Prospective Payment System (IPPS). For instance, MPFS compensates the aforementioned AI-assisted diagnosis of diabetic retinopathy under a fee-for-service model. By contrast, CMS also compensates a machine learning system for diagnosis of large vessel occlusion strokes under the IPPS—a provision paying hospitals for inpatient stays but making no allowance for individual radiologists (or other providers) [9].

The question of price—and who will pay it—is the central question of implementing AI in radiology, yet there is no consistent standard by which prices are set. Until they are able to confidently predict returns on AI investment, hospitals and other stakeholders will likely remain unwilling to bear the direct costs of implementation. Alternative modes of payment may be part of the solution. As Chen et al. note, "Payment for AI in the current fee-for-service environment may be challenging, and sustained adoption of AI may not occur within the framework of the IPPS and Physician Fee Schedule" [9].

Alternative payment models (APMs), such as those contained in the CMS Quality Payment Program, may be useful for investigating sustainable methods of reimbursing AI in radiology [9]. Understanding best practices in reimbursement will furnish stakeholders with the pricing estimates needed to inform cost–benefit analyses of adopting AI in healthcare.

The Merit-based Improvement Payment System (MIPS) is a program of the CMS Quality Payment Program that adjusts provider FFS rates (up or down by up to 9% in 2022) based on performance in four categories: quality, clinical practice improvement, resource use, and advancing care information [7]. While APMs are not widespread in US practice, programs such as MIPS are useful proving grounds for new payment models that may better reflect the costs of adopting new technologies in clinical care [8]. Critically, MIPS incentivizes the adoption of new technologies independent of specific reimbursement codes [6].

APMs will not, by themselves, make the implementation of many AI products in radiology financially feasible. AI must also demonstrate significant efficiency gains that allow providers to more rapidly achieve "break even." If, as seems probable, AI makes radiology workflows more efficient, the gains in efficiency may lead directly to lower practice costs and, under certain payment models like MIPS, potentially higher CMS reimbursement (as a consequence of achieving improved resource use, for instance). While such APMs remain rare, the promise of upward-adjusted reimbursement rates may spur hospitals and providers on the leading edge toward early implementation of AI in radiology.

THE EFFICIENCY QUESTION

It is probable that machine learning algorithms will make radiology workflows more efficient. This is the expectation of all concerned stakeholders. Without significant gains in efficiency and the attendant savings in financial costs and clinical hours, the promise of AI in medicine will likely be overshadowed by the significant upfront and sustained investment such technologies require.

Given the rapidly increasing costs in the US healthcare system and the growing tendency for payers to pass on such costs to patients, it is imperative that AI in radiology demonstrates cost and work hour savings for it to be deployed in widespread clinical use [7]. For radiologists, AI algorithms that automate all or part of unskilled yet time-consuming tasks like segmentation, lesion measurement, data mining in the EHR, and other quantification tasks will represent an important first step toward realizing this promise [10].

By automating such tasks, for which highly compensated physicians are now typically used, hospitals will reduce the marginal costs—the production cost of an additional product or service— nearly to zero [11]. Of course, automation may bring about another cost-saving measure, though one that radiologists fear: fewer radiology jobs. AI will not eliminate the need for radiologists in the near-to-mid future but may moderately diminish the number of radiology positions available [10]. However, the relationship between labor automation and job elimination is complex: it is frequently observed during historical periods of technological transformation that automation enables workers to shift to other, higher-order tasks, rather than eliminating the need for their professional input altogether [11].

However, in a future in which an AI algorithm is equal in performance to radiologists across all tasks, the economics tilt in favor of AI. It is unlikely AI will ever fully replace all or even most radiologists, as the need for human oversight remains a medical and ethical priority. Yet radiologists will feel significant pressure from an AI that is able to perform equally in many of the tasks they are trained in. At that distant point in the future, the only overhead required to maintain a radiological practice are the marginal operating costs of operating the AI itself, which, as has been addressed elsewhere, are rapidly decreasing as computing power increases for ever lower costs. As Jha notes, "AI could achieve unprecedented economies of scale, at negligible costs" [11]. As the prior example of IDx-DR makes clear, however, theoretical efficiency gains may not be sufficient for achieving financial feasibility in the short term.

HIDDEN COSTS IN THE PURSUIT OF EFFICIENCY

While the promise of efficiency is compelling, payers and providers must also ask themselves whether they are fully assessing the impact of AI in clinical radiology on physician workloads. Perhaps the AI-enabled increase in clinical information at radiologist's disposal may make cases more time-consuming to resolve [7]. As FDA-approved AI devices and algorithms proliferate, the training and education costs for radiographers, nurses, medical assistants, radiologists, and others may become burdensome. And perhaps radiologists become overburdened by as-yet-undefined regulatory documentation requirements secondary to the introduction of AI in their practice. It is not inconceivable that AI-assisted radiology workflows may become less efficient than traditional workflows in the short term.

The EHR is a useful consideration here. The EHR was initially heralded as a time-saving boon to providers that enabled patient health data to travel easily between practice settings. Studies have found, however, that providers spend more time charting in an electronic setting than they did previously [12]. Moreover, EHR documentation requirements are becoming both objectively (in terms of minutes spent per patient encounter) and subjectively (expressed in provider surveys) more burdensome [12]. The more time providers spend charting, the less time they spend on patient care. This takes a toll not only on provider satisfaction, but also on both the provider and hospital bottom line—more time attending to administrative tasks means less time attending to medical (and billable) ones. Implementation cost estimates from AI efficiency gains should be attenuated by an awareness of the ways administrative and regulatory obligations may inadvertently make physician workflows less efficient.

CONCLUSION

While AI is already changing the practice of radiology, there are financial considerations governing AI's use that remain unresolved. In particular, until CMS compensates the true value of AI algorithms and medical devices—capturing the costs of equipment, development, and personnel training—the adoption of AI may be slowed by poor return on investment. And while it is probable that AI will spur many efficiencies, stakeholders should also carefully monitor for the risk of administrative bloat and other non-value-added tasks that can sometimes accompany technological transformations in the practice of radiology.

REFERENCES

1. Hosny A, Parmar C, Quackenbush J, Schwartz LH, Aerts HJWL. Nature Reviews Cancer (2018). https://doi.org/10.1038/s41568-018-0016-5

2. Chang PJ. Radiology (2019). DOI: 10.1148/radiol.2019192527

3. Becker's Hospital Review (2017). https://www.beckershospitalreview.com/ehrs/10-ehr-implementations-with-the-biggest-price-tags-in-2017.html

4. Wolff J, Pauling J, Keck A, Baumbach J. Journal of Medical Internet Research. (2020) DOI:10.2196/16866

5. Digital Diagnostics (2021). https://www.digitaldiagnostics.com/products/eye-disease/idx-dr/

6. Chen EM, Chen D, Chilakamarri P, Lopez R, Parikh R. Ophthalmology (2021). DOI: 10.1016/j.ophtha.2020.07.043

7. Allen B, Gish R, Dreyer K. (2019). The role of an artificial intelligence ecosystem in radiology. In: Ranschaert E., Morozov S., Algra P. (eds.) *Artificial Intelligence in Medical Imaging.* https://doi.org/10.1007/978-3-319-94878-2_19

8. Howard DH, Torres MA. JAMA (2019). DOI: 10.1001/jama.2019.15888

9. Chen MM, Golding LP, Nicola GN. Radiology: Artificial Intelligence (2021). DOI: 10.1148/ryai.2021210030

10. Recht M, Bryan RN. Journal of American College of Radiology (2017). DOI: 10.1016/j.jacr.2017.07.007

11. Jha S. Academic Radiology (2019). https://doi.org/10.1016/j.acra.2019.10.027

12. Overhage JM, McCallie Jr D. Annals of Intern Medicine (2020). DOI: 10.7326/M18-3684

PART IX
TECHNOLOGY DETERMINISM

33 Clinical View

Suely Fazio Ferraciolli, Edson Saito, Eduardo Farina, Léo Max Feuerschuette Neto, Osvaldo Landi, and Felipe Campos Kitamura

TECHNICOLOGICAL DETERMINISM DEFINITION

Technological determinism is a reductionist theory that assumes that a society's technology progresses by following its own internal logic of efficiency, while determining the development of the social structure and cultural values [1]. The term was coined by Thorstein Veblen, an American social scientist and economist. Karl Marx, a German philosopher and economist, elaborated the first major technological determinist view of socioeconomic development. He believed that technological progress creates newer ways of production in a society, which ultimately influence the cultural, political, and economic aspects of a society, thereby inevitably changing society itself. A criticism of technological determinism is that members of society are never compelled by technology itself. Society creates technology and chooses to utilize it. The responsibility for using technology and facing its outcomes is in the hands of individuals. In a less rigid interpretation of technological determinism, it is proposed that technology acts as a mediator, influencing cultural changes but ultimately subject to human control [2].

In history, there are numerous examples of this type of technological determinism. The invention of gunpowder changed the way battles are fought [2, 3]. With the discovery of nuclear energy, wars in the future could be waged using nuclear arsenals [2]. For technological determinists, the need to use technology to gain an advantage in a competition, whether between companies or countries, ensures that technology will be used in an exploitative manner [3]. Gunpowder or nuclear energy was not invented to subjugate people, but the invention of these technologies allowed for this type of use [3]. In healthcare, recent news about the use of algorithms to deny healthcare services has been reported in the United States [6, 7]. Although as in the previous example, algorithmic techniques were not developed with the goal of denying medical care to people.

In medicine, technology has undeniably improved healthcare by introducing new diagnostic methods, new drugs, and expanding physicians' capabilities in the healthcare process. But technology can lead to loss of existing knowledge [3]. As we observe in the emergency departments of several hospitals, radiology exams are replacing medical examinations as the primary diagnostic tool. This shift has resulted in a gradual loss of knowledge related to traditional methods of medical examination [4]. Therefore, technology is also influencing the level of knowledge within a society.

As has already been said, TD has been criticized for underestimating the influence of humans in shaping the path and choosing the tools to interact with technology. This idea can be reinforced by authors like Edward A. Lee (UC Berkeley, USA), in Part I (Artificial Intelligence, Humans, and Control) of the recently published open-access book *Perspectives on Digital Humanism*. He argues that engineers like himself are agents of mutation and the success and further development of their product is determined as much or more by society and the cultural milieu into which they launch their "creation." He says that digital technology co-evolved with humans, and gave the example of Facebook, as it changes its users, who then change this social media enterprise.

The customer-centric feature increasingly adopted by technology companies can be understood as a response to the theory of technological determinism. The customer-centric approach to healthcare, popularized by Apple, Fitbit, and others, can bring innovation as it can generate engagement. Healthcare is a historically regulated area and societal-driven and industry-driven innovation has the potential to create new use cases and consequently new regulatory needs for entities like the FDA. In our view, this balance between usability, medical accuracy, and affordability of these gadgets, associated with better acceptance by doctors and the population, can be the engine to make precision medicine a reality.

However, with the emergence of large language models (LLMs) in artificial intelligence, the balance between the deterministic view of technology on society and how humans use these technologies for various purposes may be altered. Remington Tonar, Co-founder and CGO of the commerce and logistics unicorn Cart.com, in his article "A.I. and technological determinism: Will A.I. exacerbate inequality or destroy it? And do we have a say?" [3], discusses how the increasing implementation of AI has the potential to exacerbate social inequalities. This article addressed only specialized AI algorithms, better known as "weak AI," long before the boom of generative

DOI: 10.1201/9781003095279-42

AI. With the advent of large language models (LLMs), which are becoming increasingly powerful and multimodal, the involvement of society, government, and major technology companies in the discussion about the evolutionary path of AI, use cases, and the way these algorithms are implemented has become increasingly important.

TECHNICAL DETERMINISM WORLDWIDE

We generally seek increased productivity, with apps managing all aspects of our lives. It turns out that in recent years that the number of things to be managed has increased dramatically. Every day we swim in a sea of information, useful or not, and our brain suffers because of it.

It is through the behavior of a society that humanity evolves. Scientific advances appear all the time, but they only fulfill their role in evolution when society as a whole embraces the technology in question. In recent decades, knowledge exchange has become more accessible, exponentially accelerating the impact of technology on our lives.

Today, we communicate with anyone, anytime. In contrast, individuals used to have geographical barriers when they needed to talk to each other face to face. The cost to deliver a message was much higher than today, considering factors such as transportation and time.

It is still possible for a person to leave their home and visit a family member who lives far away, just to say good morning. It may sound weird at best, but it is possible. The freedom of choice present in this case is very clear. It turns out that there is practically a consensus that this communication would be better through some technological device. Whether because of the obvious ease or the need for a sense of belonging, we change our habits following the current technological trends at the time.

It is quite clear that the time savings provided by technology are a spectacular benefit. However, it is necessary to emphasize that the aim for greater productivity is no longer achievable when technology starts to consume most of our time.

Artificial intelligence (AI) presents itself as a centerpiece on the board when we confront freedom and technological determinism. AI has acted as a catalyst for change in diverse industries like aviation, agriculture, economy, and health. Protagonism must be accompanied by responsibility, which is the role of everyone involved in the creation, use, and monitoring of these technologies.

AI models allow us to complete tasks impossible for the human brain. With the inevitable advance, an increasing distance must be created between our capabilities and computers. This advantageous performance will justify the expansion of the scope affected by AI, going far beyond what we see today.

AI and technology as a whole seek to filter out all repetitive or low-cognitive–value tasks, leaving humans the final say in what they do. The thought brings comfort, but is it immutable? What if a set of these little tasks is enough to distort our wills or even guide our actions?

AI models bring nuances, like biases, that often go unnoticed precisely because they take a massive amount of data into account and involve a significant number of parameters. Some of these nuances are dangerous, as in reinforcing and exacerbating gender, ethnicity, and income inequality.

The functioning of an algorithm depends, among other factors, on the dataset from which it has been trained. The tool will likely not work appropriately on populations not represented in the initial database. Thus, algorithms can accentuate the inequality present in society [8, 9].

There are a few ways to solve this problem. It is necessary to fully understand how algorithms reproduce the differences between people so that adjustments and optimization plans can occur. The data used in training is an essential part. They need to be as representative as possible. Finally, we need constant monitoring and standardized approaches to evaluating the performance of algorithms across different groups and minorities [8, 9].

TECHNOLOGICAL DETERMINISM IN MEDICINE

Technology and digital health have changed and are on a pace to dramatically change the patient journey and experience through the healthcare system, especially regarding the physician–patient relationship. A partnership between healthcare workers and patients is necessary to improve outcomes related to patient treatment. For example, studies show that monitoring sensors can improve the management of HbA1c in diabetic patients [10].

Despite these digital health techs being promising, we need to be careful when implementing them and thinking about how to regulate them. With increased availability of information, misinformation has also increased, and all the apps developed for certain types of patients need to be well examined to check if they do not contain misinformation that can lead to a worsened outcome.

If regulation and implementation of those healthcare technologies can be made safely, it will be possible to add more value to the healthcare system and to more closely monitor patients with the Internet of Things embedded in medical devices. It will also allow for solutions that alert patients regarding the time to take medication and organize screening for patients who need the recommended screening exams such as a low-dose chest CT scan for smokers or a colonoscopy for patients 50 years of age or older. During a consultation with a doctor, digital health tech can improve the physician–patient relationship, as audio-based electronic health records have shown to be an alternative [11] to the classic ones, and the doctor would have more time interacting with patient. Also, with the rise of telehealth, new devices will be created for facilitating medical examination, and wearables IoT devices will record more data into patient's electronic health record than a doctor can record in a 30-minute consultation every three months, improving that consultation for a better patient–doctor relationship and maybe facilitating treatment adoption [12].

However, besides all of these cited techologies being useful in the healthcare environment, most of them were not created with the purpose of improving healthcare. Many digital health applications with different embedded technologies were designed to solve an issue that often is not a real deal for those working in the medical environment. One example is that half of the digital health start-ups will fail in two years [13], which happens due to different issues, but one of them is not taking account added burden on healthcare workers when creating the business and only assuming a ready willingness to use new techologies.

In an AI favorable view, radiologists and emergency physicians did not foresee the usefulness of an automatic Alberta stroke program early CT scoring (ASPECTS) tool. Nevertheless, companies developed such a scoring tool that is proving to be beneficial clinically.

TECHNICAL DETERMINISM IN RADIOLOGY

We cannot think about the evolution of radiology without associating it with technological development. Along with computational development came computerized tomography and, later, magnetic resonance, two imaging tools so powerful that they revolutionized the way radiologists diagnose and also treat diseases with interventional procedures.

So, facing the theory of technological determinism, have these technologies determined the medical processes and structures that would be organized from their creation? Did they change the way people saw radiology (even how they look at medicine in general) and the organizational culture involved?

Nowadays, there seems to be a need for imaging-based diagnosis: usually, most clinicians send the patient home from the ER only after an imaging exam, and most patients believe that their diseases are better investigated when an imaging procedure is performed. So, could we infer that these technologies have influenced the way medicine itself is practiced and the basis of physician–patient relations? How often have we witnessed a physician being pressed by the patient to request an imaging exam to relieve their fears?

Seeing our reality on this side of the prism, we would believe that these new technologies are driving society's response and desires for health, and by this organizational culture, the processes and structures are tailored. So, by this theory, the cornerstone for everything is technology: it creates the need from which social structure develops and cultural values are built, determining their outcome [14].

Another example is the assimilation of medical devices and artificial intelligence in radiology. Not only has technological development changed the way these data can be acquired and analyzed, but also the acceptance of and the need to acquire these new tools are growing day by day. People not only want to be diagnosed with diseases: they want to interact, manipulate and share information with their doctors; they long for not only disease prevention but also health promotion; they want faster schedules for their exams and intuitive apps to do so, with the integration of all their exams; they want all of the convenience and positive impacts that these technologies could add to their health.

On the other hand, there are also some cases where technologies have been developed based on pre-existing problems: for example, when nephrogenic sclerosis came along with the use of gadolinium, macrocyclic contrast agents were created to avoid this disease from happening. As you can see, the theory of technical determinism is very robust, but there are some events (in our case, a clinical issue) that initiated the research and development of a new technology (or medicament/ drug, in this case). Another point that may also be taken into account is serendipity, such as the famous discovery of penicillin, since things can happen without the linear relation of cause and effect but merely by fruitful chance.

After this discussion, do you think the development of society brought these new technologies to radiology, or has the technological development molded and determined the outcome of the processes, structure, and social organization in this field?

Technology has progressed a long way from the radiographs and barium of past decades. Although each day brings us closer to the reality of precision medicine and radiogenomics, one question that remains is how we can utilize these technologies to promote worldwide equity. Given that most societies are based on capitalism, there is still much to do to ensure equality of access to these new technologies.

TECHNOLOGICAL DETERMINISM
Final Considerations and Future Perspectives

New digital technologies have emerged in the last decades that promise immense potential to improve healthcare delivery and accessibility worldwide [15, 16]. However, wide adoption of such tools faces several challenges, such as cost, workflow integration, and appropriate scientific validation [17]. Healthcare expenditure is one of the most challenging barriers given the rising share of global income spent on health, in part driven by technology [18]. Innovators will have to develop sustainable business models for these new applications.

Workflow integration is also an essential element for adoption as system integration has been considered critical to physicians [19]. Many studies have pointed out how user-friendliness is an essential factor for physician adoption and also how poor user experience increases workload and contributes to professional burnout [20].

Unlike traditional medical devices, which undergo many different implementation constraints like manufacturing and distribution, software-based digital health products have a faster development cycle, which may outpace our ability to rigorously validate them for safe use [21].

Regulatory institutions will have to develop and apply innovative approaches to regulate these new digital medical products and overcome this critical barrier to adoption, like the efforts recently made for the regulation of artificial intelligence as medical devices by the FDA [22, 23].

This healthcare transformation has been not only technological but also cultural. With the accelerated development of new technologies and ready availability of information on the web, patients are turning to healthcare professionals expecting them to answer their questions about it.

In this new dynamic, it is vital that patients are empowered to become more involved in the available options regarding their health, with shared decision-making and responsibility. Therefore, physicians are transitioning from authoritative figures to being healthcare guides, using their expertise to steer and safeguard patients through all the different technologies and information now available [10].

The speed of innovation can overwhelm patients and healthcare professionals. Consequently, developers must focus on the human aspect when designing new solutions [24]. Stakeholders will need to work together to ensure that the necessary information is presented in a clear and friendly manner for professionals and patients to help make the optimal use of these technologies.

REFERENCES

1. https://en.wikipedia.org/wiki/Technological_determinism#:~:text=Technological%20dete rminism%20is%20a%20reductionist,social%20structure%20and%20cultural%20values

2. https://www.communicationtheory.org/technological-determinism/

3. https://remingtontonar.medium.com/a-i-and-technological-determinism-will-a-i -c294a12abc0a

4. https://www.statnews.com/2017/10/16/artificial-intelligence-physicians/

5. https://www.fastcompany.com/90688912/the-case-for-making-hearing-aids-and-insulin -monitors-as-sexy-as-an-apple-product

6. https://www.statnews.com/2023/12/12/humana-algorithm-medicare-advantage-patients -lawsuit/

7. https://www.statnews.com/2023/11/14/unitedhealth-class-action-lawsuit-algorithm-medi- care-advantage/

8. Sorin V, Klang E. Artificial intelligence and health care disparities in radiology. *Radiology.* 2021 Sept 21;301(3). https://doi.org/10.1148/radiol.2021210566

9. Waite S, Scott J, Colombo D. Narrowing the gap: Imaging disparities in radiology. *Radiology.* 2021 Feb 9;299(1). https://doi.org/10.1148/radiol.2021203742

10. Meskó B, Drobni Z, Bényei É, Gergely B, Győrffy Z. Digital health is a cultural transformation of traditional healthcare. *Mhealth.* 2017 Sep 14;3:38. doi: 10.21037/mhealth.2017.08.07

11. Payne TH, Alonso WD, Markiel JA, Lybarger K, White AA. Using voice to create hospital progress notes: Description of a mobile application and supporting system integrated with a commercial electronic health record. *Journal of Biomedical Informatics.* 2018;77:91–96.

12. Messinger AI, Luo G, Deterding RR. The doctor will see you now: How machine learning and artificial intelligence can extend our understanding and treatment of asthma. *Journal of Allergy and Clinical Immunology.* 2020;145(2):476–478.

13. Francis J. Half of digital health start-ups will fail within two years of launch accenture finds. *Newsroom.* 2015. https://newsroom.accenture.com/news/half-of-healthcare-it-start-ups-will -fail-within-two-years-of-launch-accenture-finds.htm> [Accessed Jan 10, 2022]

14. Technology Determinis. https://shaunad00186041.wordpress.com/2018/05/28/technology -determinism/amp/

15. Solomon, DH, Rudin, RS. Digital health technologies: Opportunities and challenges in rheu- matology. *Nature Reviews Rheumatology.* 2020;16:525–535. https://doi.org/10.1038/s41584-020 -0461-x

16. Manyazewal T, Woldeamanuel Y, Blumberg HM. et al. The potential use of digital health technologies in the African context: A systematic review of evidence from Ethiopia. *NPJ Digital Medicine.* 2021;4:125. https://doi.org/10.1038/s41746-021-00487-

17. Dorsey ER. The new platforms of health care. *NPJ Digital Medicine.* 2021;4:112. https://doi.org /10.1038/s41746-021-00478-5

18. Baltagi BH, Lagravinese R, Moscone F, Tosetti E. Health care expenditure and income: A global perspective. *Health Economics.* 2017 Jul;26(7):863–874. doi: 10.1002/hec.3424.

19. de Grood C, Raissi A, Kwon Y, Santana MJ. Adoption of e-health technology by physicians: A scoping review. *Journal of Multidisciplinary Healthcare.* 2016 Aug 1;9:335–344. doi: 10.2147/ JMDH.S103881.

20. Tajirian T, Stergiopoulos V, Strudwick G, Sequeira L, Sanches M, Kemp J, Ramamoorthi K, Zhang T, Jankowicz D. The influence of electronic health record use on physician burn- out: Cross-sectional survey. *Journal of Medical Internet Research.* 2020 Jul 15;22(7):e19274. doi: 10.2196/19274.

21. Mathews SC, McShea MJ, Hanley CL, Ravitz A, Labrique AB, Cohen AB. Digital health: a path to validation. *NPJ Digital Medicine.* 2019 May 13;2:38. doi: 10.1038/s41746-019-0111-3.

22. Artificial Intelligence/Machine Learning (AI/ML)-Based: Software as a Medical Device (SaMD) Action Plan. 2021 Jan. U.S. Food and Drug Administration Center for Devices & Radiological Health.

23. Good Machine Learning Practice for Medical Device Development Guiding Principles. Oct 2021. U.S. Food and Drug Administration.

24. Ayers DJ, Menachemi N, Ramamonjiarivelo Z, et al. Adoption of electronic medical records: The role of network effects. *Journal of Product & Brand Management* 2009;18:127–35.

34 Technological View

Ronit Gupta, Heena Shah, Ribhav Gupta and Rajiv Dharnipragada

TECHNOLOGY AND TECHNOLOGICAL DETERMINISM

Technological determinism puts forward the notion that technology affects a multitude of aspects within our lives, society, and culture, thereby facilitating further innovation. An illustrative example of this is the invention of the telephone and our society's current dependence on it, which has grown to include a reliance on accessory applications attached to this device. Stepping back, technology itself has a more generalizable definition: "the application of scientific knowledge to the practical aims of human life" [1]. According to this definition, anything from a hair comb to a fighter jet is considered technology. As advances continue to address societal needs, we will further build upon current technology's scaffolding and explore solutions to modern problems. One of the fundamental axioms of technological determinism is that future technological innovations will undoubtedly pose unpredictable consequences, which will guide humanity to adapt to the requirements of these new technologies [2]. On the other hand, some critics of this school of thought suggest that technology is merely a tool and we, as users, can decide how to apply it rather than remain subjects of the technologies or be restricted to the requirements necessitated for their use.

TECHNOLOGICAL AUTONOMY AND THE TECHNOLOGICAL IMPERATIVE

Another facet central to technological determinism is autonomy. Technological autonomy refers to any technology that is both independent of human control and capable of continuous independent development. This idea is closely tied to technological determinism since it assumes a parallel perspective [3]. In other words, technological autonomy assumes that the impacts technology and society have on each other cannot be bidirectional; only technology has the ability to affect society [3]. This does not necessarily imply that technology acts alone, but rather that the catalyst for new applications is derived from self-logic rather than guided by human societal advancement [4].

Relating to this idea is the technological imperative—the assumption that technological impacts on society are inevitable and that users must adapt to contemporary circumstances [2, 5]. Although some argue that individuals can actively choose to avoid societal influences, the technological imperative remains an important principle of technological determinism. It sheds light onto another perspective as to how technology shapes society and human life. In this way, technological autonomy and the technological imperative have both played a profound role in the growth of artificial intelligence (AI) through the technical advancements and scope of opportunities generated across fields, including radiology.

AI AND TECHNOLOGICAL DETERMINISM

Advances in AI have directly influenced the innovation and production of new technologies that have holistically improved healthcare. This calls for continuous development of AI-derived algorithms that can further improve care. Technology firm Jvion, for instance, offers an AI program that provides healthcare workers with information about patient risk for certain procedures and the expected responses to the treatment protocols [6]. Jvion's work is only one branch of the vast research that is being done to reach the full breadth of potential that AI can provide.

Further innovations have compelled us to rethink the ways in which we analyze and execute based upon information. We can see this impact in nearly every industry, from transportation to customer service, with the healthcare sector serving as one of the largest implementers and future markets for AI [7]. Today, diseases can be identified and diagnosed more efficiently, drugs discovered more quickly, and patients monitored according to their specific needs without constant supervision [6]. A systematic review previously conducted found that properly implemented AI also has the potential to enhance patient safety and outcomes by stratifying patients, managing drugs, and detecting errors to a higher degree of accuracy [8]. These examples are just the tip of the iceberg for the wide potential of AI to transform healthcare.

AI may revolutionize the future of technological innovation itself via two primary mechanisms: first, the creation of a wider domain of applications within fields and second, the alteration of the innovation process itself [9, 10]. The first aspect introduces the concept that AI may be categorized as a "general purpose technology" (GPT) that can be applied across facets of society for a variety

DOI: 10.1201/9781003095279-43

of purposes. An example of this phenomenon is the potential for AI-based autonomous vehicles to transform various sectors [11]. This same technology may be employed by the food industry to automate deliveries or by the transportation sector to more sustainably expand their services. The use of GPTs paves the way for broader future applications of the technology. This is the cyclical dynamic of technological determinism in action [9]. Second, AI can be used to develop technologies that can streamline tasks in the workforce to create new methods for how work can be carried out. This cycle affects many labor sectors and can have a much larger impact on society than just one new AI technology [9, 10]. This is a clear example of technology determinism as AI affects society and sets the precedent for future technologies to come.

AI has great potential in being widely applicable, particularly in situations where innovation is hindered by data processing constraints [12]. AI can be utilized to address these current technological limitations through automated decision-making, wherein decisions can be made without human involvement, rather than using other means of computing that are not as precise. Thus, this could unlock a limitless array of possibilities which humans may not independently be capable of. Many impacts have already been made across society and existing technologies and will surely continue to be made due to the ever-expanding scope of AI.

AI AND DETERMINISM IN THE SCOPE OF RADIOLOGY
Present Times

The latest advancements in AI in medicine have allowed for many new opportunities to advance patient care. Radiology is one specialty in medicine that demonstrates an extensive current use, and greater yet potential, for AI. The presence of AI in radiology can be seen across the frontiers of research, such as a deep learning algorithm for reading chest X-rays developed by a Stanford University team [13]. This, among other models, will allow radiologists to analyze images and scans more efficiently, gain a better understanding of their patient's diagnosis, and perhaps identify details that may otherwise be overlooked, ultimately increasing radiologists' performance and clinical role [14].

Despite such promising potential to improve patient care, in recent years, projections of the precise role of radiologists have oscillated widely, largely influenced by increasing implementation of AI in daily clinical practice. The harrowing deterministic belief that AI has the possibility of replacing radiologists is largely a misconception. Rather, AI has the potential to aid radiologists in their diagnostic capabilities and expand the boundaries of this field [15]. In fact, radiologists who can integrate new AI technologies into their workstream to augment their clinical skill sets are preferred over others [16]. More broadly, such radiologists will demonstrate higher value to healthcare systems given their ability to divert attention to more complex cases which AI cannot yet support [16]. There are many nuances of medical imaging and rare pathologies that AI is not yet able to recognize, thus proving the need for experienced radiologists to continue manually making and confirming diagnoses. Hence, AI may not yet be equipped for rare diseases, abnormal presentations, and unique clinical case studies, thus requiring the expert work-up of a radiologist. With AI already in early use across radiology, dependence on AI will likely continue to grow. In this sense, the current state of AI is paving the way for novel innovations that will meet current technologic gaps, which will ultimately deepen the impact of innovation on society, as called for by technology determinism.

TELERADIOLOGY AND THE PROSPECTS OF AI

Teleradiology, the transfer of patient diagnostic images between locations to facilitate remote work and professional collaboration, is central to the modern practice of radiology. Not unlike other areas of medicine, this routine functionality relies on AI to carry out many of its tasks. Specifically, AI can help do so by providing healthcare workers with more guidance to make accurate and better decisions. From a financial perspective, innovations in image sharing have decreased costs by reducing the need for physical documentation required across multiple collaborating institutions [17]. Remote access has also improved diagnostic medicine in settings where access to advanced technology or experienced physicians may not yet be readily available, such as rural areas. This technology has yielded significant impacts on patient care and the medical field, particularly during the COVID-19 pandemic. Teleradiology facilitated the remote work of radiologists to both ensure compliance with social distancing guidelines while promoting the development of large networks across physicians in different hospital systems to collaborate in the critical evaluation of COVID-19 complications [18]. This is simply one illustrative example to demonstrate the technological advancements that have been created to grow the field of medicine through collaboration.

AI's performance abilities within teleradiology have grown immensely within the past decade. The marriage of AI and teleradiology pushes medical practice into a new territory of endless possibilities. Looking forward, developments in AI are likely to complement the growth of teleradiology markets with this field serving as a launchpad for AI developers to improve their technologies, distribute their technology to the public, and promote improved patient management and diagnosis [19]. AI in teleradiology is still in its infancy when we consider the potential magnitude of the impact that can be achieved by harmonizing both. As this technology continues to be put into practice, newfound applications will arise, increasing its need in society and the rate of further advancements.

Current limitations of AI within radiology have shaped the trajectory of technological innovation by shedding light on existing gaps that can be explored to improve future medical care. Some of these prospective changes include adding AI automation to image segmentation, radiology report generation, semantic error detection, research data mining, and business intelligence [20]. Such areas of radiology currently take up disproportionately large quantities of time to complete manually and are therefore more prone to human error. To best implement these changes, the ever-evolving role of radiologists will likely require routine training in novel AI tools, as a part of continuing medical education. Serving as consultants to AI developers, the role of radiologists will likely be greatly expanded, and their value accordingly increased. Overall, the expanded use of AI in radiology will promote new niches for radiologists and healthcare professionals, research, and other branches of radiology, such as radiogenomics (the use of medical imaging to infer genetic information), to ultimately increase the value of radiology in medicine and society [20].

Technological determinism and AI will be crucial concepts interlinked with the future of radiology and innovation in the field. Current technologies impact the environment around us and influence the decisions we make and how we administer them. These actions then allow us to explore new concepts and gaps in society, lending way to future innovations. Continuous evolution along this cycle will enable medicine to reach its maximum potential through implementing new technologies and adapting to the needs of the patient and society.

REFERENCES

1. Encyclopedia Britannica (2022). https://www.britannica.com/technology/technology

2. Winner L. Bulletin of Science, Technology & Society (1997). https://doi.org/10.1177/027046769701700202

3. Dusek V. Philosophy of technology: An introduction (2006). https://doi.org/10.1002/9781444310795.ch22

4. Feenberg A. Technosystem: The social life of reason (2017). https://doi.org/10.4159/9780674982109-003

5. McOmber JB. Journal of Communication (2006). https://doi.org/10.1111/j.1460-2466.1999.tb02809.x

6. Davenport T, Kalakota R. Future Healthc J (2019). https://doi.org/10.7861/futurehosp.6-2-94

7. Bohr A, Memarzadeh K. Artificial Intelligence in Healthcare (2020). https://doi.org/10.1016/B978-0-12-818438-7.00002-2

8. Choudhury A, Asan O. JMIR Med Inform (2020). https://doi.org/10.2196/18599

9. Cockburn I, Henderson R, Stern S. The Economics of AI (2019). https://doi.org/10.7208/chicago/9780226613475.003.0004

10. Hutchinson P. IEEE Transactions on Engineering Management (2021). https://doi.org/10.1109/TEM.2020.2977222

11. Brynjolfsson E, Rock D, Syverson C. National Bureau of Economic Research (2017). https://doi.org/10.3386/w24001

12. Allam S. International Journal of Creative Research Thoughts (2016). ISSN: 2320-2882 https://papers.ssrn.com/sol3/papers.cfm?abstract_id=3821173.

13. Rajpurkur P, Irvin J, Ball RL, Zhu K, et al. PLOS Medicine (2018). https://doi.org/10.1371/journal.pmed.1002686

14. Mohan C. Indian Journal of Radiology and Imaging (2018). https://doi.org/10.4103/ijri.IJRI_256_18

15. Gampala S, Vankeshwaram V, Gadula SSP. Cureus (2020). DOI:10.7759/cureus.11137

16. Ng D, Du H, Yao MM-S, Kosik RO, Chan WP, Feng M. Quantitative Imaging in Medicine and Surgery (2021). https://doi.org/10.21037/qims-20-1083

17. Martín-Noguerol T, Lopez-Ortega R, Ros P, Luna A. European Radiology (2020). https://doi.org/10.1007/s00330-020-07205-w

18. Corapli M, Oktay C, Cil E, Corapli G, Bulut HT. Cerrahpasa Medical Journal (2021). https://doi.org/10.5152/cjm.2021.21028

19. Kalyanpur A. Academic Radiology (2020). https://doi.org/10.1016/j.acra.2019.04.011

20. Liew C. European Journal of Radiology (2018). https://doi.org/10.1016/j.ejrad.2018.03.019

35 Societal View

Lalitha S. Denduluri, Ronit Gupta, Heena Shah, Ribhav Gupta, and Rajiv Dharnipragada

OVERVIEW OF TECHNOLOGICAL DETERMINISM

Throughout history there have been key innovations that have spurred on momentous change in the foundations of society. Some of the more notable developments include the printing press, automobile, telephone, and computer. Each technology has fundamentally changed everyday life, from the way people communicate, to traveling across the globe, to ways of accessing information. Technological advancements produce direct and measurable impacts on society. These empirical examples are broadly summarized through the concept of technological determinism—a theory that technology evolves independently over time and acts as the driving force behind creating structure and culture in society [1, 2].

Medicine is not exempt from technology's evolving effects on society. From anesthesia to vaccines to drug development, technological advances have improved the quality of life and drastically increased life expectancies [3], becoming ever more integral to medicine. Artificial intelligence (AI) has rapidly risen to prominence as a technological domain with potential for widespread application and influence across specialties in medicine [4]. In recent years, AI has become increasingly intertwined with radiology [5, 6]. A 2019 study of PubMed search results with the query "Artificial Intelligence" demonstrated the prevalence of AI research in radiology, with nearly 5,000 of 82,000 artificial intelligence publications being specific to radiology [7]. The concept of technological determinism supports the development of AI as an independent force that continues to advance radiology.

APPLYING AI TO RADIOLOGY

Medicine will, undoubtedly, be a fruitful industry for AI integration, with the specialty of radiology being an early adopter [5, 6, 8]. In fact, in many ways, AI has already altered the field of radiology in recent years. Medical imaging has become increasingly advanced, and its utilization in clinical practice has led to new techniques of disease visualization and diagnosis. Recent efforts to integrate novel features and improve precision using AI have allowed traditional imaging methods like computed tomography (CT), magnetic resonance imaging (MRI), ultrasound (US), single-photon emission computed tomography (SPECT), and positron emission tomography (PET), which have been in use for decades, to provide increased patient benefit [9]. For example, AI-based tools have aided with tasks such as lesion detection for brain hemorrhage or lung nodules [9] and assessment of response to treatment, which have enabled radiologists to diagnose pathologies more quickly and effectively, facilitating patient treatment [7]. Furthermore, a recent study demonstrated how a deep learning (DL) network architecture performed equivalently to radiologists in detailed 2D breast cancer segmentation using routine MRI [10], freeing up radiologists from relatively tedious work and allowing for increased radiologist focus on more clinically relevant tasks. Radiology is not alone. In orthopedic surgery, despite several limitations, some robots have been proven to be better at pedicle screw accuracy and reducing radiation exposure than standard surgery [11–13], and in psychiatry, recent machine learning (ML) models have outperformed psychiatrists in suicide prediction [14–16]. Remarkable progress has been made since the discovery of X-rays in 1895, and under the purview of technology determinism, computers will continue to grow in their functionality, thereby reinforcing the potential for future progress [9].

Despite initial successes across specialties, it is important to acknowledge that many of the studies demonstrating AI successes are still preliminary and will require further investigation and the hard work of clinical translation before such tools can become widely adopted. A multinational diagnostic accuracy study examined over 300,000 mammograms from more than 150,000 women and found that no singular AI model currently performed better than US community benchmarks for radiologists [17]. Importantly, only when combining an automated algorithm with physicians' assessments did such models demonstrate a slightly higher specificity rate than that of unaided radiologists. Much progress is needed before fully automated algorithms in patient care can be trusted, and radiologists will still need to remain thoroughly involved in the diagnostic process.

DOI: 10.1201/9781003095279-44

HOW AI WILL AFFECT RADIOLOGIC PROFESSIONALS

With increasing implementation, AI will have progressively vital implications for those involved in radiology. Today's radiologists have greater workloads than ever before, with some studies suggesting an average radiologist must interpret an image every three to four seconds during an eight-hour workday [5]. AI has the potential to be integrated within radiological practices to automate tasks with minimal human input, thereby increasing efficiency and potentially reducing human-factor–related errors [6]. Within radiology, radiographers obtain medical imaging for patients, after which radiologists visually assess and interpret these scans to propose diagnoses or assess treatment effects [18]. Of note, a recent study showed radiographers and radiologists having varying responses to the integration of AI within their work. Whereas both groups agree that the advent of AI will be beneficial for workforce shortages, radiographers appeared to have an overall less favorable view than radiologists [18]. Radiographers had greater concern about how AI would impact their roles, viewing it more as a potential threat to their jobs and skill development, and were less confident regarding how they would respond to the changing field [18]. On the other hand, radiologists tended to view AI as a tool that would eliminate more repetitive tasks, giving them time for other more challenging and time-consuming work. In addition, radiologists were less concerned with AI curtailing their professional autonomy as, on average, they viewed innovation as an auxiliary aid to their primary position and were assured in the continued need for their skillset [18]. Even within this one specialty, we can see how different professional groups have differing opinions on the topic, thus highlighting the need to understand multiple viewpoints.

On a general note, many hold the concern that the rapid advancement of AI in society will replace human workers with machines. Undeniably, although global automation has historically spurred growth in many job sectors, automation has led to job loss in other fields. The World Economic Forum (WEF) estimates that by 2025, nearly 85 million jobs will be replaced by machines [19]. While this may be the case for certain industries, radiology does not appear to be a field where professionals will be replaced with AI for the foreseeable future. In fact, AI will enable routine tasks to be performed faster and more accurately than by humans. Despite these benefits, it will be important to acknowledge that with the greater implementation of AI in radiology, the roles of radiologic professionals will evolve [5, 7]. As a result, those that work in tandem with AI will more likely continue their practice, whereas those that do not embrace these changes may be eventually replaced [20].

CRITIQUES TO TECHNOLOGICAL DETERMINISM

Technological determinism faces perceived flaws and expected criticism. Many consider the theory that technology independently develops over time and drives its own integration into daily life [1, 2] as putting undue weight and influence on technology. Some question whether technology has the power to single-handedly shape and influence society independent of human involvement. Such theorists posit that even if technology's progression is a given, it is humans who must consciously and intentionally choose to implement technologies into industries like communication, transportation, and medicine [18]. Some argue that technological and societal developments take place in tandem, and that both are intertwined as partners, rather than engaging in a unidirectional relationship [1]. According to these criticisms, AI's introduction into radiology critically depends on the efforts of computer scientists, radiology professionals, and hospital administrators—and not because AI's path is predetermined and involuntary.

IMPORTANT SOCIETAL CONSIDERATIONS

Beyond the technological developments that would result through the symbiosis of AI and radiology, there are several pertinent questions that emerge regarding accountability, regulation, equity, data, and privacy with the increased use of AI.

One major consideration regarding the use of AI in medical practice is that of accountability. Who would be considered at fault for a mistake or a misdiagnosis? Would the responsible party be the AI system, or the humans that contributed to its operations? Furthermore, if it is determined that the AI system is at fault, how would the situation be rectified? On the other hand, if individuals are found to be responsible, then would they be the doctors, the technicians, or the developers of the systems? The crux of these questions is whether AI systems can be evaluated to the standards of responsible moral entities, which is defined as having complete autonomy and sentience, akin to humans [21]. While AI systems have not yet achieved the status of being responsible moral entities [21, 22], at the very least, before implementation, AI systems need to be trained and tested against the highest standards to be trusted by medical professionals and patients. Moreover, developers of these systems should be responsible for delivering high-quality algorithms and for providing training to medical professionals using their products [23]. Beyond that, additional responsibility may need to be determined case-by-case, based on the situation at hand.

Regulation will be a driving factor of the ethical and responsible implementation of AI in medicine and radiology. Clear definitions and oversight measures must be outlined by regulatory agencies such as the US Food and Drug Administration (FDA), the International Medical Device Regulators' Forum (IMDRF), the World Health Organization (WHO), and other bodies to ensure these medical AI systems are adequately managed and critically evaluated in the case of a mishap [23]. A recent analysis devised a framework for the conscientious implementation of AI in medicine, in which a reliable and practical algorithm must have 1) real-time monitoring, 2) the ability to identify the cause of potential failures, 3) the capability for rapid correction of possible failures, and 4) the ability to be efficient and continually refined [23]. These requisites help to build confidence in the use of AI systems in medicine.

Within the status quo of the healthcare system, equity and bias are two of the biggest concerns. Under technological determinism, if AI algorithms are implemented poorly and without thorough testing, health inequities may be exacerbated, and so it is crucial to address this issue head on. The United States exhibits greater levels of socioeconomic health disparities when compared to many other developed countries, particularly affecting people of color and lower-income populations [24]. A wealth of research demonstrates how people within these populations experience worse health outcomes and less access to medical care compared to more privileged members of society. This disparity has been demonstrated in the application of AI in dermatology, where although some propose that AI will eventually serve as the primary diagnostician for skin malignancies instead of dermatologists, some current AI algorithms have been shown to produce significantly more errors when diagnosing patients with darker skin compared to counterparts. These errors are primarily due to these models not having been trained using balanced data including sufficient numbers of darker skin subjects, and thus not being able to accurately analyze samples from these underrepresented patient populations [14, 25]. This, and other examples, call attention to the critical need for population-representative datasets to avoid further accentuation of existing inequities [14, 23, 25].

This relates to another significant issue within the realm of AI in medicine: data bias. While there are oceans of medical data that can be analyzed, vital questions remain as to how data can be accessed, transformed, input, and stored in an effective and efficient manner for AI systems. One of the key issues centers around data curation and data pre-processing, an example of which is the segmentation of certain objects (e.g., lesions) within images [5]. Existing data must be transformed into a usable format for algorithms to garner any accurate insights, and while some of this curation work, like segmentation, can be automated, it often requires the efforts of trained experts for validation. Furthermore, data transparency is a significant point of consideration. ML and DL systems within AI have been labeled as "black boxes," due to how much of the internal processing that occurs between the input of data and the output of classification is unknown, or otherwise incomprehensible, to both users and developers [21]. While DL algorithms can be trained and tested to validate their accuracy, it can give cause for caution when the inner workings of the algorithm are not transparent. This lack of transparency can raise medical, ethical, and legal questions, and can diminish the trust between patients and medical professionals.

Technological determinism as a theory predicts that newly developed AI would be implemented in medicine whether we like it or not. Being mindful of valid counterarguments against the concept, an important consideration remains regarding patient privacy. Patient privacy has been a central tenet of modern healthcare practices, particularly in recent years given the widespread use of electronic health records (EHR) which has resulted in more stored information being potentially accessible to malicious actors online. The COVID-19 pandemic saw increased cyberattacks on the medical industry, where medical tasks, communication, and other practices have increasingly migrated to virtual spaces [26]. While the Health Insurance Portability and Accountability Act (HIPAA) established strict standards for protecting patient privacy, it could not fully prevent cybersecurity attacks from occurring. A study found that the primary driver of patient hesitancy to allow providers to share their medical data was concerns regarding health information privacy [27, 28]. As AI and medicine become more intertwined, it is critical to have security defenses in place that anticipate, prevent, and respond to cybersecurity threats and attacks on patient privacy.

CONCLUSION AND FUTURE OUTLOOK

At the core of technology determinism is a belief in the perpetual nature of technology-driven societal advancement. Through this lens, we can see how AI has the potential to revolutionize medicine and bring forth a new era of technological innovation in radiology. In doing so, AI will change the fabric of the field and the roles of radiologic professionals. Concurrently, there will be a plethora of questions raised regarding safety, ethics, and the future. While the technological determinism framework has notable critiques, under its assumptions we can expect the continued development of AI to lead to further improvements in the accuracy and efficiency of diagnostic radiology, and therefore sustained reliance on AI. As a result, societal considerations such as accountability, equity, and privacy, will need to be mindfully addressed. The proliferation of AI in the everchanging landscape of radiology will continue to drive advancements in medicine and society moving forward, requiring a deeper understanding of the relationship and future prospects between AI and medicine.

REFERENCES

1. Hallström, J. "Embodying the past, designing the future: Technological determinism reconsidered in technology education." *Int J Technol Des Educ* 32, 17–31 (2022). https://doi.org/10.1007/s10798-020-09600-2.

2. Mezentsev, S. "Technological determinism: Breakthrough into the future." In *Communicative Strategies of Information Society, vol 80. European Proceedings of Social and Behavioural Sciences* (pp. 240–248). European Publisher, 2020. https://doi.org/10.15405/epsbs.2020.03.02.29.

3. Crimmins, Eileen M. "Lifespan and Healthspan: Past, Present, and promise." *The Gerontologist* 55(6), 901–911 (2015). https://doi.org/10.1093/geront/gnv130.

4. Tai, Michael Cheng-Tek. "The impact of artificial intelligence on human society and bioethics." *Tzu Chi Med J* 32(4), 339–343 (2020). https://doi.org/10.4103/tcmj.tcmj_71_20.

5. Hosny, Ahmed et al. "Artificial intelligence in radiology." *Nat Rev Cancer* 18(8), 500–510 (2018). https://doi.org/10.1038/s41568-018-0016-5.

6. Oren, Ohad et al. "Artificial intelligence in medical imaging: Switching from radiographic pathological data to clinically meaningful endpoints." *The Lancet* 2(9), E486–E488 (2020). https://doi.org/10.1016/S2589-7500(20)30160-6.

7. European Society of Radiology (ESR). "What the radiologist should know about artificial intelligence – an ESR white paper." *Insights Imaging* 10, 44 (2019). https://doi.org/10.1186/s13244-019-0738-2.

8. Briganti, G., and Le Moine, O. "Artificial intelligence in medicine: Today and tomorrow." *Front Med* 7, 27 (2020). https://doi.org/10.3389/fmed.2020.00027.

9. Bercovich, Eyal, and Javitt, Marcia C. "Medical imaging: From roentgen to the digital revolution, and beyond." *Rambam Maimonides Med J* 9(4), e0034 (2018). https://doi.org/10.5041/RMMJ.10355.

10. Hirsch, Lukas et al. "Radiologist-level performance by using deep learning for segmentation of breast cancers on MRI scans." *Radiol Artif Intell* 4(1) (2021). https://doi.org/10.1148/ryai.200231.

11. Fogel, Alexander L., and Kvedar, Joseph C. "Artificial intelligence powers digital medicine." *NPJ Digit Med* 1, 5 (2018). https://doi.org/10.1038/s41746-017-0012-2.

12. Han, X. G., and Tian, W. "Artificial intelligence in orthopedic surgery: Current state and future perspective." *Chin Med J* 132(21), 2521–2523 (2019). https://doi.org/10.1097/CM9.0000000000000479.

13. Kochanski, Ryan B., Lombardi, Joseph M., Laratta, Joseph L., et al. "Image-guided navigation and robotics in spine surgery." *Neurosurgery* 84(6), 1179–1189 (2019). https://doi.org/10.1093/neuros/nyy630.

14. Arnold, M. H. "Teasing out artificial intelligence in medicine: An ethical critique of artificial intelligence and machine learning in medicine." *Bioethic Inq* 18, 121–139 (2021). https://doi.org/10.1007/s11673-020-10080-1.

15. Passos, Ives Cavalcante et al. "Identifying a clinical signature of suicidality among patients with mood disorders: A pilot study using a machine learning approach." *J Affect Disord* 193, 109–116 (2016). https://doi.org/10.1016/j.jad.2015.12.066.

16. Walsh, Colin G., et al. "Predicting risk of suicide attempts over time through machine learning." *Clin Psychol Sci* 5(3), 457–469 (2017). https://doi.org/10.1177/2167702617691560.

17. Schaffter, T., Buist, D. S. M., Lee, C. I., et al. "Evaluation of combined artificial intelligence and radiologist assessment to interpret screening mammograms." *JAMA Netw Open* 3(3), e200265 (2020). https://doi.org/10.1001/jamanetworkopen.2020.0265.

18. Chen, Yaru et al. "Professionals' responses to the introduction of AI innovations in radiology and their implications for future adoption: A qualitative study." *BMC Health Serv Res* 21(1), 813 (2021). https://doi.org/10.1186/s12913-021-06861-y.

19. "The future of jobs report." *World Economic Forum* (2020). https://www.weforum.org/reports/the-future-of-jobs-report-2020/in-full.

20. Langlotz, Curtis. "Will artificial intelligence replace radiologists?" *Radiol Artif Intell* 1, 3 (2019). https://doi.org/10.1148/ryai.2019190058.

21. Véliz, Carissa. "Moral zombies: Why algorithms are not moral agents." *AI & Soc* 36, 487–497 (2021). https://doi.org/10.1007/s00146-021-01189-x.

22. Verdicchio, M., and Perin, A. "When doctors and AI interact: On human responsibility for artificial risks." *Philos Technol* 35, 11 (2022). https://doi.org/10.1007/s13347-022-00506-6.

23. Bazoukis, George et al. "The inclusion of augmented intelligence in medicine: A framework for successful implementation." *Cell Rep Med* 3(1), 100485 (2022), ISSN 2666–3791. https://doi.org/10.1016/j.xcrm.2021.100485.

24. Lavizzo-Mourey, Risa J et al. "Understanding and mitigating health inequities - past, current, and future directions." *N Engl J Med* 384, 1681–1684 (2021). https://doi.org/10.1056/NEJMp2008628.

25. Adamson, Adewole S., and Smith, Avery. "Machine learning and health care disparities in dermatology." *JAMA Dermatol* 154(11), 1247–1248 (2018). https://doi.org/10.1001/jamadermatol.2018.2348.

26. He Ying et al. "Health care cybersecurity challenges and solutions under the climate of COVID-19: Scoping review." *J Med Internet Res* 23(4), e21747 (2021). https://doi.org/10.2196/21747.

27. Keshta, Ismail, and Odeh, Ammar. "Security and privacy of electronic health records: Concerns and challenges." *Egypt Inform J* 22(2), 177–183 (2021), ISSN 1110-8665. https://doi.org/10.1016/j.eij.2020.07.003.

28. Ermakova, Tatiana et al. "Security and privacy system requirements for adopting cloud computing in healthcare data sharing scenarios." Proceedings of the Nineteenth Americas Conference on Information Systems (2013). https://www.academia.edu/download/51770500/Security_and_Privacy_System_Requirements20170212-15531-67pnmb.pdf.

36 Financial View

Megan Kollitz, Kate Dembny, Madeline Ahern, Mina Estafanos, and David J. H. Wu

From the perspective of technological determinism, AI represents an inevitable force in directing the future of radiology. As such, it will impact the value of imaging and the radiologists who perform it. A financial view of AI in radiology through a deterministic lens examines the anticipated financial impacts of establishing AI as an inexorable force in radiology. The analysis includes financial impacts on the value of radiologists and their training; impacts on different payment models; and cost of implementation, including associated costs such as cybersecurity.

Through a deterministic lens, the newest advances in AI in radiology will inevitably become the "gold standard" where implementation is the only feasible choice, rather than an optional method that may or may not be applied. Where implementation is feasible, implementation *will* occur. AI in radiology will come to signify progress, and methods that exclude available AI applications will become obsolete. This is in keeping with Schumpeter's theory of creative destruction, a key tenet of technological determinism [1]. Therefore, practices and individual radiologists who do not continue to implement the most recent AI developments will diminish in value. Furthermore, radiologists and computer scientists alike expect that radiologist salaries will remain the same or increase with the incorporation of AI [2].

AI AND RADIOLOGIST VALUE WITHIN THE SCOPE OF TECHNOLOGICAL DETERMINISM

A particular manifestation of the financial impact of AI on the value of radiologists is that those radiologists who do not keep up with an increasing workload will lose value, whereas those who *do* will maintain or increase their value. The inevitable application of AI into radiology will result in increased diagnostic efficiency, leaving those radiologists who use it better equipped to keep up with the increasing workload associated with the increased demand of an aging population. This effect also indirectly increases a radiologist's value because less time will be required for more routine diagnosis, allowing for increased time devoted to challenging cases where the narrow scope of an algorithm is unlikely to yield an accurate diagnosis. Additionally, radiologists are likely to develop other skills, such as more advanced imaging techniques and interventions, and play a more direct role in aspects of patient care [3].

The rise of AI in radiology will also result in an increased value placed on advancing the training of junior radiologists more rapidly. These physicians can use AI algorithms at work as a "second opinion" to help confirm diagnoses, and as a virtual learning environment to advance training. Furthermore, while enhancing their own skills, this engagement with the algorithms will allow for quality control. A feedback loop will develop in which radiologists generate new labels, particularly with complex cases, and machines will continue to become better at assisting radiologists with image interpretation [3].

AI IN RADIOLOGY IN VARIOUS PAYMENT MODELS FROM THE STANDPOINT OF TECHNOLOGICAL DETERMINISM

It is apparent that AI will impact the value of radiologists. When discussing the deterministic impacts of AI on the financial future of healthcare, it is also important to consider how its implementation will necessarily impact different payment methods that currently exist in insurance reimbursement, including fee-for-service, value-based care, and capitation.

In a fee-for-service model, reimbursement is determined by Current Procedural Terminology (CPT) codes that are created and maintained by the American Medical Association. The codes are regularly updated, with new codes added, existing codes revised, and codes for defunct protocols removed [4]. While the CPT code does not designate a price for the service, the price charged for each code is negotiated between insurance companies and providers. An article by Golding and Nicola recognizes that many AI algorithms are often developed for atypical patient populations, and thus creating CPT codes for each of these individual niche algorithms becomes both impractical and cumbersome. For this reason, some predict that AI cannot be reimbursed in the same way as more universal components of imaging, like simply performing a head CT [5]. Despite this, through the lens of technological determinism, it would follow that the necessary incorporation of AI into radiological imaging in a fee-for-service model would determine a new system of reimbursing radiologists for the additional services resulting from working with AI.

DOI: 10.1201/9781003095279-45

In a value-based care model, physicians are paid more for better outcomes, rather than based on the number of patients they see or tests they run, with the goal of increasing quality of care while decreasing costs. Physician reimbursement is based on a score called the Merit-Based Incentive Payment System (MIPS). The score is calculated based on performance in four categories, including quality, cost, promoting interoperability, and improvement activities. However, most radiologists are exempt from cost and interoperability measures, and as a result their score is calculated primarily using quality and improvement measures [5]. While artificial intelligence may have less of a role in the determination of cost of care in this model, it can likely play a greater role in determining scoring and reimbursement for physicians. This mechanism of reimbursement became more frequently used by both Medicare and Medicaid following implementation of the Affordable Care Act [6], and reporting to these systems is often done with automated reporting and electronic data captures. This massive amount of data on provider decisions, treatment choices, and patient outcomes would be incorporated into the latest AI applications to generate new data-based best practices and improve evaluation of the quality of care provided by radiologists.

Finally, AI will lead to cost savings in capitation systems. In this model, a physician is paid a specified amount to care for a patient for a period of time, regardless of the care sought or treatment provided. These systems often have measures to account for patient health status, such as increased compensation for an elderly patient with a number of health conditions than for a young and generally healthy individual. This creates an incentive to care for as many patients as possible. With the addition of AI to radiology services and the potential to pre-screen images with these tools, radiologists will be able to evaluate more images in a shorter period of time, increasing payment from capitation, and decreasing healthcare costs [7]. This may drive more healthcare systems toward capitation payment models. Current criticisms of capitation include that physicians could cherry-pick profitable patient groups. Utilization of AI to project healthcare costs for individual patients could be used to reduce the risk of cherry-picking by reducing the value placed on selecting for more profitable groups. And of course, within all of these payment systems, AI will reduce costs by increasing the likelihood of early detection of abnormalities and promoting early intervention [7].

COST OF AI IN RADIOLOGY FROM THE STANDPOINT OF TECHNOLOGICAL DETERMINISM

From a deterministic standpoint, regardless of the payment model used, it is anticipated that the initial financial cost of each major development of AI in radiology will skyrocket before a sharp decline in price and subsequent plateau. One might anticipate this outcome from the standpoint of technological determinism because the inevitability of implementation without competition will result in high costs. This trend is already evident in real-world examples of AI development. For instance, as interest in machine learning increased sharply, in 2015 IBM purchased $5Bn USD worth of healthcare IT companies. One of these companies, Merge Healthcare, came with a massive radiologic dataset that would allow for training deep learning algorithms. In 2013 Google spent $5M USD on a GPU chip with 21 teraflops of processing power. Three years later in 2016, Nvidia came out with a 21-teraflop GPU chip as well, and by 2017, eight of them were available in a supercomputer for just over $100,000 [8]. Today, Nvidia's 110-teraflop Titan V can be purchased for a personal computer for $2,999 [9]. It may be surmised that the latest advances in AI and radiology will follow similar trends, with initially steep costs that taper over time.

One of the necessary theories of technological determinism is the "unintended consequences" account, which asserts that the effects that technology has may be outside the control of society [10]. Certainly, one of the most important financial consequences of the extended use of artificial intelligence is the critical need for an increase in cybersecurity. The use of artificial intelligence in medical practice has come to mean that large amounts of patient data are stored and shared among often unregulated individuals and organizations.

A reimagining of the landscape of medical record privacy is at hand, as data must not only be blinded to practitioners but must be protected from the looming threat of cybersecurity attacks. In 2016, it is estimated that around 16 million patient records were compromised, resulting in only 6,000 claims and $4.8M USD in payouts. In addition, the publicity surrounding a data breach can cause a patient turnover rate of 5–6%, further impacting profit margins [11]. While navigating the legal waters of data privacy can be complicated, governing bodies are making strides to reduce difficulty. The United Nations has been a fast actor in legislative updating, implementing the General Data Protection Regulation (GDPR) in 2018, which increases consumer protections and prevents opt-out data use [12]. The United States has been slower on the uptake, although Supreme Court cases such as *Sorrell v. IMS Health Inc.* (which concerned the use of health data as a marketing tactic for drug companies) are creating a strong basis for increased regulation in this sector [13]. Ultimately, the field of radiology, like all other medical specialties, must contend with the unintended consequences of the use of artificial intelligence in order to protect patient health and privacy. One method currently being employed to increase data privacy is the use of blockchain technology.

Blockchain is a specific type of database for the storage and retrieval of any electronic information. Data is collected and stored in "blocks" which are then attached to each other in chronological order, or "chain." Blockchain constitutes a decentralized data set with limited accessibility, making it a promising candidate for use in a medical setting [14].

The use of artificial intelligence in radiology is synonymous with the use of large data sets. Similar to the financial sector (where blockchain technology got its start), this data must be kept secure for the good of the users. While Electronic Health Record (EHR) data comprises the majority of forwarding movement in this subject, there is promising evidence for blockchain's use in the prevention of medical billing–related fraud [15]. Inherently, technology such as blockchain has been developed to lower labor costs and decrease unnecessary spending. Specifically, this technology has been guided to lower overall operational costs. Worth mentioning is the cost of employing blockchain technology, which varies widely, but runs in the range of $100,000 [16]. Blockchain has been cited as a possible method of reducing credentialing time for physicians. It is estimated that $7,500 is lost each day that a physician's credentials have not yet been approved and the physician is on contract [15]. While this makes up a low overall percentage of expenditures in any health system, it makes an easy target for repair.

Another opportunity for decreasing cost is by lowering the rate of malpractice lawsuits in radiological practices. According to a recent study published in the *New England Journal of Medicine*, more than 17,000 medical malpractice lawsuits are filed each year [17]. The use of blockchain technology in radiology will help secure patient data via the decentralization of data storage. This decentralization makes hacking more difficult because having multiple interconnected servers ensures that any changes cannot be isolated to one server. This technology can enable patients to have their own credentials to access health information and participate in their healthcare decision-making. This feature will make it impossible for individuals to change any information without being identified, reducing malpractice and subsequent lawsuits. The benefits here include reductions in patient harm due to malpractice and suits unduly brought against physicians, which are each associated with reduced financial costs to patients and the healthcare system.

Finally, in order to benefit from the security and savings in time and money resulting from blockchain, auditing must play a role to ensure accurate data input. This will present the blockchain system with new challenges because the entity that controls blockchain must be subject to continuous auditing from healthcare institutions and the government [18]. This auditing challenge will add to the cost of the blockchain. The risk/benefit ratio of blockchain must be evaluated carefully before implementation into a healthcare system.

Looking at the future of AI in radiology from a deterministic standpoint reveals predictable impacts on the value of radiologists, effects within different payment methods, and trends in the cost of implementation and associated costs. This lens reveals greater value placed on radiologists who keep up with increasing workloads, as well as advancing the training of junior radiologists more rapidly. Within a fee-for-service system, it necessitates a new system of reimbursement. Within the value-based care model, quality of care will be more easily ascertained, and defects addressed. The increased number of images reviewed may drive more interest in a capitation system, while reducing cherry-picking. The cost of implementing AI in radiology will result in an initially high cost that will decrease over time with new developments and increased competition. Finally, AI in radiology will come with associated costs such as blockchain storage and auditing, but these advances come with their own benefits and financial savings as well.

REFERENCES

1. Sato T. Technological Frame and Best Praxis in the Age of Artificial Intelligence. In: Kazeroony H.H., Tsang D. (eds.). *Management Education and Automation*. Abingdon, Oxon: Routledge; 2021:36–37.

2. Eltorai AE, Bratt AK, Guo HH. Thoracic Radiologists' Versus Computer Scientists' Perspectives on the Future of Artificial Intelligence in Radiology. *J Thorac Imaging*. 2020;35(4):255–259. doi:10.1097/RTI.0000000000000453.

3. Ng D, Du H, Yao MM-S, Kosik RO, Chan WP, Feng M. Today's Radiologists Meet Tomorrow's AI: The Promises, Pitfalls, and Unbridled Potential. *Quant Imaging Med Surg*. 2021;11(6):2775–2779. doi:10.21037/qims-20-1083.

4. The CPT® code process. American Medical Association. https://www.ama-assn.org/about/cpt-editorial-panel/cpt-code-process#:~:text=The%20CPT%C2%AE%20Editorial%20Panel,is%20composed%20of%2017%20members. Published 2021.

5. Golding LP, Nicola GN. A Business Case for Artificial Intelligence Tools: The Currency of Improved Quality and Reduced Cost. *J Am Coll Radiol*. 2019 Sep;16(9 Pt B):1357–1361. doi:10.1016/j.jacr.2019.05.004. PMID: 31492415.

6. Evaluating Medicaid Value-Based Care Models. Ama-assn.org. https://www.ama-assn.org/system/files/2019-04/medicaid-value-based-care-models.pdf. Published 2021.

7. Van Leeuwen K, de Rooij M, Schalekamp S, van Ginneken B, Rutten M. How Does Artificial Intelligence in Radiology Improve Efficiency and Health Outcomes? *Pediatr Radiol*. 2021. doi:10.1007/s00247-021-05114-8

8. Dreyer KJ, Geis JR. When Machines Think: Radiology's Next Frontier. *Radiology*. 2017;285(3):713–718. doi:10.1148/radiol.2017171183

9. Nvidia Titan V. https://www.nvidia.com/en-us/titan/titan-v/#:~:text=GROUNDBREAKING%20CAPABILITY,NVIDIA%20CUDA%20for%20maximum%20results.&text=NVIDIA%20TITAN%20users%20now%20have,software%20on%20NVIDIA%20GPU%20Cloud.

10. Kline RR. Technological Determinism. In: Smelser NJ, Baltes PB, Editors, *International Encyclopedia of the Social & Behavioral Sciences*. Pergamon; 2001:15495–15498, ISBN 9780080430768. https://doi.org/10.1016/B0-08-043076-7/03167-3.

11. Campbell N. Managing to Succeed: Radiology's Cybersecurity Landscape. *Radiol Today*. 2018;19(7):8.

12. Pesapane F, Volonté C, Codari M, Sardanelli F. Artificial Intelligence as a Medical Device in Radiology: Ethical and Regulatory Issues in Europe and the United States. *Insights into Imaging*. 2018;9(5):745–753. doi:10.1007/s13244-018-0645-y.

13. Cartwright-Smith L, Lopez N. Sorrell v. IMS HEALTH Inc. Data Mining of PHARMACY Records and Drug Marketing as Free Speech. *Public Health Rep*. 2013;128(1):64–66. doi:10.1177/003335491312800109.

14. Verde F, Stanzione A, Romeo V, Cuocolo R, Maurea S, Brunetti A. Could Blockchain Technology Empower Patients, Improve Education, and Boost Research in Radiology Departments? An Open Question for Future Applications. *J Digit Imaging*. 2019;32(6):1112–5.

15. Abdullah S, Rothenberg S, Siegel E, Kim W. School of Block–Review of Blockchain for the Radiologists. *Acad Radiol*. 2020;27(1):47–57. doi:10.1016/j.acra.2019.06.025.

16. Ahluwalia S, Mahto RV, Guerrero M. Blockchain Technology and Startup FINANCING: A Transaction Cost Economics Perspective. *Technol Forecast Soc. Change*. 2020;151:119854. doi:10.1016/j.techfore.2019.119854.

17. Jena AB, Seabury S, Lakdawalla D, Chandra A. Malpractice Risk According to Physician Specialty. *N Engl J Med*. 2011;365(7):629–636. https://doi.org/10.1056/NEJMsa1012370.

18. Bible W, Raphael J, Riviello M, Taylor P, Valiente IO. *Blockchain Technology and its Potential Impact on the Audit and Assurance Profession.* CPA Canada, AICPA; 2017:1–28. https://www.aicpa.org/content/dam/aicpa/interestareas/frc/assuranceadvisoryservices/downloadabledocuments/blockchain-technology-and-its-potential-impact-on-the-audit-and-assurance-profession.pdf.

Index

Printed in the United States
by Baker & Taylor Publisher Services